11 - 19 PROGRESSION

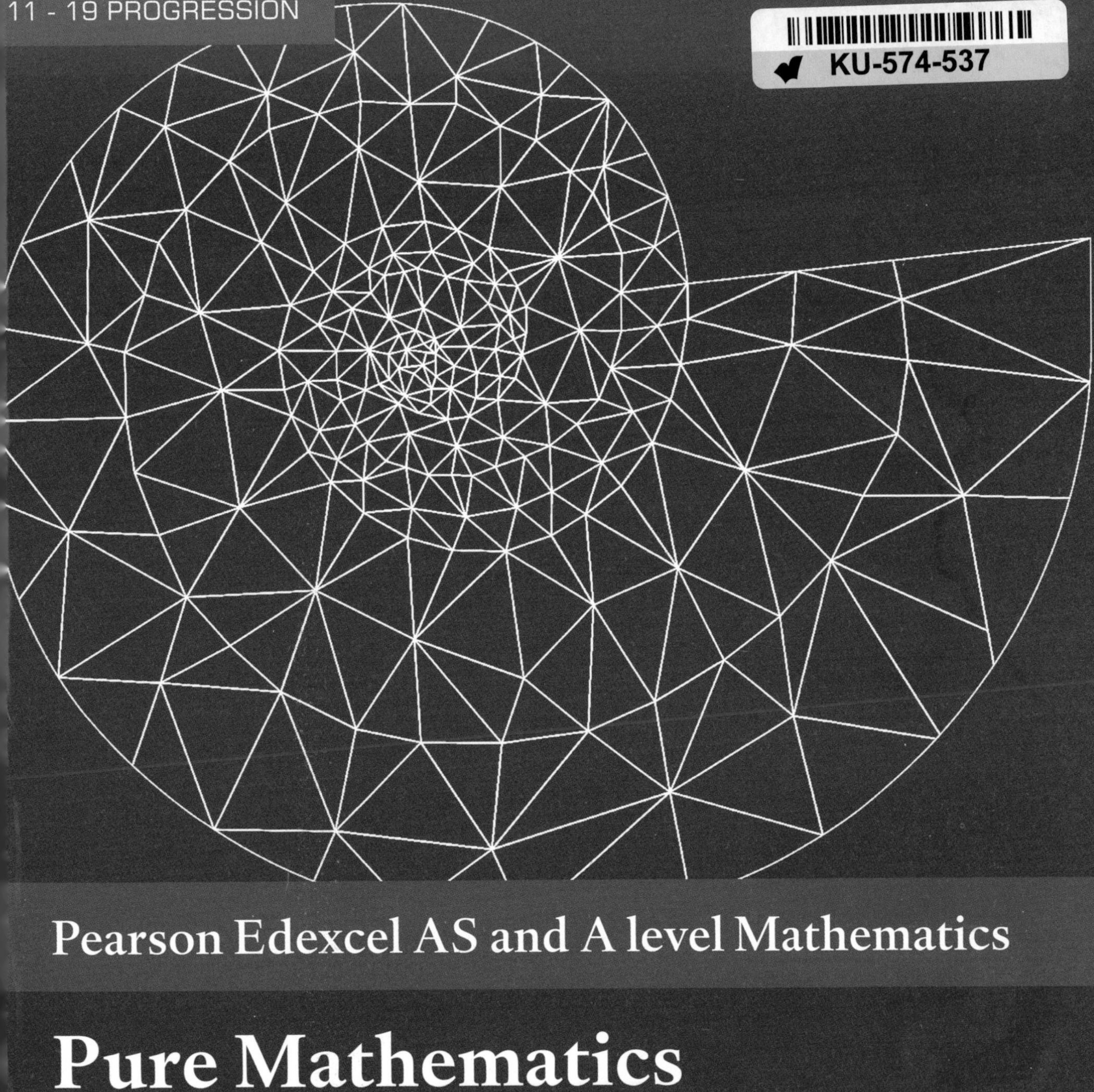

Pearson Edexcel AS and A level Mathematics

Pure Mathematics

Year 1/AS

Practice Book

Series Editor: Harry Smith

Authors: Jack Barraclough, Tara Doyle, Su Nicholson

Pearson

Contents

How to use this book

The Pure Mathematics Year 1/AS Practice Book is designed to be used alongside your Pearson Edexcel AS and A level Mathematics Pure Year 1/AS textbook. It provides additional practice, including problem-solving and exam-style questions, to help make sure you are ready for your exam.

- The chapters and exercises in this practice book match the chapters and sections in your textbook, so you can easily locate additional practice for any section in the textbook.
- Each chapter finishes with two sets of problem-solving practice questions at three different difficulty levels.
- An Exam question bank at the end of the book provides mixed exam-style questions to help you practise selecting the correct mathematical skills and techniques.

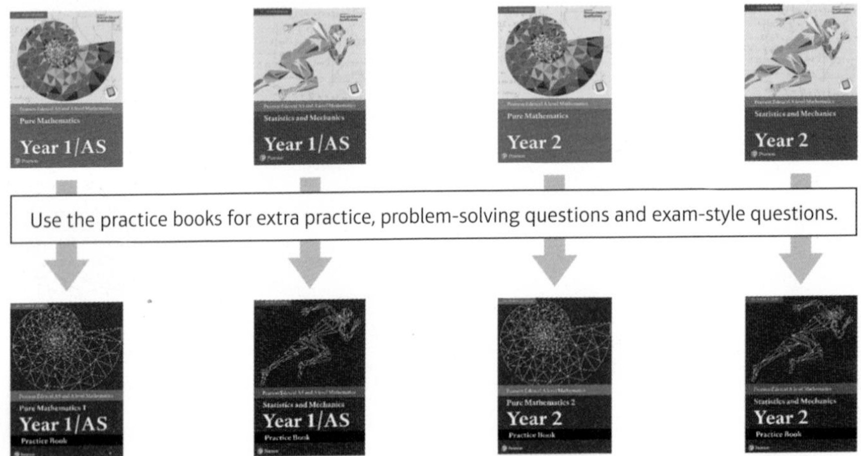

Finding your way around the book

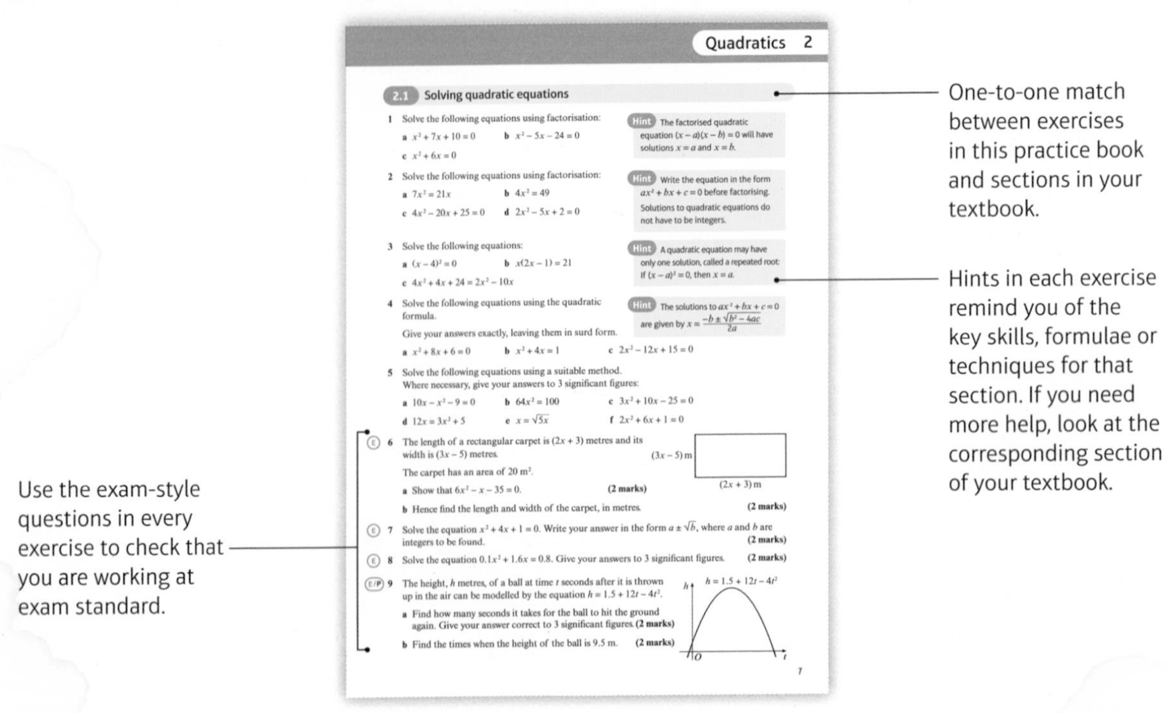

Exam-style questions are flagged with Ⓔ and have marks allocated to them.

Problem-solving questions are flagged with Ⓟ

Bronze questions might have more steps to lead you through the technique, or require a more straightforward application of the skills from that chapter.

Silver questions are more challenging, and provide less scaffolding. If you're struggling with the Silver question, try the Bronze question first.

You can find more exam-style questions on this chapter in the Exam question bank.

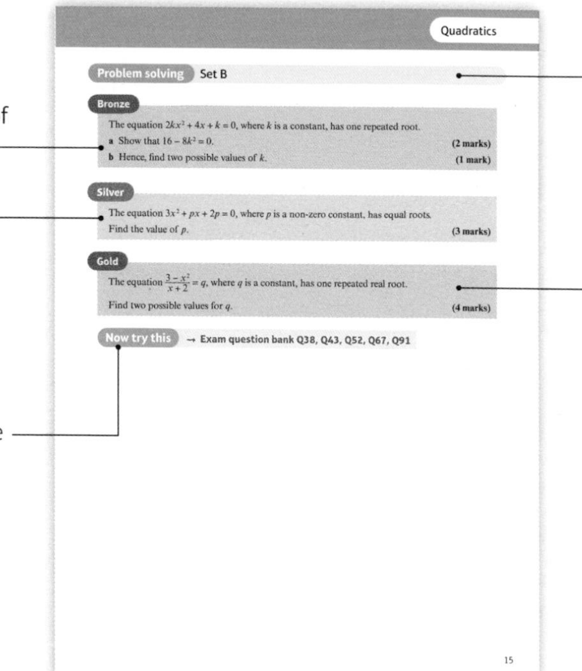

Each chapter ends with two sets of exam-style problem-solving questions which draw on material from throughout the chapter and from earlier chapters.

Gold questions involve tricky problem-solving elements, or might require you to think more creatively. If you can answer the Gold questions then you can be confident that you are ready to tackle the hardest exam questions.

One challenge of the exam is that you aren't usually told which techniques or strategies you need to apply to a particular question. The questions in the Exam question bank are not ordered by topic, so you need to choose the appropriate mathematical skills.

There are a lot more questions in the Exam question bank than there will be on your exam paper. Don't try and tackle them all at once, but make sure you try some of the trickier questions from the end of the question bank.

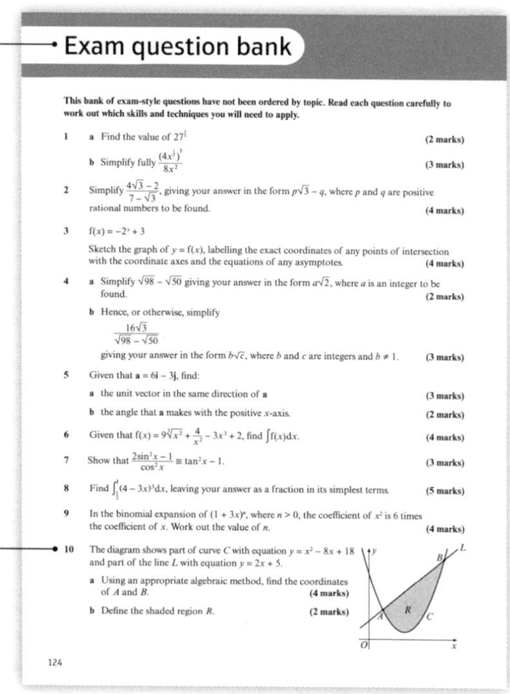

Published by Pearson Education Limited, 80 Strand, London, WC2R 0RL.

www.pearsonschoolsandfecolleges.co.uk

Series editor Harry Smith
Edited by Haremi Ltd
Typeset by York Publishing Solutions Pvt. Ltd., INDIA

First published 2019

22 21 20 19
10 9 8 7 6 5 4 3 2 1

British Library Cataloguing in Publication Data
A catalogue record for this book is available from the British Library

ISBN 9781292274683

Printed in Italy by L.E.G.O S.p.A

Note from the publisher
Pearson has robust editorial processes, including answer and fact checks, to ensure the accuracy of the content in this publication, and every effort is made to ensure this publication is free of errors. We are, however, only human, and occasionally errors do occur. Pearson is not liable for any misunderstandings that arise as a result of errors in this publication, but it is our priority to ensure that the content is accurate. If you spot an error, please do contact us at resourcescorrections@pearson.com so we can make sure it is corrected.

1.1 Index laws

1 Simplify these expressions:

a $b^3 \times b^4 \times b^2$ **b** $\frac{a^5}{a^3}$

c $5x^7 \times 3x^2$ **d** $(2x^2)^3 \div 4x^5$

Hint Use the laws of indices to simplify powers of the same base:

$a^m \times a^n = a^{m+n}$ $(a^m)^n = a^{mn}$

$a^m \div a^n = a^{m-n}$ $(ab)^n = a^n b^n$

2 Expand these expressions and simplify if possible:

a $x^3(4x^2 - 7) + 2x^5$ **b** $-5x^2(3 - 8x)$

c $x(3x + 4) - 7(5x - 2)$

Hint A minus sign before a bracket changes the signs of the terms inside the brackets.

3 Simplify these expressions:

a $\frac{x^9 - x^5 - x^7}{x^3}$ **b** $\frac{12x^3 + 8x^7}{4x}$

c $\frac{6x^4 + 12x^3 - (4x^5)^2}{2x^3}$

Hint Divide each term in the numerator by the denominator.

Remember that $a^1 = a$ and $a^0 = 1$

4 Simplify these expressions:

a $3x^4y^2 \times 7xy^3$ **b** $\frac{25x^{12}y^2}{5x^3y}$

Hint Simplify the numbers first and then use the laws of indices to simplify each letter.

E 5 Simplify these expressions:

a $\frac{27r^5s^7}{3r^3}$ **(2 marks)**

b $\frac{2(4a^4b^2)^3}{8ab^2}$ **(2 marks)**

E/P 6 **a** Given that $32 = 2^a$, find the value of a. **(1 mark)**

b Given that $81 = 3^b$, find the value of b. **(1 mark)**

E/P 7 Given that $\frac{33x^6 - 3(xy)^4 + (3x^2)^3}{6x^2}$ can be written in the form $-10x^p - \frac{1}{2}x^qy^r$, find the values of p, q and r. **(3 marks)**

E 8 Write down the value of:

a $(125x^3)^{\frac{1}{3}}$ **(2 marks)**

b $\frac{20x^{\frac{5}{4}}}{5x^{\frac{1}{4}}}$ **(2 marks)**

c $(xy)^0$ **(1 mark)**

1.2 Expanding brackets

1 Expand these expressions and simplify if possible:

a $(x+3)(x+4)$ b $(x+7)(x-7)$

c $(x-6)(x-3)$

Hint Multiply each term in one expression by each term in the other expression and then simplify.

A minus sign in front of a term will change the sign of the term it is multiplied with.

2 Expand these expressions and simplify if possible:

a $(x+y)^2$ b $(x-4)^2$ c $(2x-5y)^2$

Hint $(a+b)^2 = (a+b)(a+b)$

3 Expand these expressions and simplify if possible:

a $(5x+3y)(4x-7y)$ b $(x-9y)(x^2-1)$

c $(x+y)(x-3y+2)$

Hint Multiply each term in one expression by each term in the other expression and then simplify. A minus sign in front of a term will change the sign of the term it is multiplied with.

4 Expand these expressions and simplify if possible:

a $x(2x-4y)(5x-3y-9)$

b $(3x-2y)^3$

c $(2x-1)(x+3)(4x-2)$

Hint You can expand three brackets by expanding and simplifying one pair of brackets first, then multiplying your expanded expression by every term in the third bracket.

(E) 5 Expand and simplify $(a+b)^3$ **(2 marks)**

(E/P) 6 Given that $(2x-7)(ax+b) = 6x^2 - 13x - 28$, find the values of the constants a and b. **(2 marks)**

(E) 7 The length of each edge of a cube is $(2x+1)$ cm. Find an expression in terms of x for:

a the area of one face of the cube **(2 marks)**

b the volume of the cube. **(2 marks)**

(E/P) 8 The dimensions of a patio are shown, with lengths given in metres. The six edges are straight lines.

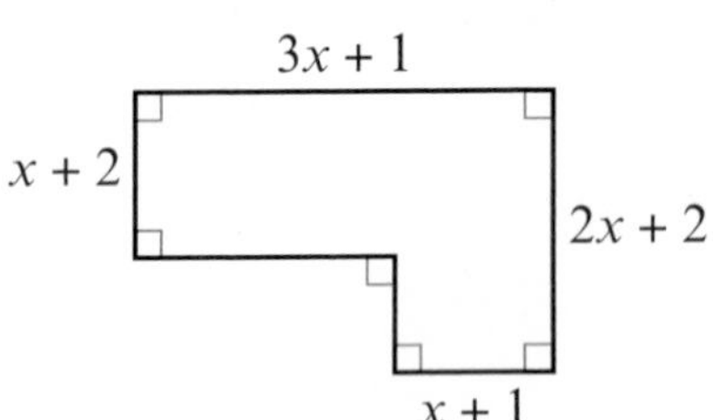

a Find an expression for the area of the patio in terms of x. **(3 marks)**

b Given that the area is 62 m^2, find the value of x. **(3 marks)**

(E/P) 9 Given that $(5a-b)(3a+2b)(2a-b) = pa^3 - qa^2b - rab^2 + sb^3$, find the values of the constants p, q, r and s. **(2 marks)**

1.3 Factorising

1 Factorise these expressions completely:

a $6x + 18$ **b** $27x^2 - 9x$

c $x^3 - 4x^2$

Hint Find the highest common factor of the terms and write this factor outside the brackets. You can check your answers by expanding the brackets.

2 Factorise these expressions completely:

a $7ab^2 + 21a^2b^2$ **b** $8ab - 64b^3$

c $5ab^4 + 20a^3b^2c + 15b^5c^2$

Hint Find the highest common numerical factor, the highest common power of a and the highest common power of b. Then write these outside the brackets.

3 Factorise:

a $x^2 + 5x + 6$ **b** $x^2 - 3x - 10$

c $x^2 - 49$ **d** $y^2 - 16x^2$

Hint These are all quadratic expressions. They can be factorised into two brackets. For parts **c** and **d** you can use the rule for the difference of two squares: $x^2 - y^2 = (x + y)(x - y)$

4 Factorise completely:

a $3x^4 + 6x^3 - 18x^2$ **b** $4x^2 + 28x + 48$

c $2x^2 + x - 3$

Hint Start by writing part **b** as $4(x^2 + 7x + 12)$, then factorise the expression inside the brackets.

E **5** Factorise completely $x - 25x^3$ **(2 marks)**

E **6** Factorise completely $36x - 16x^3$ **(2 marks)**

E **7** Factorise completely $8x^3 + 20x^2 + 8x$ **(2 marks)**

/P **8** Simplify $\dfrac{2x^3 + 6x^2 + 4x}{x^2 + 4x + 3}$ **(3 marks)**

1.4 Negative and fractional indices

1 Simplify:

a $x^{-4} \times x^{-1}$ **b** $\dfrac{x^5}{x^6}$ **c** $3x^{-10} \div 6x^3$

Hint Use the rules $a^m \times a^n = a^{m+n}$ and $a^m \div a^n = a^{m-n}$

2 Simplify:

a $2x^{\frac{1}{4}} \times 5x^{\frac{2}{3}}$ **b** $9x^{\frac{5}{2}} \div 3x^{\frac{2}{3}}$

c $(8x^3)^{\frac{2}{3}}$ **d** $7x^{-0.25} \times 2x^{0.5}$

Hint You can use the laws of indices with fractional powers.

3 Simplify:

a $\sqrt[5]{x} \times \sqrt{x}$ **b** $\sqrt[4]{x} \div (\sqrt{x})^3$ **c** $(\sqrt[3]{x})^2 \times (\sqrt{x})^7$

Hint Rewrite the questions with fractional powers using $a^{\frac{1}{m}} = \sqrt[m]{a}$ and $a^{\frac{n}{m}} = \sqrt[m]{a^n}$, and then simplify.

4 Evaluate:

a $16^{\frac{3}{2}}$ **b** $(-4)^{-3}$ **c** $\left(\dfrac{125}{27}\right)^{-\frac{2}{3}}$

Hint For parts **b** and **c**, you can use $a^{-m} = \dfrac{1}{a^m}$ You can use your calculator to evaluate negative and fractional powers.

5 Simplify:

a $\left(\dfrac{16}{25}x^3\right)^{\frac{5}{2}}$ **b** $\dfrac{x^5 + 8x^3}{x^{10}}$ **c** $\dfrac{30x^2 - 18x^5}{6x^7}$

Hint For part **a**, apply the power to both the fraction and x^3.

(E) 6 a Evaluate $64^{\frac{3}{2}}$ **(2 marks)**

b Simplify fully $x^3\left(2x^{\frac{1}{3}}\right)^6$ **(2 marks)**

(E) 7 a Find the value of $25^{-\frac{1}{2}}$ **(2 marks)**

b Simplify $x\left(\frac{3}{2}x^{-\frac{1}{4}}\right)^8$ **(2 marks)**

(E/P) 8 a Evaluate $81^{\frac{3}{4}}$, giving your answer as an integer. **(2 marks)**

b Simplify fully $\dfrac{\left(3x^{\frac{1}{2}}\right)^3}{9x^2}$ **(3 marks)**

(E/P) 9 Given that $y = \frac{1}{9}x^2$, express each of the following in the form kx^n, where k and n are constants.

a $y^{\frac{1}{2}}$ **(2 marks)**

b $2y^{-1}$ **(2 marks)**

(E/P) 10 Express 25^{2x-5} in the form 5^y, where $y = ax + b$ for some constants a and b to be determined. **(2 marks)**

1.5 Surds

1 Simplify:

a $\sqrt{18}$ b $\sqrt{63}$ c $\sqrt{250}$

Hint You can manipulate surds using the rules $\sqrt{ab} = \sqrt{a} \times \sqrt{b}$ and $\sqrt{\frac{a}{b}} = \frac{\sqrt{a}}{\sqrt{b}}$
To simplify, look for factors of each number that are square numbers.

2 Simplify:

a $\frac{\sqrt{27}}{3}$ b $\frac{\sqrt{98}}{7}$ c $\frac{\sqrt{24}}{2}$

Hint Simplify the numerator first, then check whether the denominator is a factor of the numerator. If so, you can divide through by it.

3 Simplify:

a $4\sqrt{32} - 3\sqrt{8}$ b $\sqrt{75} + 2\sqrt{12} - \sqrt{27}$

c $\sqrt{200} + \sqrt{18} - 2\sqrt{72}$

Hint Simplify each surd, then collect like terms.

4 Expand and simplify if possible:

a $\sqrt{3}(\sqrt{27} - 1)$ b $(1 + \sqrt{2})(3 - 2\sqrt{2})$

c $(4 - \sqrt{3})(6 - \sqrt{7})$

Hint Multiply each term in one expression by each term in the other expression.

(E) 5 a Simplify $\sqrt{20} + \sqrt{45}$, giving your answer in the form $a\sqrt{b}$ where a and b are integers. **(2 marks)**

b Express $\sqrt{112}$ in the form $a\sqrt{7}$, where a is an integer. **(1 mark)**

(E/P) 6 Solve the equation $x - \sqrt{60} = 2\sqrt{3} - x$, giving your answer in the form $\sqrt{a} + \sqrt{b}$, where a and b are integers. **(2 marks)**

E **7** Expand and simplify:

a $(2-\sqrt{7})(\sqrt{7}-1)$ **(3 marks)**

b $(2\sqrt{7}+3)^2$ **(2 marks)**

E **8** Simplify:

a $(3\sqrt{11})^2$ **(1 mark)**

b $(7+\sqrt{3})(2-\sqrt{3})$ **(3 marks)**

E/P **9** Given that $243\sqrt{3} = 3^a$, find a. **(2 marks)**

1.6 Rationalising denominators

1 Simplify:

a $\frac{1}{\sqrt{3}}$ **b** $\frac{35}{\sqrt{5}}$ **c** $\frac{9}{3\sqrt{3}}$

Hint For fractions in the form $\frac{1}{\sqrt{a}}$ multiply numerator and denominator by $\sqrt{a}$

2 Simplify:

a $\frac{\sqrt{7}}{\sqrt{21}}$ **b** $\frac{\sqrt{12}}{\sqrt{72}}$ **c** $\frac{\sqrt{75}}{\sqrt{125}}$

Hint Multiply numerator and denominator by the surd denominator and then simplify.

3 Rationalise the denominator and simplify:

a $\frac{1}{1+\sqrt{2}}$ **b** $\frac{2}{\sqrt{7}-1}$ **c** $\frac{3}{\sqrt{6}-\sqrt{5}}$

Hint For fractions in the form $\frac{1}{a+\sqrt{b}}$ multiply the numerator and denominator by $a-\sqrt{b}$

For fractions in the form $\frac{1}{a-\sqrt{b}}$ multiply the numerator and denominator by $a+\sqrt{b}$

4 Rationalise the denominator and simplify:

a $\frac{\sqrt{3}+\sqrt{2}}{\sqrt{3}-\sqrt{2}}$ **b** $\frac{3-\sqrt{5}}{\sqrt{5}+5}$ **c** $\frac{3}{(2\sqrt{2}+3)^2}$

Hint For part **c**, start by expanding and simplifying the denominator.

E **5** Express $\frac{2}{2-\sqrt{3}}$ in the form $a+b\sqrt{3}$, where a and b are integers to be found. **(2 marks)**

E **6** Write each expression in the form $a+b\sqrt{5}$, where a and b are integers to be found:

a $\frac{1+\sqrt{5}}{\sqrt{5}-2}$ **(3 marks)**

b $\frac{7+\sqrt{5}}{3+\sqrt{5}}$ **(3 marks)**

E/P **7** $\frac{3x^5 - x^{\frac{5}{2}}}{\sqrt{x}}$ can be written in the form $3x^p - x^q$. Write down the values of p and q. **(2 marks)**

E/P **8** Show that $\frac{(7-2\sqrt{x})^2}{\sqrt{x}}$ can be written in the form $Ax^{-\frac{1}{2}} + Bx^{\frac{1}{2}} - C$ where A, B and C are constants to be found. **(3 marks)**

E/P **9** The diagram shows a rectangle with sides of length $(\sqrt{14}-2)$ cm and x cm. The area of the rectangle is 5 cm². Find x, giving your answer as simply as possible in surd form. **(3 marks)**

$\sqrt{14}-2$

x

Problem solving Set A

Bronze

a Simplify $\sqrt{147} - \sqrt{75}$ giving your answer in the form $a\sqrt{3}$, where a is an integer. **(2 marks)**

b Hence, or otherwise, simplify $\dfrac{24\sqrt{2}}{\sqrt{147} - \sqrt{75}}$ giving your answer in the form $b\sqrt{6}$, where b is an integer. **(2 marks)**

Silver

a Expand and simplify $(11 + \sqrt{5})(\sqrt{5} + 1)$ giving your answer in the form $a + b\sqrt{5}$, where a and b are integers. **(2 marks)**

b Hence, or otherwise, simplify $\dfrac{11 + \sqrt{5}}{\sqrt{5} - 1}$ giving your answer in the form $c + d\sqrt{5}$, where c and d are integers. **(2 marks)**

Gold

Simplify $\dfrac{6\sqrt{3} - 4}{5 - \sqrt{3}}$, giving your answer in the form $p\sqrt{3} - q$, where p and q are positive rational numbers. **(4 marks)**

Problem solving Set B

Bronze

a Express 4^{x+2} in the form 2^y, stating y in terms of x. **(2 marks)**

b Hence, or otherwise, solve the equation $4^{x+2} = 32$. **(2 marks)**

Silver

a Rewrite the equation $(4^{x-1})^2 - 6(4^{x-1}) + 8 = 0$ in the form $y^2 + by + c = 0$, where $y = 4^{x-1}$ and b and c are constants to be found. **(1 mark)**

b Factorise your equation from part **a** and hence, or otherwise, solve for x. You must show each step of your working. **(3 marks)**

Gold

Solve the following equation, showing each step of your working:

$(9^{x-1})^2 - 30(9^{x-1}) + 81 = 0$ **(5 marks)**

Now try this → **Exam question bank Q1, Q2, Q4, Q24, Q36, Q46, Q64**

2.1 Solving quadratic equations

1 Solve the following equations using factorisation:

a $x^2 + 7x + 10 = 0$ **b** $x^2 - 5x - 24 = 0$

c $x^2 + 6x = 0$

Hint The factorised quadratic equation $(x - a)(x - b) = 0$ will have solutions $x = a$ and $x = b$.

2 Solve the following equations using factorisation:

a $7x^2 = 21x$ **b** $4x^2 = 49$

c $4x^2 - 20x + 25 = 0$ **d** $2x^2 - 5x + 2 = 0$

Hint Write the equation in the form $ax^2 + bx + c = 0$ before factorising. Solutions to quadratic equations do not have to be integers.

3 Solve the following equations:

a $(x - 4)^2 = 0$ **b** $x(2x - 1) = 21$

c $4x^2 + 4x + 24 = 2x^2 - 10x$

Hint A quadratic equation may have only one solution, called a repeated root: If $(x - a)^2 = 0$, then $x = a$.

4 Solve the following equations using the quadratic formula.

Give your answers exactly, leaving them in surd form.

a $x^2 + 8x + 6 = 0$ **b** $x^2 + 4x = 1$ **c** $2x^2 - 12x + 15 = 0$

Hint The solutions to $ax^2 + bx + c = 0$ are given by $x = \frac{-b \pm \sqrt{b^2 - 4ac}}{2a}$

5 Solve the following equations using a suitable method. Where necessary, give your answers to 3 significant figures:

a $10x - x^2 - 9 = 0$ **b** $64x^2 = 100$ **c** $3x^2 + 10x - 25 = 0$

d $12x = 3x^2 + 5$ **e** $x = \sqrt{5x}$ **f** $2x^2 + 6x + 1 = 0$

E **6** The length of a rectangular carpet is $(2x + 3)$ metres and its width is $(3x - 5)$ metres.

$(3x - 5)$ m

$(2x + 3)$ m

The carpet has an area of 20 m^2.

a Show that $6x^2 - x - 35 = 0$. **(2 marks)**

b Hence find the length and width of the carpet, in metres. **(2 marks)**

E **7** Solve the equation $x^2 + 4x + 1 = 0$. Write your answer in the form $a \pm \sqrt{b}$, where a and b are integers to be found. **(2 marks)**

E **8** Solve the equation $0.1x^2 + 1.6x = 0.8$. Give your answers to 3 significant figures. **(2 marks)**

E/P **9** The height, h metres, of a ball at time t seconds after it is thrown up in the air can be modelled by the equation $h = 1.5 + 12t - 4t^2$.

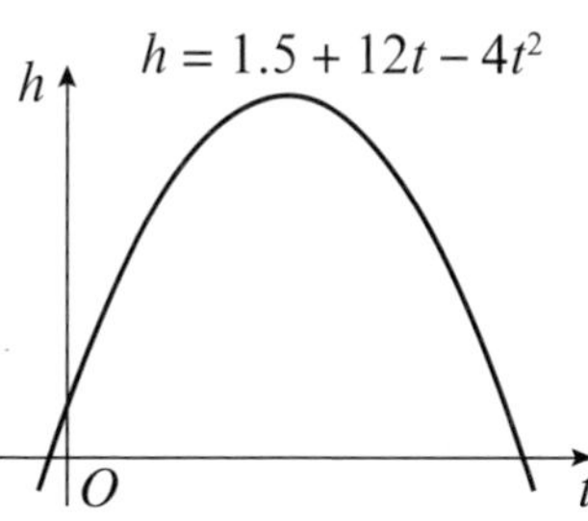

a Find how many seconds it takes for the ball to hit the ground again. Give your answer correct to 3 significant figures. **(2 marks)**

b Find the times when the height of the ball is 9.5 m. **(2 marks)**

2.2 Completing the square

1 Complete the square for these expressions:

a $x^2 + 2x$ **b** $x^2 - 8x$

c $x^2 - 12x$

Hint $x^2 + bx = \left(x + \frac{b}{2}\right)^2 - \left(\frac{b}{2}\right)^2$

2 Complete the square for these expressions:

a $x^2 - 8x + 12$ **b** $x^2 - x - 12$

c $x^2 - 3x - 4$

Hint When a quadratic expression is in completed square form, x only appears once. The constants in your answer do not have to be integers.

3 Write each of these expressions in the form $a(x + b)^2 + c$, where a, b and c are constants to be found:

a $3x^2 - 12x + 17$ **b** $5x^2 - 10x + 12$ **c** $3x^2 - 7x + 2$

Hint Start by factorising the first two terms of each expression.

4 Solve these quadratic equations by completing the square. Leave your answers in surd form.

a $x^2 - 2x - 10 = 0$ **b** $5x^2 + 12x + 6 = 0$

c $3x^2 - 7x - 2 = 0$

Hint Write the left-hand side of each equation in completed square form. Then use inverse operations to make x the subject of the equation. Remember to include ± when you take square roots of both sides of the equation.

E/P **5** $f(x) = x^2 - 7x - 2, x \in \mathbb{R}$

a Express $f(x)$ in the form $(x - a)^2 - b$, where a and b are constants. **(2 marks)**

b Hence write down the minimum value of $f(x)$. **(1 mark)**

E **6** $h(x) = 4 - 2x - 3x^2, x \in \mathbb{R}$

a Express $h(x)$ in the form $a(x + b)^2 + c$, where a, b and c are constants. **(2 marks)**

b Hence, or otherwise, find the exact solutions to $h(x) = 0$. **(2 marks)**

E **7** $6x - 2 - x^2 \equiv q - (x + p)^2$, where p and q are integers.

a Find the value of p and the value of q. **(3 marks)**

b Hence, or otherwise, solve the equation $6x - 2 - x^2 = 0$. **(2 marks)**

2.3 Functions

1 Using the functions $f(x) = x^2 - 5x + 6$ and $g(x) = |\sqrt{x + 3}|$, find the values of:

a $f(3)$ **b** $g(22)$

c $f(-1.5)$ **d** $f(4) + g(6)$

Hint To find the value of $f(a)$, substitute a for x in the expression for $f(x)$.

2 The functions r and s are given by
$r(x) = x^2 + 2x - 3$ and $s(x) = 2x + 1, x \in \mathbb{R}$.
Find the two values of x for which $r(x) = s(x)$.

Hint You can check your answer by evaluating $r(x)$ and $s(x)$ for each of your values of x.

3 Find all roots of the following functions. Give any non-integer roots in exact form.

a $f(x) = (x + 2)(x + 7)$ **b** $f(x) = 81 - x^2$
c $f(x) = x^6 + 7x^3 + 12$ **d** $f(x) = x^3 - 4x^2 - 21x$

Hint The roots of a function are the values of x for which $f(x) = 0$.

4 The function $h(x)$ is defined as $h(x) = x^2 + 6x + 2, x \in \mathbb{R}$.

a Write $h(x)$ in the form $(x + p)^2 + q$.

b Hence, or otherwise, find the roots of $h(x)$, leaving your answers in surd form.

Hint The expression $(x + p)^2$, where x is a real number and p is a constant, must always be greater than or equal to 0.

c Write down the minimum value of $f(x)$ and state the value of x for which it occurs.

E/P 5 The function f is defined as $f(x) = x^2 + 6x + 13, x \in \mathbb{R}$.

a Write $f(x)$ in the form $(x + p)^2 + q$, where p and q are constants to be found. **(2 marks)**

b Hence, or otherwise, explain why $f(x) > 0$ for all values of x and find the minimum value of $f(x)$. **(2 marks)**

E/P 6 Find all real roots of the following functions:

a $f(x) = x^8 + 6x^4 - 7$ **(2 marks)**

b $f(x) = 6x^{10} - 5x^5 + 1$ **(2 marks)**

c $f(x) = x^{\frac{1}{4}} - 3x^{\frac{1}{2}} + 2$ **(2 marks)**

E/P 7 The function f is defined as $f(x) = 2^{2x} - 6(2^x) + 8, x \in \mathbb{R}$.

a Write $f(x)$ in the form $(2^x - a)(2^x - b)$, where a and b are real constants. **(2 marks)**

b Hence find the two roots of $f(x)$. **(2 marks)**

2.4 Quadratic graphs

1 Sketch graphs of each of the following equations, showing the coordinates of the points where the graph crosses the coordinate axes.

a $y = x^2 + 11x + 18$ **b** $y = 4x^2 - 16$
c $y = -6x^2 + 2x$

Hint Factorise each equation to find the points where $y = 0$. These are the values of x at the points where the graph crosses the x-axis. To find the y-intercept, substitute $x = 0$ into the equation.

2 Find the coordinates of the turning point on each of these graphs:

a $y = (x - 1)^2 + 9$ **b** $y = x^2 + x - 6$
c $y = -x^2 - 13x - 42$

Hint You can find coordinates of the turning point on a quadratic curve by completing the square. The curve with equation $y = (x - a)^2 + b$ will have a turning point at (a, b).

3 Sketch the graphs of the following equations. For each graph, indicate where the graph crosses the coordinate axes, and write down the coordinates of the turning point and the equation of the line of symmetry.

Hint A quadratic graph has a vertical line of symmetry that passes through its turning point.

a $y = x^2 - 6x + 20$ **b** $y = -2x^2 - 5x - 2$ **c** $4x^2 - y = 4x + 3$

4 Sketch the graphs of the following equations. For each graph, indicate where the graph crosses the coordinate axes, leaving your answer in surd form. Write down the coordinates of the turning point and the equation of the line of symmetry.

Hint After drawing your axes, write in all the required coordinate points and then draw a smooth curve through these points.

a $y = x^2 + 7x + 5$ **b** $y = -5x^2 - 12x - 3$ **c** $y = 2x^2 + 7x + 4$

(E) 5 The expression $8x - 7 - x^2$ can be written in the form $q - (x - p)^2$, where p and q are integers.

a Find the value of p and the value of q. **(3 marks)**

b Sketch the curve with equation $y = 8x - 7 - x^2$, showing clearly the coordinates of any points where the curve crosses the coordinate axes. **(3 marks)**

(E) 6 $f(x) = x^2 + 6x + 4, x \in \mathbb{R}$

a Express $f(x)$ in the form $(x + a)^2 + b$, where a and b are constants. **(2 marks)**

The curve C with equation $y = f(x)$ crosses the y-axis at point P and has a minimum point at the point Q.

b Sketch the graph of C, showing the coordinates of points P and Q. **(3 marks)**

c Explain why the equation $f(x) = -6$ has no real solutions. **(1 mark)**

(E/P) 7 $p(x) = 3 - 2x, q(x) = x^2 - 9x - 10, x \in \mathbb{R}$

a Solve the equation $q(x) = 0$. **(2 marks)**

b Sketch the graphs of $y = p(x)$ and $y = q(x)$ on the same set of axes. Label all points where the curves intersect the coordinate axes. **(4 marks)**

(E/P) 8 The graph of $y = ax^2 + bx + c$ has a maximum at $(2, -5)$ and passes through $(3, 0)$. Find the values of a, b and c. **(4 marks)**

2.5 The discriminant

1 State the condition for which the function $f(x) = ax^2 + bx + c$ has:

Hint The discriminant of the quadratic function $f(x) = ax^2 + bx + c$ is $b^2 - 4ac$.

a 2 distinct, real roots **b** 1 repeated root

c no real roots

2 For each of these functions:

i calculate the value of the discriminant

ii write down the number of real roots of the function.

> **Hint** Be careful with negative signs. In part **a**, $a = -2$, $b = -11$ and $c = -12$.

a $f(x) = -2x^2 - 11x - 12$ **b** $f(x) = x^2 + 6x + 9$ **c** $f(x) = 3x^2 - 12x + 18$

3 Calculate the value of the discriminant for each of these functions and match them to the sketch graphs.

> **Hint** A quadratic graph with one repeated real root will have its turning point on the x-axis.

a $f(x) = x^2 - 10x + 21$ **b** $f(x) = -x^2 - 10x - 25$ **c** $f(x) = x^2 + 8x + 19$

i
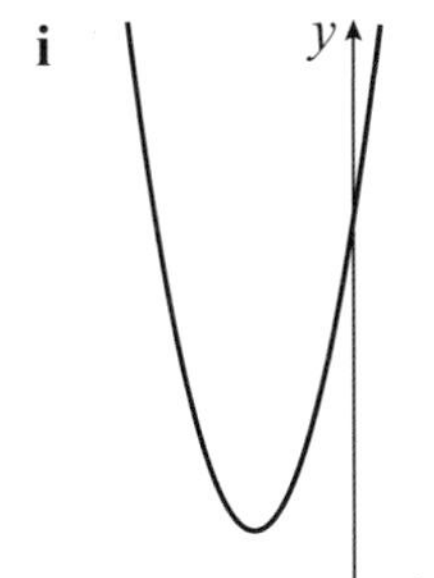

ii
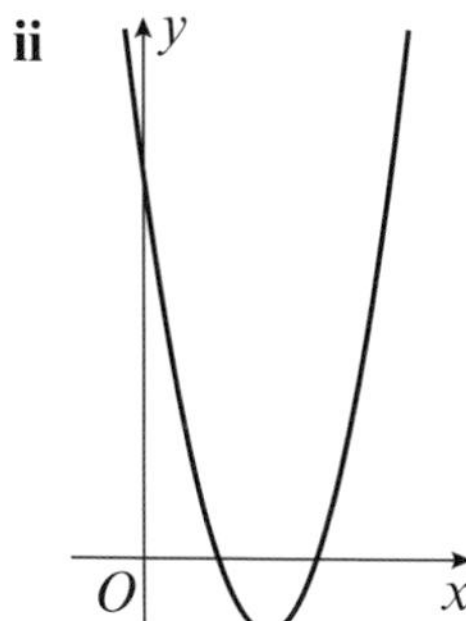

iii
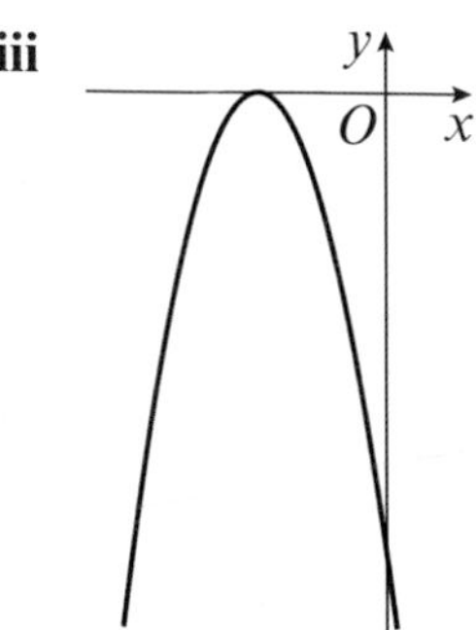

E/P **4** $f(x) = x^2 + kx + 25$, $x \in \mathbb{R}$

a Find the discriminant of $f(x)$ in terms of k. **(2 marks)**

b Given that the equation $f(x) = 0$ has one repeated root, find the possible values of k. **(2 marks)**

E/P **5** $f(x) = x^2 + (k + 3)x + k$, where k is a real constant and $x \in \mathbb{R}$.

a Find the discriminant of $f(x)$ in terms of k. **(2 marks)**

b Show that the discriminant of $f(x)$ can be expressed in the form $(k + a)^2 + b$, where a and b are constants to be found. **(2 marks)**

c Show that, for all values of k, the equation $f(x) = 0$ has distinct real roots. **(2 marks)**

E/P **6** The equation $kx^2 + 3x - 5 = 0$, where k is a constant, has two distinct real roots. Find the range of possible values of k. **(3 marks)**

E/P **7** Find the range of values of p for which the equation $2x^2 + 7x + p = 0$ has no real solutions. **(3 marks)**

2.6 Modelling with quadratics

E **1** A ball is kicked through the air from a level playing field. The path of the ball can be modelled by the equation $y = x - 0.04x^2$, where x is the horizontal distance (in metres) and y is the vertical height (in metres) from the point where it was kicked.

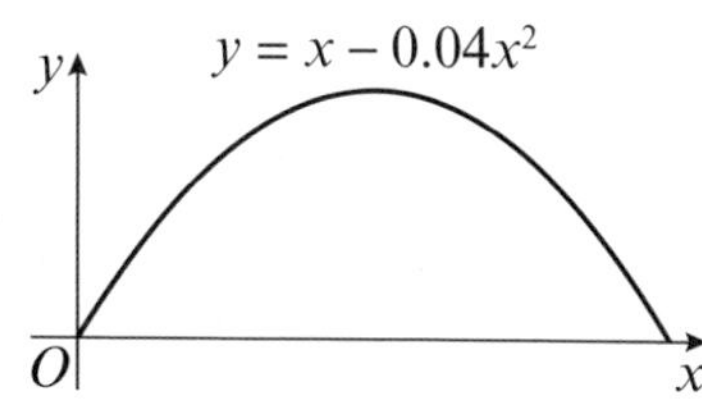

a Find the horizontal distance travelled by the ball from the point where it was kicked to the point where it lands on the playing field again. **(2 marks)**

b Show that the equation for the path of the ball can be written in the form $y = a(x - b)^2 + c$, where a, b and c are constants to be found. **(3 marks)**

c Using your answer to part **b**, or otherwise, find the maximum height the ball reaches above the playing field, and the horizontal distance at which this maximum height is reached. **(2 marks)**

Hint The maximum height of the ball will occur at the turning point of the path.

(E) **2** A company makes a particular type of phone case.
The annual profit made by the company is modelled by the equation

$$P = 73.5x - 5.25x^2 - 107.5$$

where P is the annual profit, measured in thousands of pounds, and x is the selling price of the phone case, in pounds. The company wishes to maximise its annual profit.

a Show that this model can be written in the form $P = A - 5.25(x - B)^2$, where A and B are constants to be determined. **(2 marks)**

Hint For part **a** you need to complete the square.

b Hence, or otherwise, state, according to the model:

i the maximum possible annual profit

ii the selling price of the phone case that maximises the annual profit. **(2 marks)**

(E/P) **3** The diagram shows the cross-section of a canal. O and B are points on the surface of the water, with $OB = 2.5$ m. At point A the canal is 3 m deep. The cross-section is modelled as a quadratic curve OAB. x and y are the horizontal and vertical distances from O in metres.

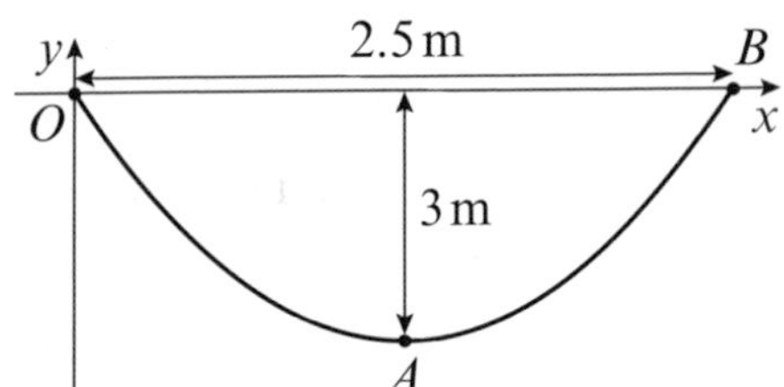

a i Write down the coordinates of point A.

ii Hence find an equation for the curve OAB. **(3 marks)**

The cross-section of a narrowboat is modelled as a rectangle of width 2.7 m, with constant depth below water of 1 m.

b Determine whether or not it is possible for the narrowboat to pass through the canal. **(2 marks)**

(E/P) **4** A car manufacturing company uses a model to determine how the number of workers employed, w, affects the number of cars produced per day, q.
The model suggests $q = 0.4w - 0.00008w^2$.

a Suggest a reason why this model does not contain a constant term. **(1 mark)**

On a particular day, 1500 workers are employed.

b Find the number of cars produced on this day. **(1 mark)**

c According to the model, what is the maximum number of cars that can be produced and how many workers are needed for this? **(3 marks)**

E/P **5** A ball is thrown from the top of a cliff.
The height, h metres, of the ball above ground level after t seconds is modelled by the function
$\text{h}(t) = 120 + 12.25t - 4.9t^2$

a Give a physical interpretation of the meaning of the constant term 120 in the model. **(1 mark)**

b Write $\text{h}(t)$ in the form $A - B(t - C)^2$, where A, B and C are constants to be found. **(3 marks)**

c Using your answer to part **b**, or otherwise, find, with justification:

i the time between the instant the ball is thrown and the instant it reaches ground level

ii the maximum height of the ball above the ground and the time at which this maximum height is reached. **(5 marks)**

6 A company makes a particular type of T-shirt.
The annual profit made by the company is modelled by the equation $P = 125x - 6.25x^2 - 465$, where P is the profit measured in thousands of pounds and x is the selling price of the T-shirt in pounds.
A sketch of P against x is shown in the diagram.

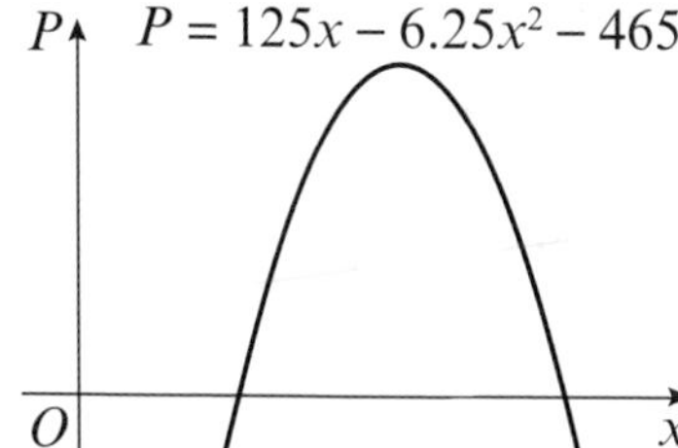

Ellie claims that this model is not valid for a selling price of £4.00, because the value of P is negative.

a Suggest a reason why Ellie may be incorrect. **(1 mark)**

b Write the model in the form $P = c - a(x - b)^2$, where a, b and c are constants to be found. **(3 marks)**

c Given that the company made an annual profit of more than £120 000 find, according to the model, the least possible selling price for the T-shirt. **(3 marks)**

The company wishes to maximise its annual profit.

d State, according to the model:

i the maximum possible annual profit

ii the selling price of the T-shirt that maximises the annual profit. **(2 marks)**

P **7** In microeconomics, the market price and quantity of an item are found by relating two functions.
Inverse supply expresses supply price $£P_s$ in terms of the quantity supplied, q thousands.
Inverse demand expresses demand price $£P_d$ in terms of the quantity demanded, q thousands.
Market equilibrium occurs when $P_s = P_d$
In this model:
Inverse supply is $P_s = 8q + 8$
Inverse demand is $P_d = q^2 - 14q + 48 \quad (0 \leqslant q \leqslant 6)$
Find the market equilibrium quantity and price given by this model.

Problem solving Set A

Bronze

The sketch shows the path of a stone that is kicked through the air from level ground.

The path of the stone can be modelled by the function $h(x) = 2x - x^2$, where x metres is the horizontal distance the stone travels from the place where it was kicked, and h metres is the vertical height of the stone above ground level.

a Write $h(x)$ in the form $h(x) = A - (x - B)^2$, where A and B are constants to be found. **(3 marks)**

b Using your answer to part **a**, or otherwise, solve $h(x) = 0$ and find the horizontal distance the stone has travelled when it lands on the ground. **(3 marks)**

Silver

A stone is thrown from the top of a cliff.

The path of the stone can be modelled by the function $h(x) = 114 + 10.4x - 5.2x^2$, where x metres is the horizontal distance the stone travels, and h metres is the vertical height of the stone above ground level.

a Give a physical interpretation of the meaning of the constant term 114 in the model. **(1 mark)**

b **i** Show that $h(x)$ can be rearranged to give $h(x) = 114 - 5.2(x^2 - 2x)$.

ii Hence, or otherwise, write $h(x)$ in the form $h(x) = A - 5.2(x - B)^2$, where A and B are constants to be found. **(3 marks)**

c Using your answer to part **b ii**, or otherwise, find, with justification:

i the horizontal distance the stone has travelled when it lands on the ground

ii the maximum height of the stone above the ground and the horizontal distance at which this maximum height is reached. **(5 marks)**

Gold

A stone is thrown from the top of a cliff.

The path of the stone can be modelled by the function $h(x) = 125 + 12.75x - 4.5x^2$, where x metres is the horizontal distance the stone travels, and h metres is the vertical height of the stone above ground level.

a Give a physical interpretation of the meaning of the constant term 125 in the model. **(1 mark)**

b Find, with justification, the maximum height of the stone above the ground. **(3 marks)**

c If measured in a straight line, what is the distance from the point where the stone is thrown to the point where it lands on the ground? **(2 marks)**

Problem solving Set B

Bronze

The equation $2kx^2 + 4x + k = 0$, where k is a constant, has one repeated root.

a Show that $16 - 8k^2 = 0$. **(2 marks)**

b Hence, find two possible values of k. **(1 mark)**

Silver

The equation $3x^2 + px + 2p = 0$, where p is a non-zero constant, has equal roots.

Find the value of p. **(3 marks)**

Gold

The equation $\frac{3 - x^2}{x + 2} = q$, where q is a constant, has one repeated real root.

Find two possible values for q. **(4 marks)**

Now try this → **Exam question bank Q38, Q43, Q52, Q67, Q91**

3 Equations and inequalities

3.1 Linear simultaneous equations

1 Solve these simultaneous equations by elimination:

a $6x + y = 9$
$4x - y = 11$

b $2x + 3y = 8$
$3x + 2y = 7$

c $4x - 3y = 2$
$5x - 7y = 9$

Hint Either the x- or y-coefficients in the pair of equations need to have the same value.

You can then add or subtract the equations to eliminate one variable and solve for the other.

You need to find the value of both variables.

2 Solve these simultaneous equations by substitution:

a $5x - 2y = 3$
$x + 4y = 5$

b $2x + 5y = 37$
$y = 11 - 2x$

c $4x + 3y = 5$
$2x - 6y = -5$

Hint Rearrange one equation to make x or y the subject and then substitute this expression into the other equation and solve.

3 Solve these simultaneous equations:

a $3(x - y) + 6 = 0$
$y + x = 8$

b $5(x - 1) = -2y$
$3x - 29 = 4y$

c $\frac{4x - y}{2} = 11$
$\frac{5 - 3x}{5} = y$

Hint Rearrange the equations into the form $ax + by = c$, where a, b and c are constants, and then solve using either elimination or substitution.

(E) 4 Solve the simultaneous equations
$x + y = 2$
$2y = 18x - 6$ **(2 marks)**

(E) 5 Solve the simultaneous equations
$3x - y = -5$
$0.5y + 2x = 4$
giving your answers in exact form. **(2 marks)**

(E/P) 6 $6ky + 9x = 12$
$ky - x = 4.5$
are simultaneous equations where k is a constant.

a Show that $x = -1$. **(2 marks)**

b Given that $y = 7$, find the value of k. **(1 mark)**

(E/P) 7 Two students are attempting to solve the simultaneous equations
$4x + 6y = 10$
$2x = 5 - 3y$
Ben says that these equations have no solutions, and Nisha says that they have infinitely many solutions. Who is correct? Explain your answer. **(2 marks)**

3.2 Quadratic simultaneous equations

1 Solve the simultaneous equations:

a $xy = 64$
$4x - y = 60$

b $x^2 + y^2 = 10$
$x + y = 4$

c $x - y + 1 = 0$
$3x^2 - 4y = 0$

Hint A quadratic equation can contain terms involving xy, x^2 and y^2.

Rearrange the linear equation to make x or y the subject, then substitute this expression into the quadratic equation and solve.

2 Solve the simultaneous equations:

a $y - x^2 + 3x + 2 = 0$
$y - 2x + 6 = 0$

b $x^2 + 2x - y = 14$
$y + 2 = x$

c $x^2 + y^2 = 5$
$y = 5 - 3x$

Hint Each set of equations will have two pairs of solutions.

You need to identify the solutions and pair them up correctly.

3 Solve these simultaneous equations, giving your answers to 2 decimal places:

a $3x - 7 = y$
$x^2 - 3x - 2 = 2y$

b $2x^2 - xy + y^2 = 8$
$x + y = 1$

c $y^2 - 5x^2 = 20$
$4x - 7 = y$

E **4** Solve the simultaneous equations

$x + y = 4$

$4y^2 - x^2 = 12$ **(4 marks)**

E **5** Solve the simultaneous equations

$y + 2x + 1 = 0$

$y^2 + 3x^2 + 2x = 0$ **(4 marks)**

E/P **6**

$mx - y - 2 = 0$

$x^2 - 2x + y^2 - 4y = 4$

where m is a real constant.

Given that these simultaneous equations have exactly one pair of solutions, find the two possible values of m, giving your answers in exact form. **(7 marks)**

E/P **7** The values of x and y satisfy the simultaneous equations

$y - 2x = 8$

$x^2 + 2ky + 4k = 0$

where k is a non-zero constant.

a Show that $x^2 + 4kx + 20k = 0$. **(2 marks)**

b Given that $x^2 + 4kx + 20k = 0$ has equal roots, find the value of k. **(3 marks)**

c For the value of k found in part **b**, solve the simultaneous equations. **(3 marks)**

3.3 Simultaneous equations on graphs

1 Sketch the graphs of each pair of equations on the same set of axes. In each case, find the coordinates of the point of intersection.

a $2x - y = 8$
$4x + 3y = 6$

b $2x - 5y = 5$
$3x - 2y = 13$

Hint You can sketch straight lines in the form $ax + by = c$ quickly by setting $x = 0$ and solving to find where the line crosses the y-axis, and similarly setting $y = 0$ and solving to find the points where the line crosses the x-axis.

2 a On the same set of axes, sketch the curve with equation $y = x^2 - 8x + 15$ and the line with equation $y = 3x - 3$.

Hint First label the points where the curve and the line cross the coordinate axes. Then sketch both graphs.

b Find the coordinates of the points where the line and the curve intersect.

c Verify your solutions to part **b** are correct by substitution.

3 a On the same set of axes, sketch the curve with equation $y = 3x^2 + 8x - 3$ and the line with equation $y = 7x - 2$.

b Find the coordinates of the points of intersection, giving your answers to 2 decimal places.

(E) 4 The diagram shows the curve with equation $y = x^2 - 4x + 16$ and the line with equation $y = x + 12$. The curve intersects the line at points A and B.

Using an appropriate algebraic method, find the coordinates of A and B. **(4 marks)**

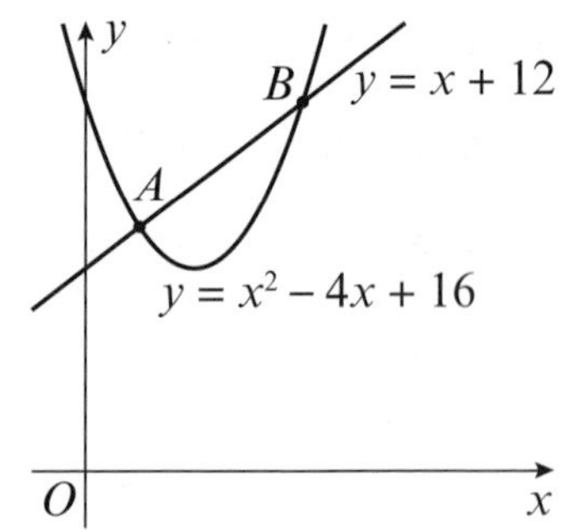

(E) 5 The diagram shows part of curve with equation $y = -x^2 - 2x + 5$ and part of the line with equation $y = -2x + 1$. The curve intersects the line at points A and B.

Using an appropriate algebraic method, find the coordinates of A and B. **(4 marks)**

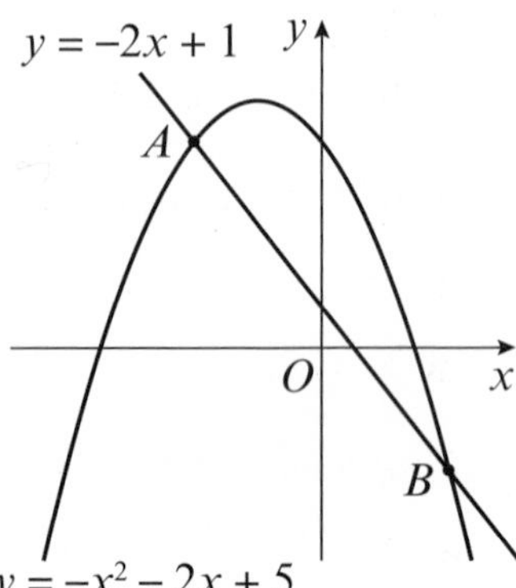

(E/P) 6 $p(x) = 1 - \frac{1}{2}x$, $q(x) = x^2 - 4x - 10$

a Sketch the graphs of $y = p(x)$ and $y = q(x)$ on the same set of axes.
Label all points where the graphs cut the coordinate axes. **(4 marks)**

b Use an algebraic method to find the coordinates of any points of intersection of the graphs $y = p(x)$ and $y = q(x)$. **(4 marks)**

E/P 7 The diagram shows a sketch of the curve C with equation $y = 2x - 3\sqrt{x}$ $(x \geqslant 0)$, and the line l with equation $y = 3x - 12$.

The line cuts the curve at point A as shown in the diagram. Using algebra, find the x-coordinate of point A. **(5 marks)**

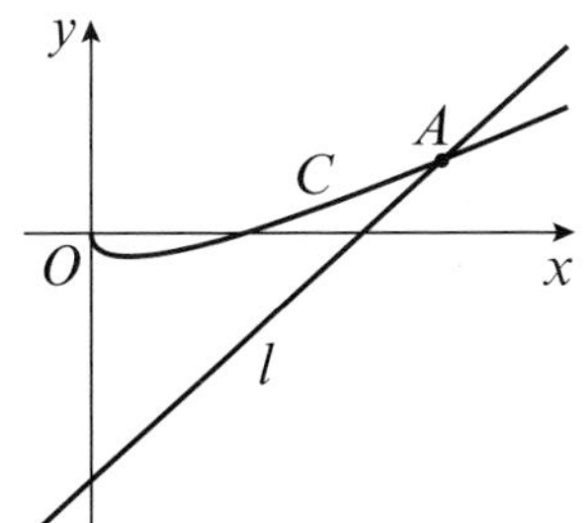

E/P 8 $f(x) = 6 - 5x$ and $g(x) = 4 - 0.5x$

a Use an algebraic method to find the exact coordinates of the point of intersection of the graphs of $y = f(x)$ and $y = g(x)$. **(3 marks)**

b Sketch these graphs on the same set of axes, showing all intersections with the coordinate axes. **(2 marks)**

3.4 Linear inequalities

1 Find the set of values of x for which:

a $x + 4 \geqslant 12$ **b** $7 - 5x < 62$

c $3(3 - 2x) \leqslant 2(3 + 2x)$

Hint You can solve inequalities like equations. However, if you divide both sides by a negative number, you need to reverse the inequality sign.

2 Find the set of values of x for which:

a $7 \leqslant 5x - 3 \leqslant 17$ **b** $\frac{x}{3} - \frac{2x}{9} < 3$

c $3x(3 + x) + x^2 \geqslant 1 + x(6 + 4x)$

Hint For part **c**, expand the brackets and simplify before solving.

3 Use set notation to describe the set of values of x for which:

a $2(1 + x) < 4 - x$ and $x - 3 < 4x + 6$

b $\frac{x}{4} + 3 \leqslant 6$ and $2(3x - 5) \geqslant 20$

c $0.5(4x + 3) < 2.5$ or $\frac{x - 3}{5} > 7$

Hint In set notation $x < 3$ and $x > -1$ is written as $\{x : -1 < x < 3\}$, or alternatively $\{x : x > -1\} \cap \{x : x < 3\}$

E 4 Find the set of values of x for which:

a $2x - 6 > 3 - x$ **(2 marks)**

b $4x - 5 < 3 + 2x$ **(2 marks)**

E 5 Use set notation to write the set of values of x for which:

a $20 - 2x > 15 - 7x$ **(2 marks)**

b $x - 5 \leqslant 3 - 9x$ **(2 marks)**

E 6 Use set notation to write the set of values of x for which:

a $3(3x + 4) \geqslant 2 - x$ **(2 marks)**

b $7(x - 3) < 3(x - 3)$ **(2 marks)**

E/P 7 The width of a rectangular field is x metres, $x > 0$. The length of the pitch is 30 m more than its width. Given that the perimeter of the pitch must be less than 400 m,

a form a linear inequality in x **(2 marks)**

b solve your inequality. **(2 marks)**

E/P 8 Use set notation to describe the set of values of x for which:

a $3(2 + x) \geqslant 2 - x$ and $x + 7 > 5x - 1$ **(4 marks)**

b $0.25(8x + 4) < 5.5$ or $\frac{x-2}{9} > 11$ **(4 marks)**

3.5 Quadratic inequalities

1 Find the set of values of x for which,

a $x^2 < 25$ **b** $x(2x + 1) < 3$

c $0 > 1 - 3x^2 - 2x$ **d** $2x^2 \geqslant 5x$

Hint First, replace the inequality sign by an equals sign and solve the quadratic equation to find the critical values. Then use a sketch of the graph to find the required set of values.

2 Use set notation to describe the set of values of x for which:

a $x^2 - 5x + 14 < 0$ and $5x + 2 > 12$

b $2x^2 - 3x - 5 > 0$ and $x^2 + 3x - 4 < 0$

c $3x^2 + 12x - 15 < 0$ and $x^2 + 5x + 6 > 0$

Hint Draw number lines to illustrate the pairs of inequalities.

3 Given that $x \neq 0$, find the set of values of x for which:

a $36 > \frac{1}{x^2}$ **b** $3 < \frac{5}{x}$ **c** $16 + \frac{6}{x} \geqslant \frac{2}{x}$

Hint Start by multiplying both sides of the inequality by x^2.

E 4 Show that $x^2 - 6x + 11 > 0$ for all real values of x. **(3 marks)**

E/P 5 Find the set of values of x for which:

a $2x - 2 > 7 - x$ **(2 marks)**

b $x^2 - 6x \leqslant 40$ **(4 marks)**

c **both** $2x - 2 > 7 - x$ **and** $x^2 - 6x \leqslant 40$ **(1 marks)**

E/P 6 The equation $x^2 + (k - 2)x + (4 - 2k) = 0$, where k is a constant, has two distinct real roots.

a Show that k satisfies $k^2 + 4k - 12 > 0$ **(3 marks)**

b Find the set of possible values of k. **(4 marks)**

E/P 7 The diagram shows the plan of a park. The edges are straight lines and the lengths shown in the diagram are given in metres.

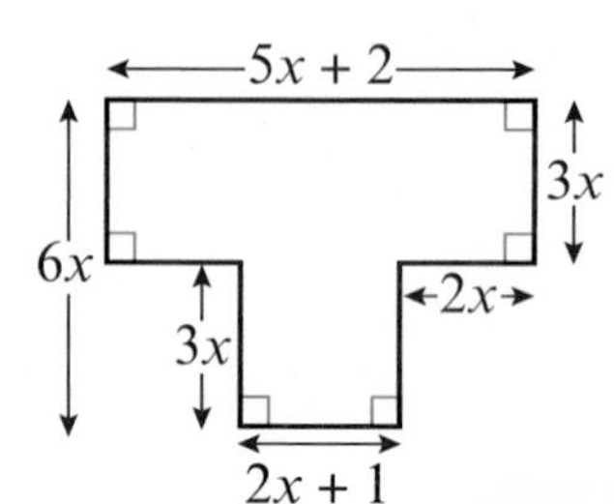

a Given that the perimeter of the park is greater than 103 m, show that $x > 4.5$. **(3 marks)**

b Given that the area of the park is less than 702 m^2, form and solve a quadratic inequality in x. **(5 marks)**

c Hence state the range of the possible values of x. **(1 mark)**

E/P **8** The equation $2x^2 - 3kx + k = 0$ (where k is a constant) has no real roots.
Find the set of possible values of k. **(4 marks)**

E/P **9** The curve C has equation $y = x^2 + 3px + 8p$, where p is a real constant.
The straight line L has equation $y = 1 - 2x$. Given that L does not intersect C, find the set of possible values of p, giving your answers in exact form. **(8 marks)**

E/P **10** Given that $x \neq 1$, find the set of values for which $\frac{3}{x-1} < 2$ **(6 marks)**

3.6 Inequalities on graphs

1 For each pair of functions,
 i sketch the graphs of $y = f(x)$ and $y = g(x)$ on the same set of axes and find the coordinates of the point of intersection
 ii write down the solution to the inequality $f(x) < g(x)$.

Hint The values of x for which the curve $y = f(x)$ is below the curve $y = g(x)$ satisfy the inequality $f(x) < g(x)$.

a $f(x) = 8 - 3x$
 $g(x) = 7x - 2$

b $f(x) = 2x - 3$
 $g(x) = 5x + 11$

c $f(x) = 6 - 2x$
 $g(x) = 9 - 5x$

2 For each pair of functions,
 i sketch the graphs of $y = f(x)$ and $y = g(x)$ on the same set of axes and find the coordinates of the points of intersection
 ii write down the solutions to the inequality $f(x) \leqslant g(x)$.

a $f(x) = x^2 + 6$
 $g(x) = -2x + 9$

b $f(x) = 4 - x^2$
 $g(x) = 3x + 4$

c $f(x) = x^2 + x - 12$
 $g(x) = -x - 9$

Hint Set $f(x) = g(x)$, rearrange into a standard quadratic form and then solve to find the points of intersection.

3 For each pair of functions,
 i sketch the graphs of $y = f(x)$ and $y = g(x)$ on the same set of axes and find the coordinates of the points of intersection
 ii use set notation to write down the solutions to the inequality $f(x) > g(x)$.

a $f(x) = 2x^2 + 2x$
 $g(x) = 5x + 2$

b $f(x) = -x^2 + 5x + 6$
 $g(x) = 3x + 6$

c $f(x) = x^2 + 3x + 2$
 $g(x) = 2 - x$

Hint The values of x for which the curve $y = f(x)$ is above the curve $y = g(x)$ satisfy the inequality $f(x) > g(x)$.

E **4** **a** Sketch the graphs of $f(x) = 3x - 7$ and $g(x) = 3 - x$ on the same set of axes. **(2 marks)**

b Find the coordinates of any points of intersection. **(2 marks)**

c Find the set of values of x for which $3x - 7 > 3 - x$. **(1 mark)**

E/P **5** **a** Sketch the graphs of $f(x) = 8x - 12 - x^2$ and $g(x) = 2x - 7$ on the same set of axes. Label all points where the curves cut the coordinate axes. **(4 marks)**

b Find the coordinates of any points of intersection. **(4 marks)**

c Find the set of values of x for which $8x - 12 - x^2 > 2x - 7$ **(1 mark)**

E/P **6** $p(x) = \frac{7}{2}x - \frac{31}{2}$, $q(x) = x^2 - 3x - 10$

a Solve the equation $q(x) = 0$. **(2 marks)**

b Sketch the graphs of $y = p(x)$ and $y = q(x)$ on the same set of axes. Label all points where the curves cut the coordinate axes. **(4 marks)**

c Use an algebraic method to find the coordinates of any points of intersection of the graphs $y = p(x)$ and $y = q(x)$. **(4 marks)**

d Write down, using set notation, the set of values of x for which $p(x) < q(x)$. **(2 marks)**

3.7 Regions

1 **a** On a coordinate grid, shade the region satisfying the inequalities

$y \geqslant -3$, $x \geqslant 0.5$ and $y \leqslant -2x + 6$

b Find the points of intersection of the lines

$y = -3$, $x = 0.5$ and $y = -2x + 6$

c Find the area of the shaded region.

Hint If $y \geqslant f(x)$ or $y \leqslant f(x)$, the graph of the function $y = f(x)$ is included in the region and is represented by a solid line.

You can substitute a coordinate point into an inequality to test whether the point lies within the required region.

P **2** **a** On a coordinate grid, shade the region that satisfies the inequalities

$y < -(x + 3)(x - 4)$, $y \leqslant x + 8$ and $y > -x + 4$

b Find the coordinates of the vertices of the shaded region.

c Briefly explain why none of these vertices lies within the shaded region.

Hint If $y > f(x)$ or $y < f(x)$, the graph of the function $y = f(x)$ is not included in the region and is represented by a dotted line.

E/P **3** The sketch shows graphs of the equations

$y = x^2 + 3x + 2$, $y = 2x^2 + 4x + 2$ and $y = x + 2$

a Find the coordinates of the points of intersection of the graphs. **(5 marks)**

b Write down a set of inequalities that define the shaded region. **(3 marks)**

c State which, if any, of the points of intersection lie within the region represented by the inequalities. **(1 mark)**

Hint Vertices lie within the region only if both intersecting lines are included. Therefore, a vertex must be between two solid lines to be considered within the region.

E/P **4** **a** On a coordinate grid, shade the region comprising all points whose coordinates satisfy the inequalities $y \leqslant 3x + 6$, $2y + x \leqslant 5$ and $y \geqslant \frac{3}{2}$ **(3 marks)**

b Work out the area of the shaded region. **(4 marks)**

E/P **5** **a** On a coordinate grid, shade the region comprising all points whose coordinates satisfy the inequalities $y \leqslant 2x + 6$, $4y + 17x \leqslant 4$ and $y \geqslant 1$. **(3 marks)**

b Work out the area of the shaded region. **(4 marks)**

E/P **6** **a** On a coordinate grid, sketch the graphs of the equations
$y = x^2 + 3x - 4$, $y = x^2 - x - 2$ and $x = 1$

Label the coordinates of all points where the graphs intersect the coordinate axes. **(6 marks)**

b Use an algebraic method to find the coordinates of any point of intersection of the graphs. **(6 marks)**

c On your sketch, shade the region that satisfies the inequalities
$y \leqslant x^2 + 3x - 4$, $y \leqslant x^2 - x - 2$ and $x \leqslant 1$ **(1 mark)**

E/P **7** The diagram shows a sketch of the curve C with equation $y = 2\sqrt{x} - 3$, $x \geqslant 0$, and the line l with equation $y = 10 - 1.28x$

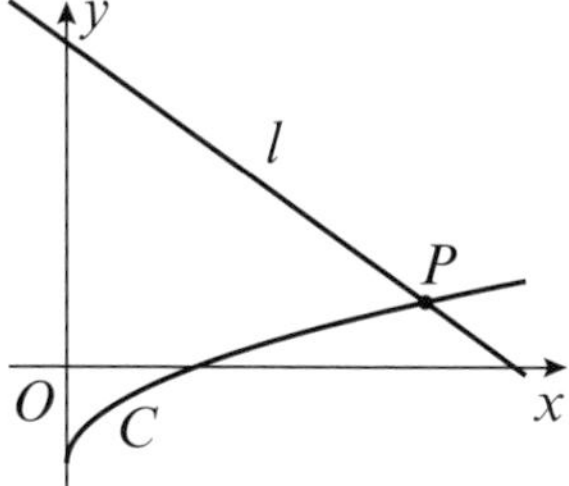

The line cuts the curve at point P.

a Using algebra, find the x-coordinate of point P. **(5 marks)**

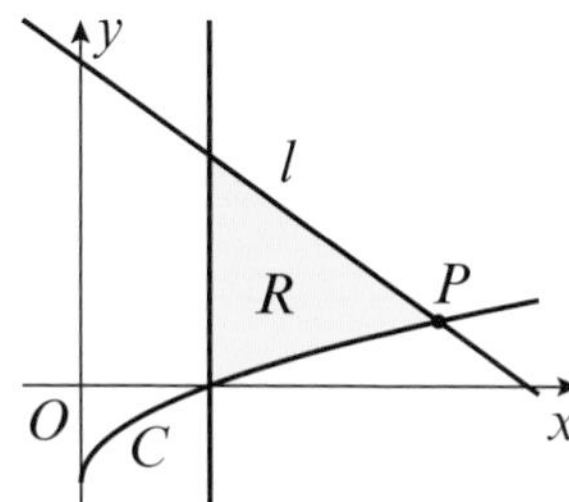

The region R is enclosed by curve C, the line l, and the vertical line through the point where C meets the x-axis.

b Identify the inequalities that define R.
The boundary lines are included in the region. **(3 marks)**

Problem solving Set A

Bronze

Given the simultaneous equations

$y = 2x + 3$

$x^2 - y + 2k = 0$

where k is a non-zero constant,

a show that $x^2 - 2x + (2k - 3) = 0$. **(2 marks)**

Given that $x^2 - 2x + (2k - 3)$ has equal roots,

b use the discriminant to form an equation, and solve this to find the value of k. **(3 marks)**

c For this value of k, solve the simultaneous equations. **(3 marks)**

Silver

Given the simultaneous equations

$y + \frac{1}{4} = 4x$

$x^2 - 4ky + 3k = 0$

where k is a non-zero constant,

a show that $x^2 - 16kx + 4k = 0$. **(2 marks)**

Given that $x^2 - 16kx + 4k$ has equal roots,

b find the solution of the simultaneous equations for this value of k. **(6 marks)**

Gold

The simultaneous equations

$y + \frac{1}{2} = 5x$

$x^2 - 4ky + 3k = 0$

where k is a non-zero constant, have exactly one pair of solutions.

Find the solution. **(8 marks)**

Problem solving Set B

Bronze

$p(x) = x + 6$ and $q(x) = x^2 + 5x + 6$

a Solve the equation $q(x) = 0$. **(2 marks)**

b Sketch the graphs of $y = p(x)$ and $y = q(x)$ on the same set of axes.
Label all points where the curves intersect the coordinate axes. **(4 marks)**

c Use an algebraic method to find the coordinates of any points of intersection of the graphs $y = p(x)$ and $y = q(x)$. **(4 marks)**

d Write the set of values of x for which $p(x) < q(x)$. **(2 marks)**

Silver

$p(x) = -x - 1.75$ and $q(x) = x^2 + 3x - 18$

a Sketch the graphs of $y = p(x)$ and $y = q(x)$ on the same set of axes. Label all points where the curves intersect the coordinate axes. **(6 marks)**

b Use an algebraic method to find the coordinates of any points of intersection of the graphs $y = p(x)$ and $y = q(x)$. **(4 marks)**

c Write the set of values of x for which $p(x) > q(x)$. **(2 marks)**

d Shade the region that satisfies the inequalities:

$p(x) < y < q(x), \quad x > 3$ **(2 marks)**

Gold

$p(x) = 3 + 2x - x^2$, $q(x) = 3x^2 + 2x - 1$

a Sketch the graphs of $y = p(x)$ and $y = q(x)$ on the same set of axes.
Label all points where the graphs intersect each other and the coordinate axes. **(10 marks)**

b Write down, using set notation, the set of values of x for which $p(x) < q(x)$. **(2 marks)**

c Shade the region on your sketch which satisfies the inequalities
$y \geqslant p(x)$, $y \geqslant q(x)$ and $x > -1.5$ **(3 marks)**

Now try this → **Exam question bank Q10, Q13, Q23, Q32, Q35, Q48, Q55, Q62, Q65, Q71**

4 Graphs and transformations

4.1 Cubic graphs

1 Sketch the following curves and indicate clearly any points of intersection with the axes.

Hint Set each factor equal to 0 and solve to find the x-coordinates of the points where the graph crosses the x-axis. Substitute $x = 0$ into the equation to find the y-intercept.

a $y = (x + 1)(x - 2)(x + 3)$

b $y = x(2x - 1)(2x + 1)$

c $y = (x - 2)(x + 4)^2$

2 Factorise each equation fully and sketch the curve:

a $y = x^3 + 2x^2 - 3x$

b $y = -x^3 - 2x^2 + 3x$

c $y = 6x^3 - 3x^2$

Hint One of the factors will be x, so the curve will pass through the origin.

3 Sketch the following curves and indicate clearly any points of intersection with the axes.

a $y = (x - 2)(x^2 + x + 3)$

b $y = (x - 3)^3$

c $y = -(x - 4)^3$

Hint The curve given by the equation $y = (x - a)^3$ is obtained when the graph of $y = x^3$ is translated a units to the right.

(E) 4 **a** Factorise completely $x^3 + 8x^2 + 16x$. **(2 marks)**

b Sketch the curve with equation $y = x^3 + 8x^2 + 16x$, showing the coordinates of the points at which the curve cuts or touches the x-axis. **(2 marks)**

(E) 5 **a** Factorise completely $25x - 4x^3$. **(3 marks)**

b Sketch the curve C with equation $y = 25x - 4x^3$.
Show on your sketch the coordinates at which the curve meets the x-axis. **(3 marks)**

(E/P) 6 The diagram shows a graph with equation $y = ax^3 + bx^2 + cx + d$, where a, b, c and d are real constants.

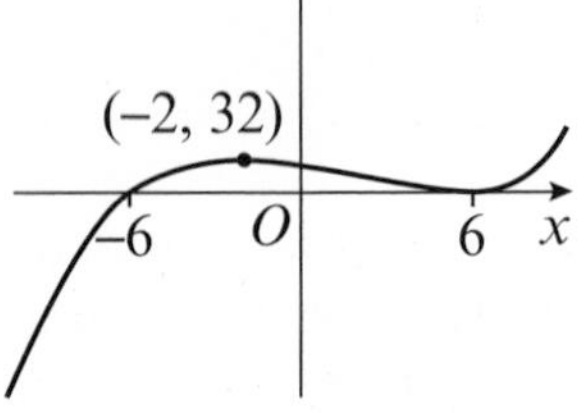

The graph passes through the points (–6, 0) and (–2, 32) and touches the x-axis at the point (6, 0).
A student attempts to find the equation of the curve, and writes the following working:

$y = (x + 6)(x - 6)^2$
$y = (x + 6)(x^2 - 12x + 36)$
$y = x^3 - 6x^2 - 36x + 216$

a Explain the mistake the student has made. **(1 mark)**

b Find the correct equation of the curve. **(4 marks)**

 **7** The diagram shows a sketch of the curve with equation $y = \mathrm{f}(x)$.

The curve touches the x-axis at the point $(1, 0)$ and passes through the point $(4, 0)$.

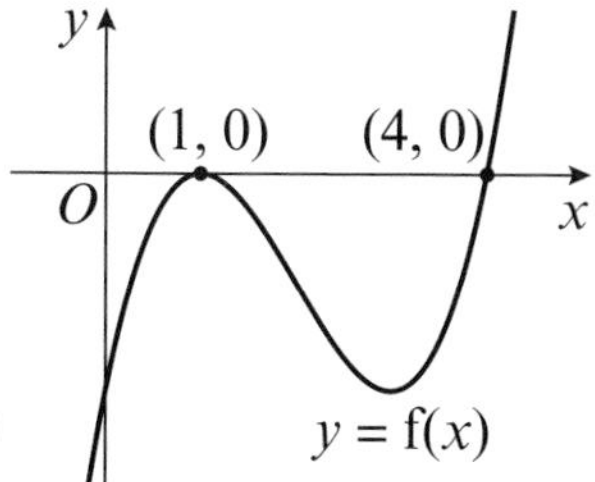

a The equation of the curve can be written in the form $y = x^3 - ax^2 + bx - c$, where a, b and c are positive integers. Calculate the values of a, b and c. **(4 marks)**

b Write down the coordinates of the point where the graph crosses the y-axis. **(1 mark)**

E/P **8** The diagram shows the graph of $y = ax^3 - bx^2 + cx + d$, where a, b, c and d are real constants.

Find the values of a, b, c and d. **(4 marks)**

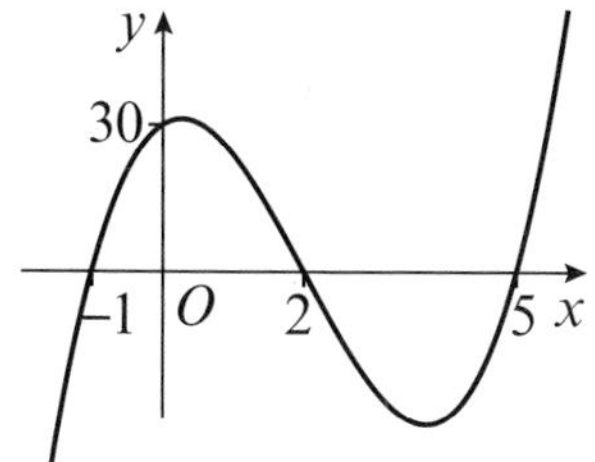

4.2 Quartic graphs

1 Sketch the following curves and indicate clearly any points of intersection with the axes.

a $y = (x + 1)(x + 3)(x - 1)(x - 3)$

b $y = -(x - 2)^2(x - 5)^2$

c $y = (2x + 1)^3(x - 3)$

Hint A quartic graph can be written in the form $y = ax^4 + bx^3 + cx^2 + dx + e$, where a, b, c, d and e are real constants and $a \neq 0$. If the right-hand side of the equation is factorised, you can determine any points of intersection with the x-axis by setting each factor equal to 0.

2 Sketch the curves and indicate clearly any points of intersection with the axes.

a $y = (x^2 - x - 2)(x^2 + x - 12)$

b $y = (x - 5)^2(x^2 - 3x + 2)$

c $y = (2x + 5)^4$

Hint Fully factorise the equations first. In part **c**, the curve only touches the x-axis at one point.

3 The graph of $y = x^4 + x^3 - cx^2 - dx + e$ is shown in the diagram, where c, d and e are real constants. The curve intersects the y-axis at the point P.

a Find the coordinates of point P.

b Find the values of c, d and e.

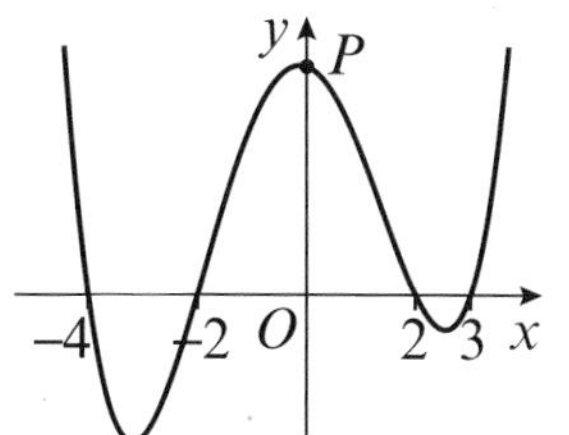

Hint Write the equation as a product of its factors, then multiply out the brackets to find the coefficients of each term.

 **4** Sketch the graph of $y = (x + 3)(x - 4)(x^2 + 7x + 10)$. **(3 marks)**

E/P **5** Sketch the graph of $y = -(x + 2)^2(x^2 - 4x + 3)$. **(3 marks)**

(E/P) 6 The graph of $y = -x^4 - bx^3 + cx^2 + dx - e$ is shown in the diagram, where b, c, d and e are real constants. The curve intersects the y-axis at the point P.

a Find the coordinates of P. **(2 marks)**

b Find the values of b, c, d and e. **(3 marks)**

(E/P) 7 The graph of $y = x^4 - bx^3 - cx^2 + dx + e$ is shown in the diagram, where b, c, d and e are real constants. The curve intersects the y-axis at the point P.

a Find the coordinates of P. **(2 marks)**

b Find the values of b, c, d and e. **(3 marks)**

4.3 Reciprocal graphs

1 Sketch on the same diagram:

a $y = \frac{2}{x}$ **b** $y = \frac{6}{x}$

Hint These are graphs of the type $y = \frac{k}{x}$ with $k > 0$. The curves will appear in the first and third quadrants, and the greater the value of k, the further the curve will be from the origin.

2 Sketch on the same diagram:

a $y = -\frac{3}{x}$ **b** $y = -\frac{7}{x}$

Hint These are graphs of the type $y = \frac{k}{x}$ with $k < 0$. The curves will appear in the second and fourth quadrants, and the smaller the value of k, the further the curve will be from the origin.

3 Sketch on the same diagram:

a $y = \frac{5}{x^2}$ **b** $y = \frac{10}{x^2}$

Hint These are graphs of the type $y = \frac{k}{x^2}$ with $k > 0$. The y-values are always positive, so the curves will appear in the first and second quadrants.

4 Sketch on the same diagram:

a $y = -\frac{1}{x^2}$ **b** $y = -\frac{9}{x^2}$

Hint These are graphs of the type $y = \frac{k}{x^2}$ with $k < 0$. The y-values are always negative, so the curves will appear in the third and fourth quadrants.

(E) 5 **a** On the same set of axes, sketch the graphs of $y = \frac{1}{x}$ and $y = \frac{4}{x^2}$ **(4 marks)**

b State the equations of the asymptotes. **(2 marks)**

(E) 6 **a** On the same set of axes, sketch the graphs of $y = -\frac{3}{x}$ and $y = -\frac{15}{x^2}$ **(4 marks)**

b State the equations of the asymptotes. **(2 marks)**

4.4 Points of intersection

1 In each case:
 i sketch the two curves on the same axes
 ii find the coordinates of the points of intersection.

Hint If the coordinates of the points of intersection are not clear from your sketch, set the equations equal and solve to find the x-coordinates. Then substitute these values to find the y-coordinates.

 a $y = x(x - 4)$ and $y = x^2(3 - x)$

 b $y = x^2 + x - 2$ and $y = x^3 + x^2 - 2x$

2 a On the same set of axes, sketch the curves with equations $y = x^3$ and $y = \frac{5}{x^2}$

 b Show that the x-coordinates at the points of intersection of these curves satisfy the equation $x^5 - 5 = 0$.

Hint For part **b**, set the equations equal to each other and rearrange to equal 0.

 c State the number of real solutions to this equation.

3 a On the same set of axes, sketch the curves with equations $y = -\frac{3}{x^2}$ and $y = -x^2(2x + 5)$

 b Show that the x-coordinates at the points of intersection of these curves satisfy the equation $x^4(2x + 5) - 3 = 0$.

 c State the number of real solutions to this equation.

Hint The number of real solutions is the same as the number of times the two curves intersect.

4 In each case:
 i sketch the curves with equations $y = \mathrm{f}(x)$ and $y = \mathrm{g}(x)$ on the same set of axes
 ii state, giving a reason, the number of real solutions to the equation $\mathrm{f}(x) = \mathrm{g}(x)$.

Hint The number of points of intersection of the curves $y = \mathrm{f}(x)$ and $y = \mathrm{g}(x)$ will be the same as the number of real solutions to the equation $\mathrm{f}(x) = \mathrm{g}(x)$.

 a $\mathrm{f}(x) = -\frac{1}{x}$ and $\mathrm{g}(x) = (x + 1)(x - 3)^2$ b $\mathrm{f}(x) = \frac{7}{x^2}$ and $\mathrm{g}(x) = x^2(4x - 9)$

E/P 5 The curve C has equation $y = \frac{2}{x} - 5$, $x \neq 0$, and the line l has equation $y = 2x - 5$.
Find the coordinates of the points of intersection of $y = \frac{2}{x} - 5$ and $y = 2x - 5$. **(5 marks)**

E/P 6 a On the same set of axes, sketch the graphs of $y = x^2(x^2 - x - 2)$ and $y = -\frac{1}{4}x^3$ **(5 marks)**

 b Write down the number of real solutions to the equation $-4x^2(x^2 - x - 2) = x^3$ **(1 mark)**

 c Show that the equation $-4x^2(x^2 - x - 2) = x^3$ can be rearranged to give $x^2(4x^2 - 3x - 8) = 0$ **(2 marks)**

 d Hence determine the x-coordinates of the points of intersection, giving your answers exactly. **(3 marks)**

E/P **7** The diagram shows a sketch of the curve C with equation $y = \frac{1}{x} + 2, x \neq 0$.

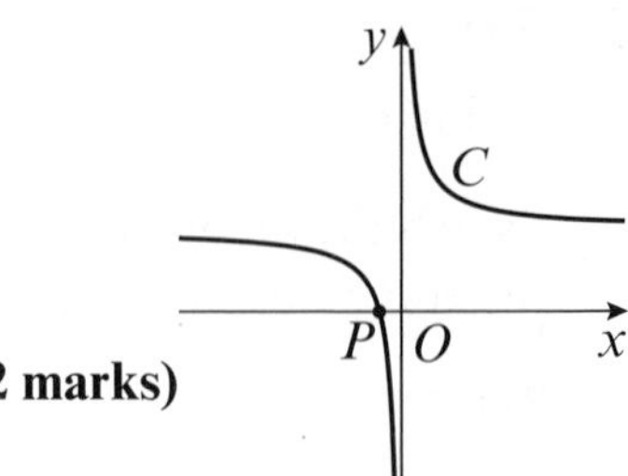

The curve crosses the x-axis at the point P.

a i State the x-coordinate of the point P.

ii State the equations of the asymptotes of the curve. **(2 marks)**

The curve D has equation $y = x^2 + 2$ for all real values of x.

b On the same set of axes, copy the graph of curve C and sketch the graph of curve D. Mark clearly where curve D crosses the y-axis. **(3 marks)**

c Find the coordinates of the point(s) of intersection of the graphs. **(2 marks)**

E/P **8** **a** On the same set of axes, sketch the graphs of:

i $y = x(x + 3)(2 - x)$ **ii** $y = -\frac{4}{x}$

In each case, show clearly the coordinates of all the points where the curves cross the coordinate axes. **(6 marks)**

b Using your sketch, state, giving a reason, the number of real solutions to the equation $x(x + 3)(2 - x) + \frac{4}{x} = 0$ **(2 marks)**

E/P **9** **a** Sketch and clearly label the graphs of $y = \frac{2}{x}$ and $y = 4x - 2$ on a single diagram. **(4 marks)**

b Find the coordinates of the points of intersection of $y = \frac{2}{x}$ and $y = 4x - 2$. **(4 marks)**

E/P **10** Find the coordinates of any points of intersection between the curve with equation $y = x^2 - 10x - 15$ and the straight line with equation $y = 10 - 10x$. **(4 marks)**

E/P **11** Two functions f(x) and g(x) are given by $f(x) = x^2 - 9x + 24$ and $g(x) = 2x + a$, where a is a real constant.

a Given that the graphs of $y = f(x)$ and $y = g(x)$ intersect at two distinct points, find the range of possible values of a. **(4 marks)**

The diagram shows the curves $y = f(x)$ and $y = g(x)$ in the case where $a = 5$.

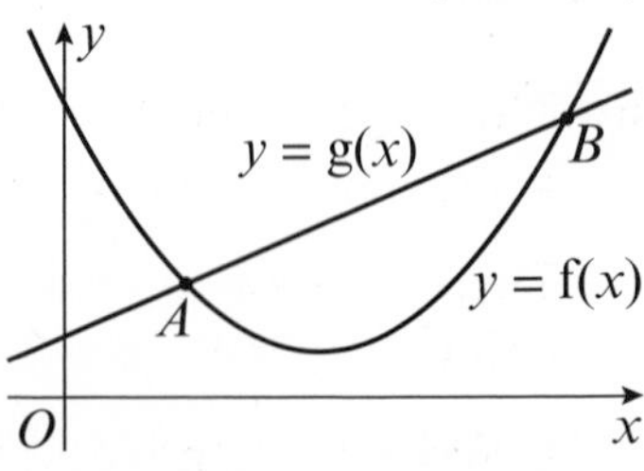

b Using an appropriate algebraic method, find the coordinates of points A and B.
Give your answers correct to 3 significant figures. **(3 marks)**

4.5 Translating graphs

1 Given that:

a $f(x) = x^2$ **b** $f(x) = x^3$ **c** $f(x) = \frac{1}{x}$

sketch each of the following graphs, stating any points where the curve crosses or touches the coordinate axes, and the equations of any asymptotes.

> **Hint** The graph of $y = f(x) + a$ is a translation of the graph $y = f(x)$ by the vector $\begin{pmatrix} 0 \\ a \end{pmatrix}$.
> The graph of $y = f(x + a)$ is a translation of the graph $y = f(x)$ by the vector $\begin{pmatrix} -a \\ 0 \end{pmatrix}$.

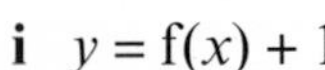

i $y = f(x) + 1$ **ii** $y = f(x - 2)$ **iii** $y = f(x + 3)$ **iv** $y = f(x) - 4$

2 $f(x) = (x + 3)^2(x + 1)(x - 2)$, $g(x) = x(x + 3)$ and $h(x) = \frac{1}{x}$

Hint Asymptotes will be translated by the same vector as the curve.

Sketch the following graphs, labelling any points where the curve crosses the axes, and stating the equations of any asymptotes.

a $y = f(x - 3)$ **b** $y = g\left(x + \frac{1}{2}\right)$ **c** $y = h(x) - 5$

3 **a** Sketch the graph of $y = f(x)$ where $f(x) = x^2(x - 2)(x + 3)$

b Sketch the curve with equation $y = f(x - 2)$

Hint Write the equation of the transformed function and substitute $x = 0$ into the equation to find the y-intercept of the curve.

E/P **4** It is given that $f(x) = 3x^3 - 9x - 6$.
The diagram shows a sketch of part of the curve with equation $y = f(x)$.

On a separate set of axes, sketch the following curves showing the new coordinates of the images of points A, B, C and D.

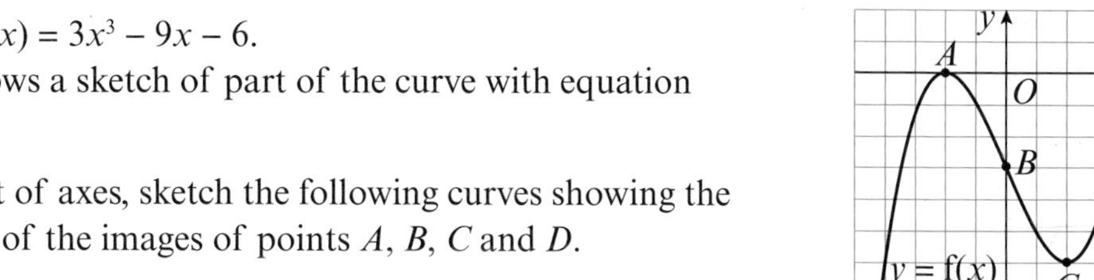

a $y = f(x) + 4$ **(2 marks)**

b $y = f(x - 3)$ **(2 marks)**

E/P **5** The diagram shows a sketch of the curve with equation $y = f(x)$.

The curve crosses the x-axis at (2, 0), touches it at (–4, 0) and crosses the y-axis at (0, –32).

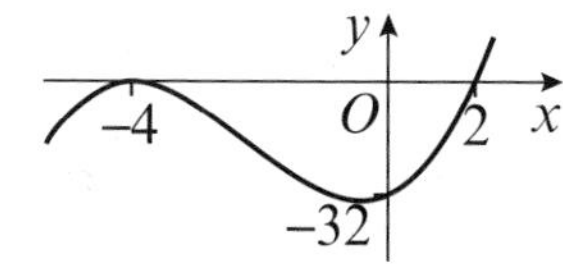

a Write down the equation of the curve in the form $y = (x + a)^2(x - b)$ where a and b are integers. **(2 marks)**

b Sketch the curve with equation $y = f(x - 2)$, clearly showing the coordinates of the points where the curve intersects with the axes. **(4 marks)**

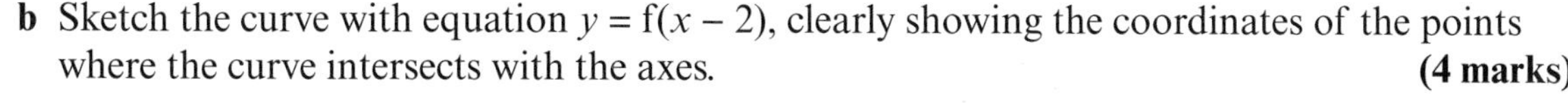

E/P **6** The diagram shows a sketch of the curve C with equation $y = f(x)$, where $f(x) = x^2(6 - 2x)$.

There is a local minimum at the origin, a local maximum at the point (2, 8) and C cuts the x-axis at the point A.

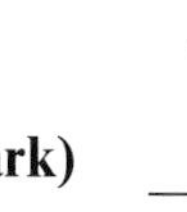

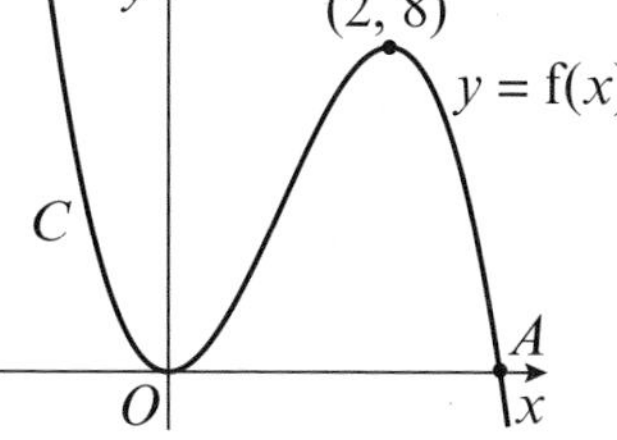

a Write down the coordinates of the point A. **(1 mark)**

The curve with equation $y = f(x) - k$, where k is a constant, has a local maximum at (2, 1).

b Write down the value of k. **(1 mark)**

E/P **7** The diagram shows the curve with equation $y = f(x)$, where $f(x) = (x + 3)(x - 2)^2$

Given that the graph of $y = f(x + a)$ passes through the origin, find the possible values of a. **(2 marks)**

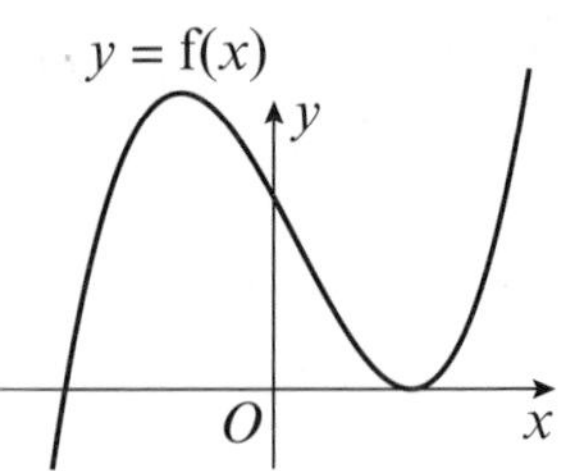

E/P **8** The sketch shows a cubic curve with equation $y = f(x)$.

The curve crosses the x-axis at P and has turning points at $Q(1, 3)$ and $R(-1, 7)$.

Given that the equation $f(x) + a = 0$, where a is a real constant, has three distinct real roots, write down the range of possible values of a.

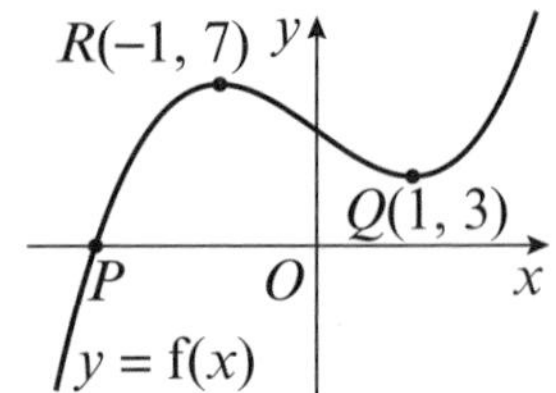

(2 marks)

4.6 Stretching graphs

1 Given that:

a $f(x) = (x - 1)^2$ **b** $f(x) = x^3 + 1$

c $f(x) = \dfrac{1}{x - 1}$

sketch each of the following graphs, stating any points where the curve crosses or touches the coordinate axes, and the equations of any asymptotes.

i $y = 3f(x)$ **ii** $y = f\left(\frac{1}{2}x\right)$ **iii** $y = -f(x)$ **iv** $y = f(-x)$

Hint The graph of $y = af(x)$ is a stretch of the graph $y = f(x)$ by a scale factor of a in the vertical direction.

The graph of $y = f(ax)$ is a stretch of the graph $y = f(x)$ by a scale factor of $\frac{1}{a}$ in the horizontal direction.

2 $f(x) = x^2 - 16$

a Sketch the curve $y = f(x)$.

b Sketch these graphs and find the equations of the curves in terms of x.

i $y = f\left(\frac{1}{2}x\right)$ **ii** $y = \frac{1}{4}f(x)$ **iii** $y = -f(x)$

c Use the equations found in part **b** to find the coordinates of the y-intercept of each curve.

Hint First find the equation of the transformed function.

Substitute $x = 0$ to find the point of intersection of the curve with the y-axis.

3 **a** Sketch the curve C with equation $y = x^2(x - 5)$.

b On the same set of axes, sketch the curve D with equation $y = (4x)^2(4x - 5)$.

c Describe the transformation from curve C to curve D.

Hint To write the transformation in terms of $f(x)$, determine what x has been multiplied by to obtain the transformed equation.

E **4** The diagram shows a sketch of part of the curve with equation $y = \mathrm{f}(x)$.

On a separate set of axes, sketch the curve with equation $y = \mathrm{f}\left(\frac{1}{2}x\right)$, showing the location and coordinates of the images of points A, B, C and D. **(2 marks)**

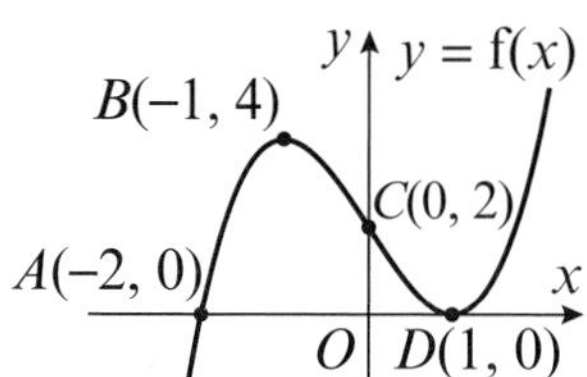

E **5** The diagram shows a sketch of part of the curve with equation $y = \mathrm{f}(x)$.

The curve has a local maximum at $A(2, 6)$, a local minimum at $B(-3, -8)$ and passes through the origin O. On separate diagrams, sketch the curves with equations:

a $y = \mathrm{f}\left(\frac{1}{2}x\right)$ **(2 marks)**

b $y = \mathrm{f}(-x)$ **(2 marks)**

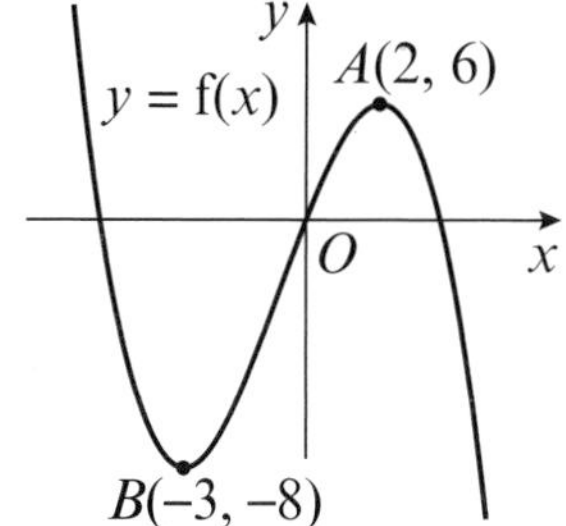

On each diagram, show clearly the locations of the y-intercept, and any local maximum and local minimum points.

E **6** The diagram shows the curve with equation $y = \mathrm{f}(x)$ where $\mathrm{f}(x) = (x + 5)^2(x - 1)$, $x \in \mathbb{R}$. The curve crosses the x-axis at $(1, 0)$, touches it at $(-5, 0)$ and crosses the y-axis at -25.

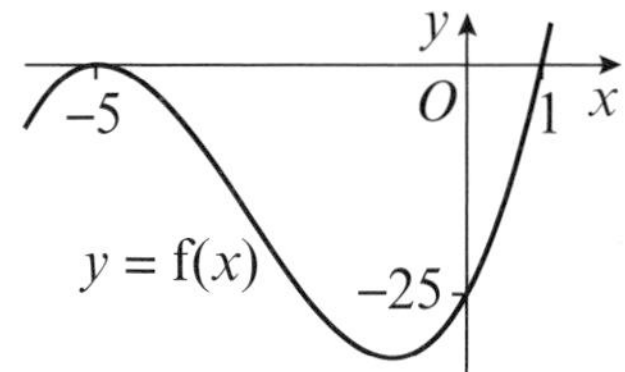

a Sketch the curve C with equation $y = 3\mathrm{f}(x)$ and state the coordinates of the points where the curve C meets the axes. **(3 marks)**

b Write down an equation of the curve C. **(1 mark)**

E/P **7** The diagram shows the graph of the curve $y = \mathrm{f}(x)$.

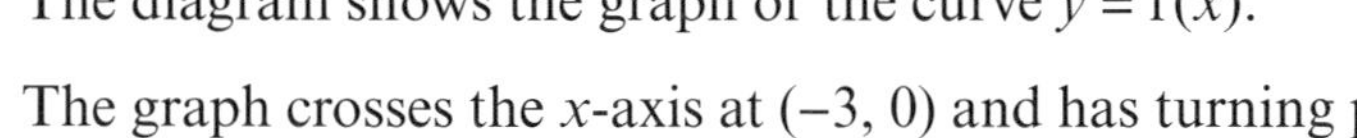

The graph crosses the x-axis at $(-3, 0)$ and has turning points at $(-2, 12)$ and $(0, 0)$.

a The graph of $y = \mathrm{f}(ax)$ has a turning point at $\left(-\frac{1}{4}, 12\right)$. Find the value of a. **(1 mark)**

b The graph of $y = b\mathrm{f}(x)$ has a turning point at $(-2, 6)$. Find the value of b. **(1 mark)**

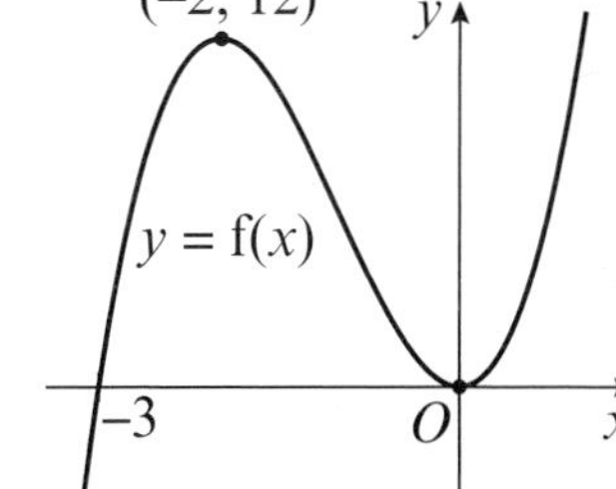

4.7 Transforming functions

1 The diagram shows a sketch of the curve $y = \mathrm{f}(x)$.

The points $A(-1, 0)$, $B(0, 4)$, $C(2, 0)$ and $D(3, 4)$ lie on the curve.

Sketch the following graphs, showing the new coordinates of A, B, C and D:

a $y = \mathrm{f}(x - 2)$ **b** $y = \mathrm{f}(x) + 3$ **c** $y = \mathrm{f}(2x)$

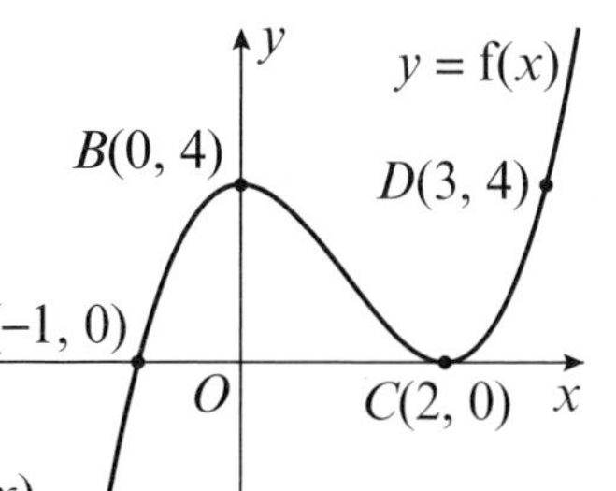

Hint Translate or stretch the points A, B, C and D and then sketch the curve through these points.

2 The diagram shows the curve $y = f(x)$ which crosses the axes at $(-2, 0)$ and $(0, 2)$, and has asymptotes at $y = 1$ and $x = -1$.

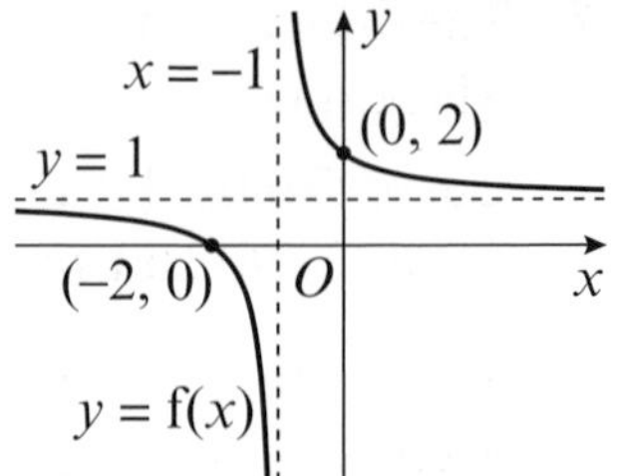

Sketch these transformed graphs, showing the coordinates of intersections with the axes and the equations of any asymptotes:

a $y = f(-x)$ **b** $y = 3f(x)$

c $y = f\left(\frac{1}{2}x\right)$

Hint Draw any transformed asymptotes first.

3 The diagram shows a sketch of the curve $y = f(x)$.

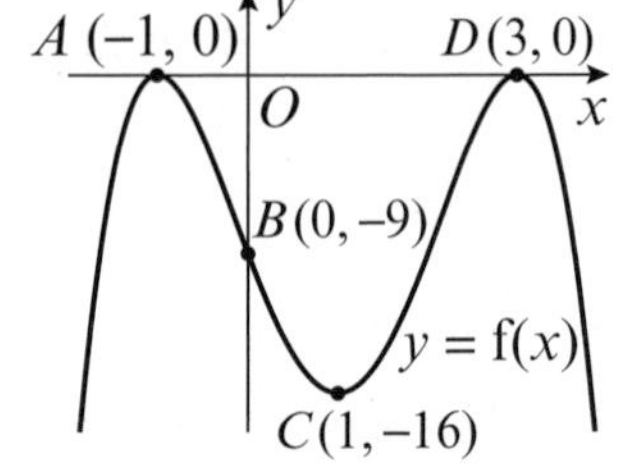

The points $A(-1, 0)$, $B(0, -9)$, $C(1, -16)$ and $D(3, 0)$ lie on the curve.

Give the coordinates of the points corresponding to A, B, C and D on each of the following transformed curves.

a $2y = f(x)$ **b** $\frac{1}{4}y = f(x)$

c $y + 1 = f(x)$

Hint Rearrange the equations into the form $y = \ldots$

E/P 4 The diagram shows a sketch of part of the curve with equation $y = f(x)$.

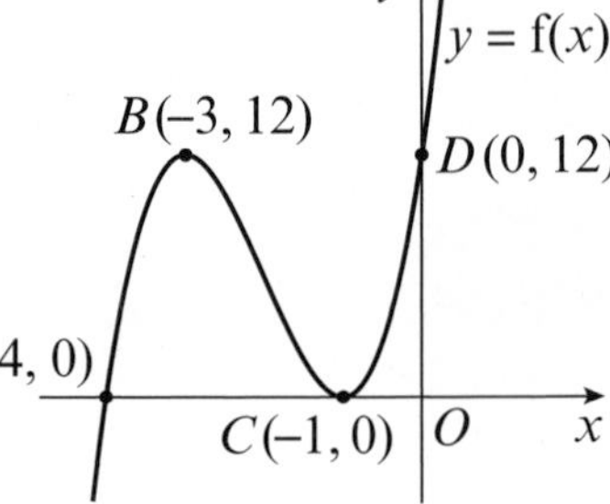

Sketch the curve with equation $y = f(x - 3)$, showing the location and coordinates of the images of points A, B, C and D. **(4 marks)**

E/P 5 The diagram shows a sketch of the curve C with equation $y = f(x)$.

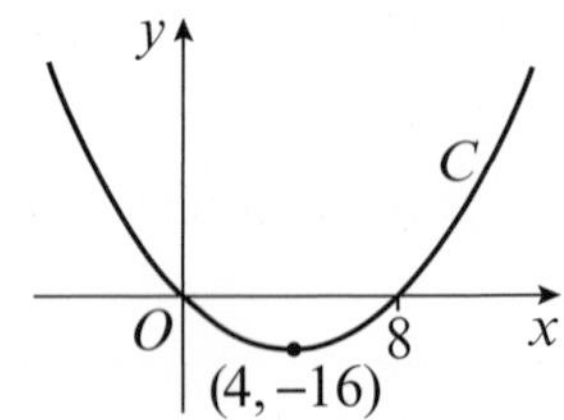

The curve C passes through the origin and the point $(8, 0)$ and has a local minimum at the point $(4, -16)$. On separate diagrams, sketch the curves with equations:

a $y = f(-x)$ **(3 marks)**

b $y = f(x - a)$, where a is a constant and $a > 0$. **(4 marks)**

On each diagram, show the locations of any points where the curve intersects the x-axis and of any local minimum or local maximum points.

E/P 6 The diagram shows a sketch of part of the curve with equation $y = f(x)$.

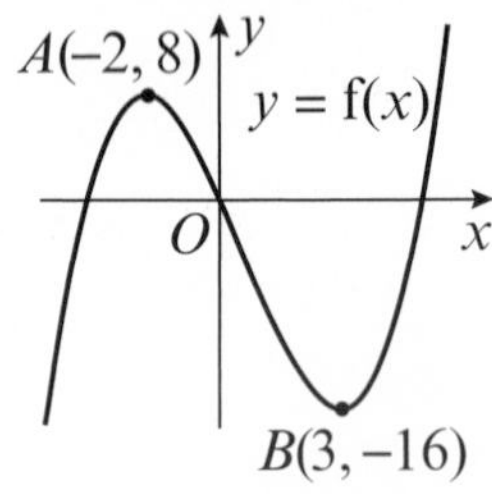

The curve has a local maximum at $A(-2, 8)$ and a local minimum at $B(3, -16)$ and passes through the origin O.
On a separate diagram, sketch the curve with equation $y = -f(2x)$, showing clearly the coordinates of the local maximum and the local minimum points and the location of the point where the curve crosses the y-axis. **(4 marks)**

E/P **7** The diagram shows a sketch of the curve C with equation $y = \mathrm{f}(x)$, where $\mathrm{f}(x) = x^2(12 - 4x)$.

Curve C has a local minimum at the origin, a local maximum at the point (2, 16) and cuts the x-axis at the point A.

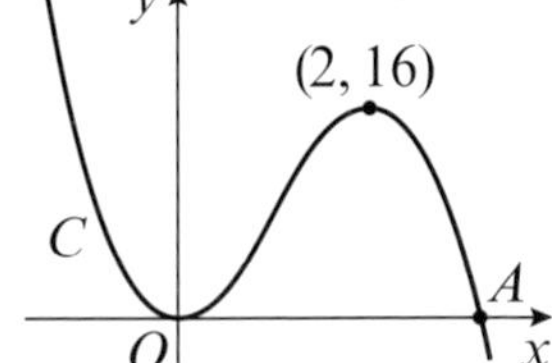

a Write down the coordinates of the point A. **(1 mark)**

b On separate diagrams sketch the curves with equations:

i $y = \mathrm{f}(x - 2)$ **ii** $y = \mathrm{f}(2x)$

On each sketch you should indicate clearly the location of the local maximum point and any points where the curve crosses or meets the coordinate axes. **(6 marks)**

c The curve with equation $y = \mathrm{f}(x) + k$, where k is a constant, has a local maximum point at (2, 5). Write down the value of k. **(1 mark)**

E/P **8** The sketch shows part of the graph of the curve $y = \mathrm{f}(x)$.

The curve has local maximum at (3, 6) and asymptote $y = 2$.
The equation $\mathrm{f}(x) + a = 0$, where a is a real constant, has exactly one real solution.
Find the range of possible values of a, giving your answer in set notation. **(3 marks)**

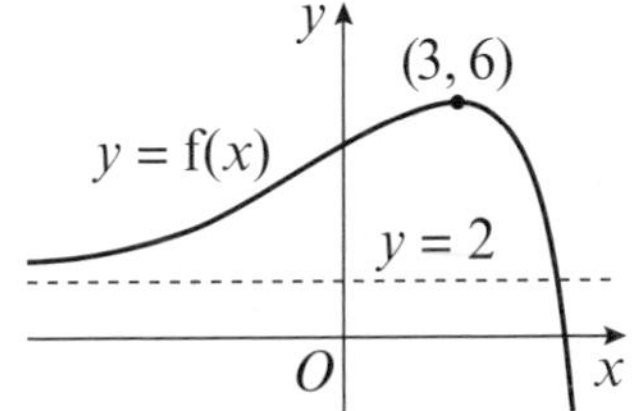

Problem solving Set A

Bronze

The graph of $y = x^3 + bx^2 + cx + d$ is shown, where b, c and d are real constants.

a Find the values of b, c and d. **(3 marks)**

b Hence find the coordinates of the point where the curve crosses the y-axis. **(1 mark)**

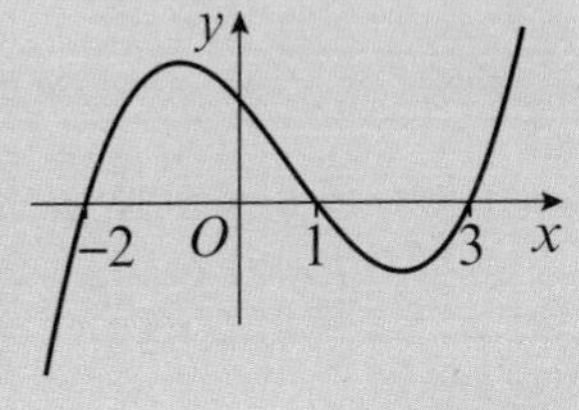

Silver

The graph of $y = x^3 + bx^2 + cx + d$ is shown, where b, c and d are real constants.

The curve crosses the x-axis at the points A, (2, 0) and (5, 0), and intersects the y-axis at (0, 5).

Use the two known roots and the y-intercept to find the values of b, c and d. **(4 marks)**

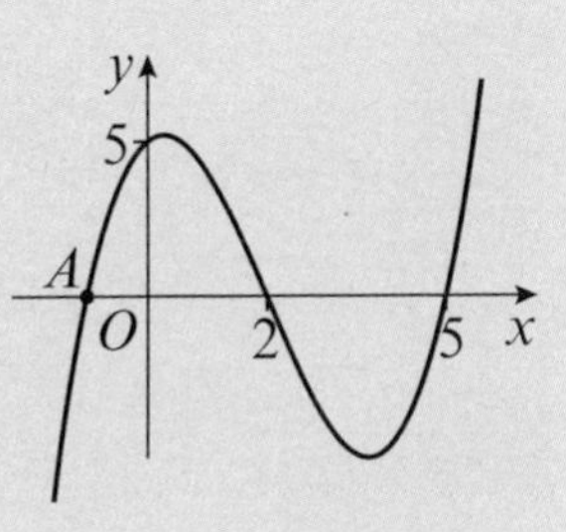

Gold

The graph of $y = \mathrm{f}(x)$ is shown, where $\mathrm{f}(x) = ax^4 + bx^3 + cx^2 + dx$ and a, b, c and d are real constants. The curve crosses the x-axis at (0, 0) and (4, 0) and touches the x-axis at (2, 0).

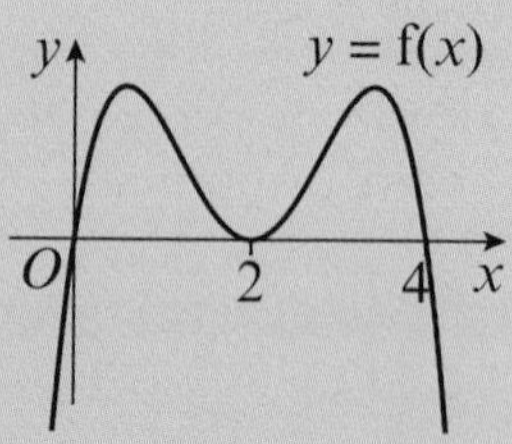

a Given that the equation of the curve can be written as $y = a(x^4 + mx^3 + nx^2 + px)$, where m, n, and p are real constants, find the values of m, n and p. **(3 marks)**

b Given further that the equation $\mathrm{f}(x) = 8$ has exactly two real solutions given by $x = 2 \pm \sqrt{2}$, find the value of a, and hence write down the equation of the curve. **(2 marks)**

Problem solving Set B

Bronze

a On the same set of axes, sketch the curves with equations $y = x^2(x - 2)$ and $y = 3x$. **(3 marks)**

b Write down the number of points of intersection of these two curves, and hence the number of real solutions to the equation $x^2(x - 2) = 3x$. **(1 mark)**

c Show that you can rearrange this equation to give $x(x + 1)(x - 3) = 0$. **(2 marks)**

d Hence determine the exact coordinates of the points of intersection. **(3 marks)**

Silver

a On the same set of axes, sketch the curves with equations $y = \frac{3}{(x-3)^2}$ and $y = \frac{6}{x}$ **(4 marks)**

b Write down the number of real solutions to the equation $\frac{3x}{(x-3)^2} = 6$ **(1 mark)**

c Determine the exact coordinates of the points of intersection of the two curves in part **a**. **(4 marks)**

Gold

The curve C has equation $y = \frac{4}{(x-2)^2}$. The line L has equation $y = 5 - x$.

a Given that the curve and the line intersect at the point (1, 4), find the coordinates of any other points of intersection. **(5 marks)**

b Hence sketch both graphs on the same set of axes, showing the coordinates of any points where the curve and line intersect each other, or either axis. **(3 marks)**

Now try this → **Exam question bank Q12, Q22, Q33, Q49, Q72, Q88**

5.1 $y = mx + c$

1 Work out the gradients of the lines joining these pairs of points:

a (4, 3), (8, 6)   b (5, 2), (7, −1)

c (3*p*, −4*p*), (8*p*, −2*p*)

Hint The gradient, m, of a line joining the points with coordinates (x_1, y_1) and (x_2, y_2) is given by $m = \frac{y_2 - y_1}{x_2 - x_1}$

2 The line *l* has gradient $\frac{1}{3}$ and passes through (0, 7). Find an equation for *l* in the form

a $y = mx + c$ b $px + qy + r = 0$

P 3 The points *A*(1, 2), *B*(4, 1) and *C*(12, *k*) are collinear. Work out the exact value of *k*.

Hint *A*, *B* and *C* are **collinear** if they all lie on the same straight line.

This means that the gradients of the line segments *AB*, *BC* and *AC* are equal.

4 For the line with equation $4x - 5y + 12 = 0$, find:

a the gradient

b the coordinates of the *y*-intercept

c the coordinates of the *x*-intercept.

Hint The equation of a straight line can be written in the forms:

- $y = mx + c$, where m is the gradient and c is the y-intercept
- $ax + by + c = 0$, where a, b and c are integers.

E/P 5 The gradient of the line joining the points (2, 3*a*) and (5*a*, −2) is −1.
Work out the value of *a*. **(2 marks)**

E/P 6 The line l_1 with gradient $-\frac{1}{5}$ passes through (0, 3). l_1 intersects the line l_2 with equation $3x - 4y + 7 = 0$ at point *P*. Find the exact coordinates of *P*. **(4 marks)**

E/P 7 The points *A*(−3*p* − 3, 2*p*), *B*(−5*p* + 1, 0) and *C*(0, 8*p*), where *p* is a constant, $p \neq 0$, are collinear. Find:

a the value of *p* **(4 marks)**

b the gradient of the line through *A*, *B* and *C*. **(2 marks)**

E/P 8 The line l_1 has gradient $\frac{2}{3}$ and passes through the point (0, −4).

a Find an equation of l_1 in the form $ax + by + c = 0$. **(3 marks)**

The line l_2 with equation $3x - ky + 25 = 0$ intersects l_1 at the point (*p*, −2) where *k* and *p* are constants. Find:

b the value of *p* **(2 marks)**

c the value of *k*. **(3 marks)**

E/P 9 The points A and B have coordinates $(-2, k + 1)$ and $(3k - 2, 6)$ where k is a constant.

Given the gradient of AB is $-\frac{1}{2}$,

a find the value of k. **(3 marks)**

The line through the points A and B passes though $(0, c)$. Find:

b the value of c **(3 marks)**

c the equation of the line in the form $ax + by + c = 0$. **(3 marks)**

5.2 Equations of straight lines

1 Find the equation of the line with gradient m that passes though the point (x_1, y_1) when:

a $m = -2$ and $(x_1, y_1) = (4, -5)$

b $m = \frac{1}{4}$ and $(x_1, y_1) = (-2, 6)$

c $m = -\frac{1}{8}$ and $(x_1, y_1) = (-3, -2)$

d $m = -\frac{3}{5}$ and $(x_1, y_1) = (1, -3)$

> **Hint** The equation of the line with gradient m that passes though the point (x_1, y_1) can be found using $y - y_1 = m(x - x_1)$

2 Find the equations of the lines that pass through these pairs of points:

a $(4, -3)$ and $(-6, 9)$ **b** $(-1, 3)$ and $(5, -2)$

c $\left(\frac{1}{2}, -\frac{1}{3}\right)$ and $\left(-\frac{3}{4}, \frac{2}{3}\right)$

> **Hint** To find the equation of the line that passes through the points (x_1, y_1) and (x_2, y_2), first find the gradient m using $m = \frac{y_2 - y_1}{x_2 - x_1}$, then use $y - y_1 = m(x - x_1)$

3 The line that passes through the points $(1, -4)$ and $(-3, 6)$ meets the x-axis at the point P. Work out the coordinates of P.

> **Hint** Find the equation of the line, then substitute in $y = 0$ and solve to find x.

4 The line that passes through the points $(-2, 5)$ and $(4, -3)$ meets the y-axis at the point Q. Work out the exact coordinates of Q.

E **5** The line l passes through the points $A(3, -6)$ and $B(-2, -10)$. Find an equation for l, giving your answer in the form $y = mx + c$. **(3 marks)**

E **6** The points $A(-6, 3)$ and $B(15, -4)$ lie on the line L. Find an equation for L in the form $ax + by + c = 0$, where a, b and c are integers. **(3 marks)**

E/P **7** The line l_1 passes through the point $(6, -3)$ and has gradient $\frac{1}{3}$. l_1 meets the line l_2 with equation $x + 2y = 10$ at the point P. Calculate the coordinates of P. **(4 marks)**

E/P **8** The line l_1 passes through the point $(5, -4)$ and has gradient $\frac{1}{4}$

a Find an equation for l_1 in the form $ax + by + c = 0$, where a, b and c are integers. **(3 marks)**

The line l_2 passes through the origin O and has gradient -5. The lines l_1 and l_2 intersect at the point P.

b Calculate the coordinates of P. **(4 marks)**

E/P 9 The line l_1 passes through the points $A(-2, 3)$ and $B(4, -1)$.

a Find an equation for l_1 in the form $ax + by + c = 0$, where a, b and c are integers. **(4 marks)**

The line l_2 with equation $3x + by - 1 = 0$ intersects l_1 at the point $P(k, k)$. Find:

b the value of k **(2 marks)**

c the value of b. **(2 marks)**

5.3 Parallel and perpendicular lines

1 Work out whether these pairs of lines are parallel.

Hint Parallel lines have the same gradient.

a $2y = 3x - 5$
$6x - 4y + 11 = 0$

b $3x - 4y + 9 = 0$
$9x + 12y - 10 = 0$

c $5x + 2y - 15 = 0$
$10x + 4y + 9 = 0$

2 Work out whether each of these pairs of lines are perpendicular.

a $3y = 2x + 7$
$4x + 6y + 1 = 0$

b $5x - 3y + 2 = 0$
$5y = 3x + 6$

c $4x - y - 5 = 0$
$2x + 8y - 15 = 0$

Hint If two lines are perpendicular, the product of their gradients is -1.
If a line l has gradient m, the gradient of a line perpendicular to l is $-\frac{1}{m}$

3 Find an equation of the line that passes through the point $(-3, 5)$ and is parallel to the line $2x - 3y + 7 = 0$.

4 Find an equation of the line that passes through the origin and is parallel to the line joining the points $(1, -6)$ and $(-2, 9)$.

5 The line l_1 has equation $4y - 8 = 3x$. The point P with x-coordinate 4 lies on l_1.
The line l_2 is perpendicular to l_1 and passes through the point P. Find an equation of l_2, giving your answer in the form $ax + by + c = 0$, where a, b and c are integers. **(4 marks)**

6 The line L has equation $2y = 1 - 3x$.

a Show that the point $P(3, -4)$ lies on L. **(1 mark)**

b Find an equation of the line perpendicular to L, which passes through P.
Give your answer in the form $ax + by + c = 0$, where a, b and c are integers. **(3 marks)**

E/P **7** The line l_1 has equation $2x - 5y + 2 = 0$.

a Find the gradient of l_1. **(2 marks)**

The line l_2 is perpendicular to l_1 and passes through the point $(3, 2)$.

b Find the equation of l_2 in the form $ax + by + c = 0$, where a, b and c are integers. **(3 marks)**

E/P **8** The line L_1 has equation $5y - 2x - k = 0$, where k is a constant.
Given that the point $A(1, 3)$ lies on L_1, find:

a the value of k **(1 mark)**

b the gradient of L_1. **(2 marks)**

The line L_2 passes through A and is perpendicular to L_1.

c Find an equation of L_2, giving your answer in the form $ax + by + c = 0$, where a, b and c are integers. **(3 marks)**

The line L_2 crosses the x-axis at the point P.

d Find the coordinates of P. **(2 marks)**

E/P **9** The line l_1 passes through the points $P(-2, 3)$ and $Q(10, 9)$.

a Find an equation for l_1 in the form $y = mx + c$, where m and c are constants. **(3 marks)**

The line l_2 passes through the point $R(12, 0)$ and is perpendicular to l_1.
The lines l_1 and l_2 intersect at the point S.

b Calculate the coordinates of S. **(5 marks)**

5.4 Length and area

1 Find the exact distance between these pairs of points:

a $(1, -2), (-5, 6)$ **b** $(-8, 4), (16, -3)$

c $(6, -5), (-10, -1)$

Hint You can find the distance d between the points (x_1, y_1) and (x_2, y_2) by using the formula $d = \sqrt{(x_2 - x_1)^2 + (y_2 - y_1)^2}$
If the number inside the square root is not a perfect square, write the surd in its simplest form.

← Section 1.5

2 Consider the points $A(-2, 5)$, $B(3, 1)$ and $C(8, -3)$. Determine whether the line segment joining the points A and B is congruent to the line segment joining the points B and C.

Hint Line segments are congruent if they are the same length.

3 The vertices of a triangle are $P(-3, 2)$, $Q(2, 5)$ and $R(4, 2)$.

Find the area of triangle PQR.

Hint Sketch the triangle and label its vertices.
Use area $= \frac{1}{2} \times$ base $\times$ perpendicular height.

P **4** The distance between the points $(-2, 11)$ and $(x, 8)$ is $\sqrt{58}$.

Find the two possible values of x.

Hint Use the formula for the distance between two points to form a quadratic equation, then solve this equation.

← Section 2.1

E 5 The line l passes through the points $A(2, -6)$ and $B(-3, 14)$. Find:

a an equation for l **(3 marks)**

b the exact length of AB. **(2 marks)**

E/P 6 The line l_1 passes through the point $A(4, 6)$ and has gradient $\frac{1}{2}$

a Find an equation of l_1, giving your answer in the form $y = mx + c$. **(3 marks)**

The point B has coordinates $(-2, 3)$.

b Show that B lies on l_1. **(1 mark)**

c Find the length of AB, giving your answer in the form $k\sqrt{5}$ where k is a constant to be found. **(2 marks)**

The point C lies on l_1 and has x-coordinate equal to p. The length of AC is 6 units.

d Show that p satisfies $5p^2 - 40p - 64 = 0$. **(3 marks)**

E 7 The point $A(-6, 1)$ and the point $B(9, -4)$ lie on the line L.

a Find an equation for L in the form $ax + by + c = 0$, where a, b and c are integers. **(3 marks)**

b Find the distance AB, giving your answer in the form $k\sqrt{10}$, where k is an integer to be found. **(2 marks)**

E/P 8 a Find an equation of the straight line passing through the points $(-3, 4)$ and $(5, -2)$. **(3 marks)**

Give your answer in the form $ax + by + c = 0$, where a, b and c are integers.
The line crosses the x-axis at point A and the y-axis at point B.

b Find the area of triangle AOB, where O is the origin. **(2 marks)**

E/P 9 The line l_1 passes through the points $P(-2, 4)$ and $Q(10, -2)$.

a Find an equation for l_1 in the form $y = mx + c$, where m and c are constants. **(3 marks)**

The line l_2 passes through the point $R(0, -7)$ and is perpendicular to l_1.
The lines l_1 and l_2 intersect at the point S.

b Calculate the coordinates of S. **(5 marks)**

c Show that the length of RS is $4\sqrt{5}$. **(2 marks)**

d Hence, or otherwise, find the exact area of triangle PQR. **(3 marks)**

5.5 Modelling with straight lines

1 For each graph:

i calculate the gradient, k, of the line

ii write down the units for the gradient

iii write a direct proportion equation connecting the two variables.

> **Hint** If two quantities, y and x, are in direct proportion they increase at the same rate.
>
> The graph of y against x is a straight line which passes through the origin.
>
> If y is directly proportional to x, then $y \propto x$ and $y = kx$ where k is a constant.
>
> The units for the gradient are $\frac{\text{units of } y\text{-axis}}{\text{units of } x\text{-axis}}$

a

b

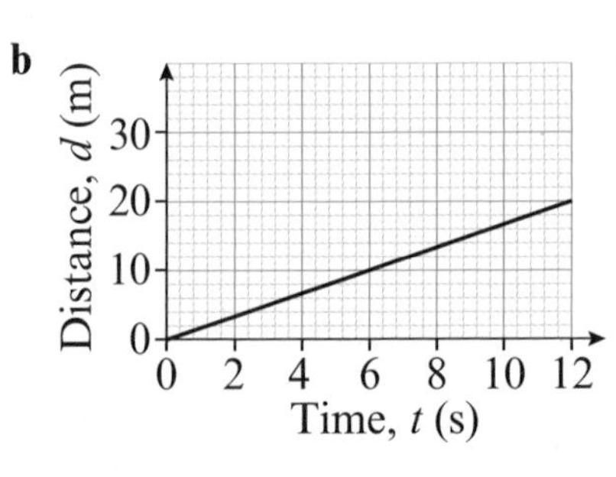

c

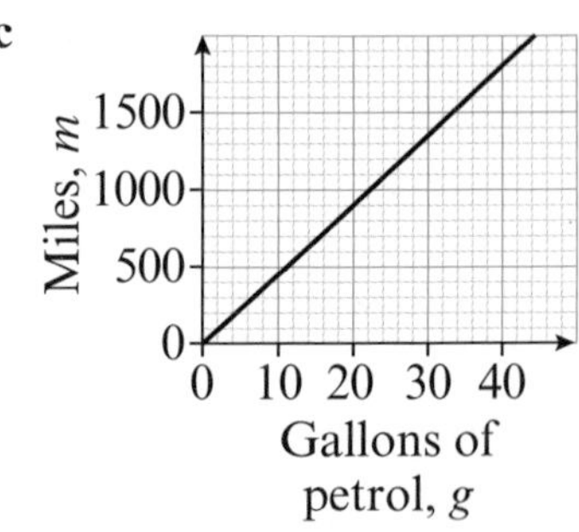

2 Draw a graph to determine whether a linear model would be appropriate for each set of data.

a

p	q
0	0
2	4
4	9
6	15
10	20
20	42

b

l	v
0	25
5	30
10	45
20	80
30	160
40	280

c

a	b
2.5	52
3.2	39
3.7	35
3.9	28
4.2	22
4.5	18

> **Hint** You can use a linear model for the relationship between two variables y and x if the points lie approximately on a straight line.

3 The graph shows the conversion between British pounds, P, and US dollars, D.

a Calculate the gradient, k, of the line.

b Write an expression for D in terms of P.

c Explain what the value of k represents in this situation.

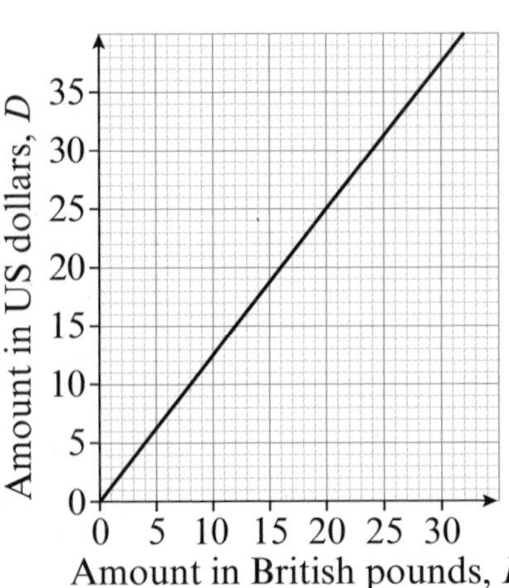

> **Hint** Use any two points on the line to calculate the gradient. As the line passes through the origin, the general form of the equation is $y = kx$.

4 Sand pours out from a 25 kg bag of sand through a hole at the bottom of the bag. The weight, w, of the bag of sand after t seconds is recorded in this table.

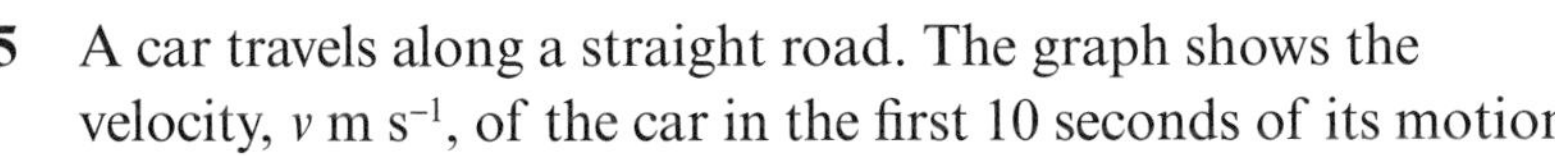

Weight, *w* kg	24.8	23.4	21.7	19.2	12.9	10.8
Time, *t* seconds	0	10	20	40	80	100

a Plot the points on a graph to determine whether a linear model is appropriate.

b Work out an equation for the line of best fit in the form $w = at + b$ where a and b are constants.

c Interpret the meaning of the constants a and b.

d Use your model to estimate the time taken for the bag of sand to empty.

e Suggest a reason your estimate may not be accurate.

Hint You can choose two points from the table to work out the gradient for your linear model.

Make sure you give your answers in the context of the question.

Identify any assumptions you have made when using your linear model which may not be realistic.

E/P **5** A car travels along a straight road. The graph shows the velocity, v m s^{-1}, of the car in the first 10 seconds of its motion.

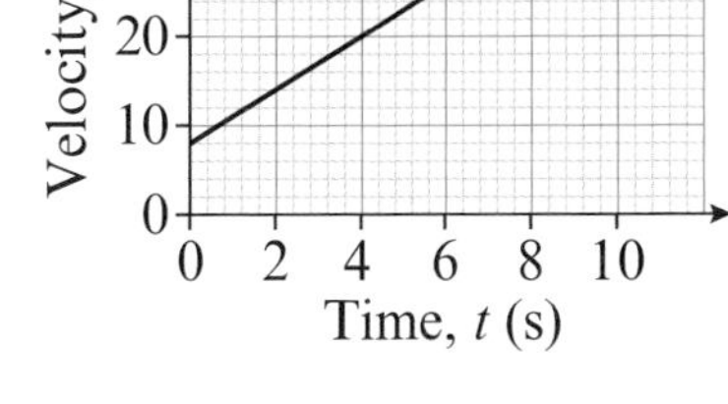

a Find a linear equation of the form $v = a + bt$ linking v and t, where t is the time in seconds. **(3 marks)**

b Interpret the values of a and b. **(2 marks)**

c Use the model to predict the velocity of the car after 15 seconds. **(1 mark)**

d Write down one reason why this might not be a realistic model. **(1 mark)**

E/P **6** The table shows the times achieved by gold medallists in the women's 100 m sprint in the Olympic Games from 1992 to 2016.

Year, *Y*	1992	1996	2004	2008	2012	2016
Time, *T* seconds	10.82	10.94	10.93	10.78	10.75	10.71

a Draw a scatter graph to show these results. **(3 marks)**

b Use the points (1996, 10.94) and (2016, 10.71) to formulate a linear model for this data. **(3 marks)**

c Use your model to predict the gold-medal winning time in the women's 100 m in the 2020 Olympic Games. **(1 mark)**

d Comment on the validity of this model for predicting winning times. **(1 mark)**

E/P **7** The charge for a taxi journey is made up of a fixed charge plus a further cost per mile travelled. A taxi company charges £20.40 for a 5-mile journey and £56.40 for a 15-mile journey.

a Write an equation linking the charge, £C, with the number of miles, m, in the form $C = am + b$ **(3 marks)**

b Interpret the values of a and b in the context of the question. **(2 marks)**

c Work out the taxi fare for a 12-mile journey. **(1 mark)**

(E/P) **8** A man buys a new motorbike for £5475. After 1 year the motorbike is worth £4380.

a Find a linear model to link the value of the motorbike, £V, with the age of the motorbike, t years. **(3 marks)**

b Interpret the gradient of your model. **(1 mark)**

c Use your model to predict the value of the motorbike after 2.5 years. **(1 mark)**

d Comment on the suitability of the model. **(1 mark)**

Problem solving Set A

Bronze

The line l_1 passes through the origin O and $A(3, 6)$ as shown in the diagram.

The length of OA is $a\sqrt{5}$.

a Find the value of a. **(2 marks)**

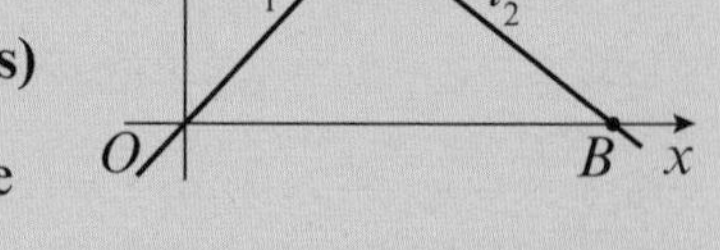

The line l_2 is perpendicular to l_1, passes through A and crosses the x-axis at the point B. Find:

b the gradient of l_2 **(2 marks)**

c an equation for l_2, giving your answer in the form $ax + by + c = 0$ **(3 marks)**

d the coordinates of B **(1 mark)**

e the area of $\triangle ABC$. **(3 marks)**

Silver

The points $A(2, 3)$ and $B(8, 0)$ lie on the line l_1, as shown in the diagram.

The line l_2 is perpendicular to l_1, passes through A and crosses the y-axis at the point C.

a Find the coordinates of C. **(5 marks)**

b Hence find the area of $\triangle ABC$. **(2 marks)**

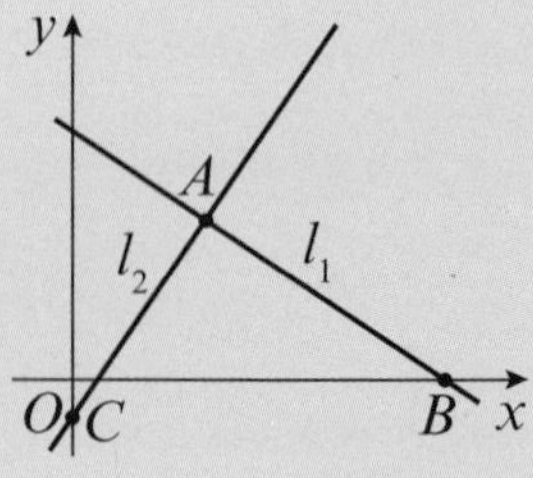

Gold

The points $A(2, 9)$ and $B(0, 5)$ lie on the line l_1, as shown in the diagram.

The line l_2 is perpendicular to l_1, passes through A and crosses the x-axis at the point C.

Find the area of $\triangle ABC$. **(8 marks)**

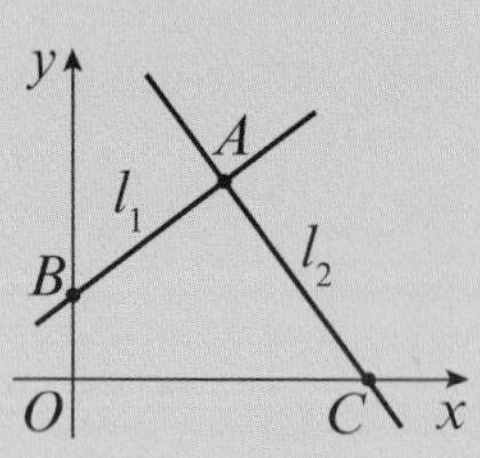

Problem solving Set B

Bronze

The point P has coordinates (7, 1). The distance from P to the point $R(10, k)$ is 4.

a Show that $k^2 - 2k - 6 = 0$. **(3 marks)**

b Given that k is positive, find the exact value of k. **(3 marks)**

Silver

The point P has coordinates (3, −4). The distance from P to the point $R(k, 2)$ is 10.

Find the two possible values of k. **(6 marks)**

Gold

The point P has a positive x-coordinate and lies on the line with equation $y = x$.

The point Q has coordinates (0, 4). Given that the distance PQ is 6, find the exact coordinates of P. **(8 marks)**

Now try this → Exam question bank Q21, Q77, Q81, Q86, Q89, Q93

6 Circles

6.1 Midpoints and perpendicular bisectors

1 Find the midpoint of the line segment joining each pair of points:

a (2, −5), (−4, 7)

b (−3, 9), (−1, −5)

c $\left(\frac{9}{5}, -\frac{3}{8}\right), \left(-\frac{3}{5}, -\frac{1}{8}\right)$

d $(2\sqrt{3}, -\sqrt{5}), (-\sqrt{3}, -3\sqrt{5})$

e $(1 + 2\sqrt{2}, 3 - \sqrt{2}), (3 + 2\sqrt{2}, 1 - 3\sqrt{2})$

> **Hint** The **midpoint** of the line segment with endpoints (x_1, y_1) and (x_2, y_2) is $\left(\frac{x_1 + x_2}{2}, \frac{y_1 + y_2}{2}\right)$

2 Find the equation of the perpendicular bisector of the line segment joining each pair of points. Give your answer in the form $ax + by + c = 0$.

a (−3, 4), (1, −8)

b (3, −5), (−7, 15)

c (2, −3), (−6, −9)

> **Hint** The **perpendicular bisector** of a line segment AB is the straight line that is perpendicular to AB and passes through the midpoint of AB. To find its equation:
>
> - find the coordinates of the midpoint (x_1, y_1)
> - find the gradient, M, of the line segment AB ← Section 5.1
> - find the gradient, m, of the perpendicular to AB using $m = -\frac{1}{M}$ ← Section 5.3
> - use $y - y_1 = m(x - x_1)$ ← Section 5.2

3 The line segment AB is a diameter of a circle, where A is (−3, 4) and B is (5, 8). Find:

a the coordinates of the centre of the circle

b the radius of the circle in the form $k\sqrt{5}$ where k is a constant to be found.

> **Hint** For part **b**, first find the diameter, d, which is the length of the line segment joining the points (x_1, y_1), (x_2, y_2), and is given by $d = \sqrt{(x_2 - x_1)^2 + (y_2 - y_1)^2}$ ← Section 5.4

4 The line segment PQ has endpoints $P(3, -5)$ and $Q(a, b)$. The midpoint of PQ is $M(6, -3)$. Find the values of a and b.

5 Points P, Q and R have coordinates (−1, −3), (5, 7) and (7, −5) respectively. Find:

a the equation of the perpendicular bisector of PQ

b the equation of the perpendicular bisector of QR

c the exact coordinates of the point of intersection of the two perpendicular bisectors.

E/P **6** The line segment PQ, where P is (3, 8) and Q is (−1, −4), is the diameter of a circle with centre C.
The line l is perpendicular to PQ and passes through C. Find an equation of l. **(7 marks)**

E/P 7 The line segment PQ is a diameter of a circle, where P is $(-3, 6)$ and Q is $(5, -2)$. Find:

a the coordinates of the centre of the circle **(2 marks)**

b the radius of the circle in the form $k\sqrt{2}$, where k is a constant to be found. **(3 marks)**

R and S are the points where the perpendicular bisector of PQ meets the circle.

c Show that $PR = RQ = QS = SP = 8$. **(4 marks)**

E/P 8 The points A and B have coordinates $(2k - 1, -3)$ and $(3, 3k + 7)$ respectively, where k is a constant. The coordinates of the midpoint of AB are $(5, p)$, where p is a constant.

a Find the value of k. **(2 marks)**

b Find the value of p. **(2 marks)**

c Find an equation of the perpendicular bisector of AB, giving your answer in the form $ax + by + c = 0$, where a, b and c are integers. **(6 marks)**

6.2 Equation of a circle

1 Write down the equation of each circle:

a Centre $(-5, 3)$, radius 8

b Centre $(7, -8)$, radius 10

c Centre $(-1, -4)$, radius $3\sqrt{2}$

Hint The equation of a circle with centre (a, b) and radius r is $(x - a)^2 + (y - b)^2 = r^2$

2 Write down the coordinates of the centre and the radius of each circle:

a $(x + 2)^2 + (y - 9)^2 = 144$ **b** $(x - 5)^2 + (y + 2)^2 = 32$ **c** $(x + 6)^2 + y^2 = 45$

3 Find the centre and the radius of the circle with the equation $x^2 + y^2 + 8x - 12y - 15 = 0$

Hint To find the coordinates of the centre and the radius, first complete the square for the terms in x and for the terms in y separately. ← **Section 2.2**

4 The line segment PQ is a diameter of a circle, where P is $(-1, 8)$ and Q is $(7, 20)$. Find the equation of the circle.

Hint Find the coordinates of the centre by finding the midpoint of PQ using $\left(\frac{x_1 + x_2}{2}, \frac{y_1 + y_2}{2}\right)$ Find the radius of the circle by finding the diameter and dividing by 2.

E/P 5 The circle with equation $(x - 1)^2 + (y - k)^2 = 50$ passes through the point $(2, 3)$. Find the possible values of k. **(5 marks)**

E/P 6 A circle has equation $x^2 + y^2 - 6x + 10y - 16 = 0$. Find:

a the coordinates of the centre of the circle **(3 marks)**

b the radius of the circle in the form $k\sqrt{2}$, where k is an integer to be found. **(3 marks)**

(E) 7 The circle C has centre (1, 6) and passes through the point $P(-3, 4)$.

a Find an equation for C. **(4 marks)**

C crosses the y-axis at the points A and B.

b Find the coordinates of the points A and B. **(3 marks)**

(E/P) 8 The midpoint of the line joining the points $P(5, 8)$ and $Q(p, q)$ is $M(-2, 3)$.

a Find the values of p and q. **(3 marks)**

PQ is the diameter of a circle.

b Find the radius of the circle. **(2 marks)**

c Find the equation of the circle, giving your answer in the form $x^2 + y^2 + ax + by + c = 0$. **(3 marks)**

(E/P) 9 The circle with equation $x^2 + y^2 + 4y - k = 0$ has radius 5.

a Find the coordinates of the centre of the circle and the value of k. **(4 marks)**

The points $P(p, 2)$ and $Q(-4, 1)$, where $p > 0$, lie on the circumference of the circle.

b Calculate the length of PQ, giving your answer in simplified surd form. **(5 marks)**

6.3 Intersections of straight lines and circles

1 Find the coordinates of the points where the circle $(x - 2)^2 + (y - 4)^2 = 52$ meets:

a the x-axis b the y-axis

Hint The circle meets the x-axis when $y = 0$ and meets the y-axis when $x = 0$.

2 The line with equation $2x + y = 7$ meets the circle $(x - 1)^2 + y^2 = 50$ at points P and Q.

Find the coordinates of P and Q.

Hint A straight line may not intersect a circle, or can intersect once (by just touching the circle), or can intersect twice.

To find the points of intersection, solve simultaneously by substituting the linear equation into the equation for the circle and solving the resulting quadratic equation.

← **Section 3.2**

3 a Show that the line $2x + y - 15 = 0$ meets the circle with equation $(x + 2)^2 + (y - 4)^2 = 45$ at exactly one point.

b Find the coordinates of the point of intersection.

Hint To find the number of points of intersection you can use the discriminant for the resulting quadratic equation $ax^2 + bx + c = 0$ where:

- $b^2 - 4ac > 0$ indicates 2 points of intersection – the line cuts the circle twice
- $b^2 - 4ac = 0$ indicates 1 point of intersection – the line is a tangent to the circle
- $b^2 - 4ac < 0$ indicates 0 points of intersection – the line does not meet the circle.

← **Section 2.5**

4 Show that the line $y = 2x + 6$ does not meet the circle $(x - 5)^2 + (y - 2)^2 = 30$.

E 5 **a** Describe completely the curve with equation $x^2 + y^2 = 100$. **(2 marks)**

b Find the coordinates of the points of intersection of the curve $x^2 + y^2 = 100$ and the line $3x + y - 10 = 0$. **(6 marks)**

E/P 6 A circle has equation $x^2 + y^2 - 6x - 23 = 0$.
A line has equation $y = x + k$, where k is a constant.

a Show that the x-coordinate of any point of intersection of the line and the circle satisfies the equation $2x^2 + 2(k - 3)x + k^2 - 23 = 0$. **(3 marks)**

b Find the values of k for which the equation $2x^2 + 2(k - 3)x + k^2 - 23 = 0$ has equal roots. **(4 marks)**

c Describe the geometrical relationship between the line and the circle when k takes either of the values found in part **b**. **(1 mark)**

E/P 7 The circle with equation $(x - 3)^2 + (y - 5)^2 = 61$ meets the straight line with equation $x + 11y = 119$ at points P and Q.

a Find the coordinates of points P and Q. **(5 marks)**

b Find an equation of the perpendicular bisector of line segment PQ. **(5 marks)**

c Show that the perpendicular bisector passes through the centre of the circle. **(2 marks)**

E/P 8 The circle C has equation $x^2 + y^2 - 10x + 8y + 25 = 0$. The line with equation $y = kx$, where k is a constant, cuts C at two distinct points. Find the range of possible values for k. **(6 marks)**

E/P 9 The line $x + 3y = k$ meets the circle with equation $(x - p)^2 + (y - 4)^2 = 45$ at the point $(4, 7)$.

a Work out the value of k. **(1 mark)**

b Work out the two possible values of p. **(5 marks)**

6.4 Use tangent and chord properties

1 The circle C has equation $x^2 + y^2 - 6x + 4y - 156 = 0$.

a Verify that the point $P(-2, 10)$ lies on C.

b Find an equation of the tangent to C at the point P, giving your answer in the form $ax + by + c = 0$.

Hint A tangent to a circle is perpendicular to the radius of the circle at the point of intersection.
For part **b**, first find the gradient, M, of the radius.
The gradient of the tangent is $m = -\frac{1}{M}$
The equation of the tangent at $P(x_1, y_1)$ can then be found using $y - y_1 = m(x - x_1)$.

2 The point $P(4, -6)$ lies on the circle with centre $(8, 2)$.

a Find the equation of the circle.

b Find an equation of the tangent to the circle at P.

Hint The radius of the circle is the distance between P and the centre.

3 The equation of a circle is $x^2 + y^2 + 6x - 8y - 16 = 0$. The points $A(2, 8)$ and $B(-7, 9)$ lie on the circle.

a Find an equation of the perpendicular bisector of the chord AB, giving your answer in the form $y = mx + c$.

b Show that the perpendicular bisector of AB passes through the centre of the circle.

Hint The perpendicular bisector of a chord of a circle always passes through the centre of the circle. Start by finding both the midpoint and gradient of the chord AB.

4 A circle has equation $(x - p)^2 + (y - 3)^2 = 20$, where $p \neq 0$ is a constant. The points $P(-2, 7)$ and $Q(-8, k)$ lie on the circle.

a Find the value of p and the two possible values of k.

b For the greater possible value of k, find the equation of the perpendicular bisector of PQ, and show that it passes through the centre of the circle.

(E) **5** The line joining the points $A(4, p)$ and $B(q, -2)$ has midpoint $M(1, -4)$.

a Find the values of p and q. **(3 marks)**

AB is the diameter of a circle.

b Find the radius of the circle. **(2 marks)**

c Find the equation of the circle in the form $x^2 + y^2 + ax + by + c = 0$. **(3 marks)**

d Find an equation of the tangent to the circle at the point $(3, -9)$, giving your answer in the form $ax + by + c = 0$. **(5 marks)**

(E/P) **6** A circle C with centre $(-5, 9)$ passes through the point $(8, 14)$.

a Show that the point $(8, 4)$ also lies on C. **(3 marks)**

The tangent to C at the point $(8, 14)$ meets the y-axis at the point A and the tangent to C at the point $(8, 4)$ meets the y-axis at the point B.

b Work out the distance AB. **(7 marks)**

(E/P) **7** The points $P(1, 10)$ and $Q(7, 8)$ lie on the circumference of a circle with centre $C(3, k)$.

a Find the equation of the perpendicular bisector of PQ. **(5 marks)**

b Find the value of k. **(2 marks)**

c Find the equation of the circle. **(3 marks)**

d Find the exact coordinates of the points of intersection of the perpendicular bisector of PQ, and the circle. **(5 marks)**

E 8 The points $P(-2, 3)$ and $Q(8, 5)$ lie on a circle with centre $(4, -1)$.

a Find the equation of the circle. **(3 marks)**

The line l passes through the centre of the circle and the midpoint of the chord PQ.

b Find an equation of l. **(5 marks)**

The line l crosses the y-axis at A and the x-axis at B.

c Find the area of triangle AOB, where O is the origin. **(4 marks)**

E/P 9 The circle C has equation $x^2 + y^2 - 2x + 6y - 30 = 0$.
The points $P(3, 3)$ and $Q(7, -5)$ lie on C.

a Find an equation for the tangent to C at Q, giving your answer in the form $ax + by + c = 0$, where a, b and c are integers. **(5 marks)**

b Find an equation of the perpendicular bisector of PQ. **(5 marks)**

c Find the coordinates of the point of intersection of the tangent and the perpendicular bisector. **(4 marks)**

6.5 Circles and triangles

1 The vertices of a triangle are $A(-1, 6)$, $B(3, 4)$ and $C(5, 2)$.

a Find the equations of the perpendicular bisectors of AB, BC and AC respectively.

b Show that the perpendicular bisectors all meet at the same point.

c Find the equation of the circle that passes through A, B and C.

Hint A unique circle, called the **circumcircle**, can be drawn through the three vertices of any triangle. The perpendicular bisectors of the three sides meet at the centre of the circle.

2 The points $P(-12, 1)$, $Q(8, 9)$ and $R(2, 15)$ lie on the circle as shown in the diagram.

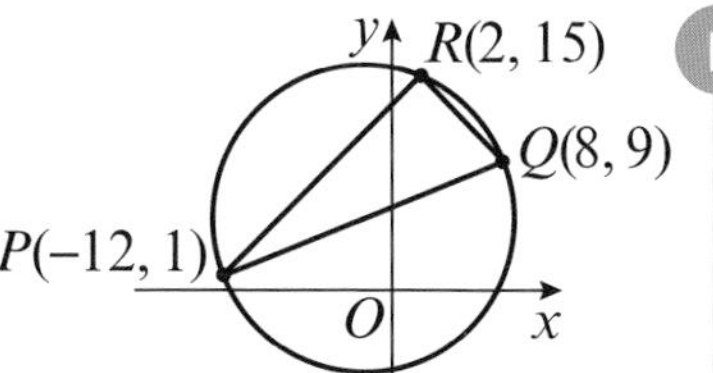

a Show that PQ is a diameter of the circle.

b Find an equation of the circle in the form $(x - a)^2 + (y - b)^2 = c$, where a, b and c are constants to be found.

Hint Use Pythagoras' theorem to determine whether a triangle is right-angled.
The angle in a semicircle is 90°, so the hypotenuse of a right-angled triangle is a diameter of the circumcircle. The centre of the circle is at the midpoint of the hypotenuse.

3 A triangle ABC has vertices $A(6, 3)$, $B(-2, 1)$ and $C(0, -7)$.

a Show that ABC is an isosceles right-angled triangle.

b Find the equation of the circumcircle of triangle ABC.

4 The points $A(6, 14)$, $B(14, 10)$ and $C(-4, 4)$ lie on the circumference of a circle. The equation of the perpendicular bisector of AC is $x + y = 10$.

a Find the coordinates of the centre of the circle.

b Find the equation of the circle.

E/P **5** A circle with centre C has equation $x^2 + y^2 - 10x + 6y = 91$.

a Find the coordinates of C. **(2 marks)**

The points $A(0, 7)$ and $B(10, 7)$ lie on the circle.

b Find the area of triangle ABC. **(4 marks)**

E/P **6** The circle C has equation $x^2 + y^2 - 4x + 8y = 33$.

a Find the centre and radius of C. **(5 marks)**

The points $P(-5, -2)$ and $Q(9, -6)$ both lie on C.

b Show that PQ is a diameter of C. **(2 marks)**

The point R lies on the positive y-axis and the angle $PRQ = 90°$.

c Find the coordinates of R. **(4 marks)**

E/P **7** A circle has equation $x^2 + y^2 + 4x - 6y - 23 = 0$.

a Find the coordinates of the centre of the circle. **(2 marks)**

b Show that the radius of the circle is 6. **(2 marks)**

The points A, B and C lie on the circle. The length of AB is 12 and the length of AC is 5.

c Find the length of BC, giving your answer to 1 decimal place. **(3 marks)**

E/P **8** The points P, Q and R have coordinates $(-2, 3)$, $(5, -1)$ and $(-3, k)$ respectively. Given that PQ is perpendicular to QR,

a find the value of k. **(4 marks)**

P, Q and R lie on the circumference of a circle.

b Find the coordinates of the centre of the circle. **(3 marks)**

E/P **9** The points $P(-4, 4)$, $Q(8, 12)$ and $R(k, 6)$ lie on the circle C as shown in the diagram.

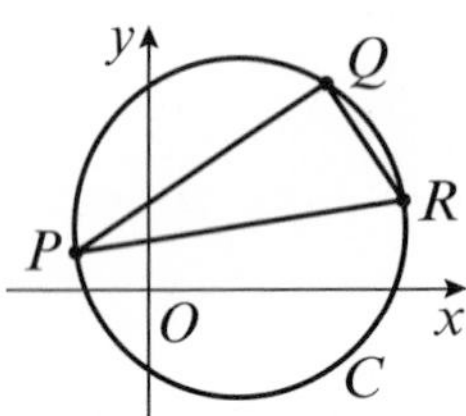

Given that PR is a diameter of C,

a find the value of k **(3 marks)**

b find an equation for C. **(5 marks)**

Problem solving Set A

Bronze

The circle C has centre (0, 5) and radius 10.

a Write down an equation of the circle. **(3 marks)**

The straight line with equation $2x + y = 1$ intersects C at the points P and Q.

b Find the coordinates of P and Q. **(6 marks)**

Silver

Show that the line with equation $4y + 3x = 44$ is a tangent to the circle with centre (5, 1) and radius 5. **(7 marks)**

Gold

The line with equation $5y - 4x = k$, where k is a constant, does not intersect the circle with equation $(x - 3)^2 + (y + 2)^2 = 41$. Find the range of possible values of k. **(9 marks)**

Problem solving Set B

Bronze

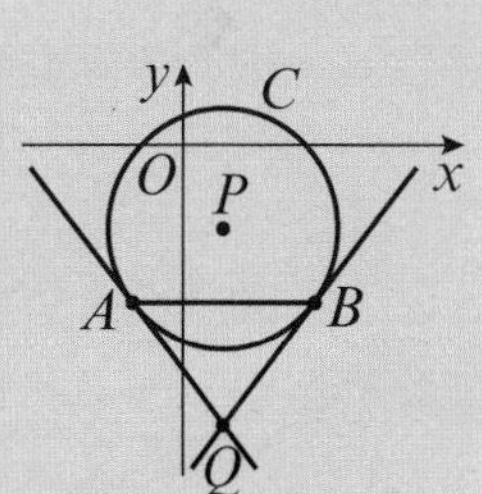

The diagram shows a sketch of the circle C with centre P.

The circle has equation $(x - 2)^2 + (y + 3)^2 = 25$.

a Write down the coordinates of P. **(2 marks)**

b State the radius of C. **(1 mark)**

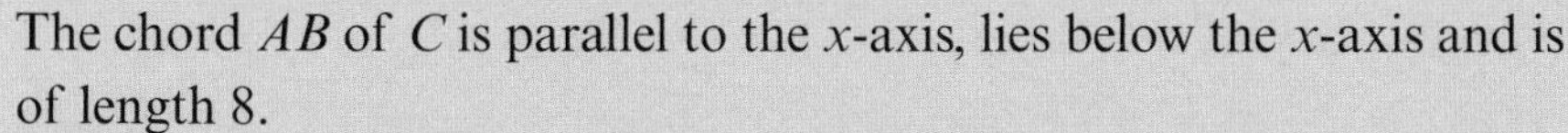

The chord AB of C is parallel to the x-axis, lies below the x-axis and is of length 8.

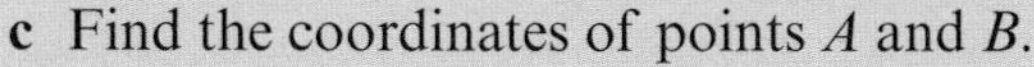

c Find the coordinates of points A and B. **(5 marks)**

d Write down the coordinates of M, the midpoint of AB. **(1 mark)**

e Show that angle $APM = 53.1°$, to one decimal place. **(2 marks)**

The tangents to C at the points A and B meet at the point Q.

f Find the length of AQ, giving your answer to 2 significant figures. **(2 marks)**

Silver

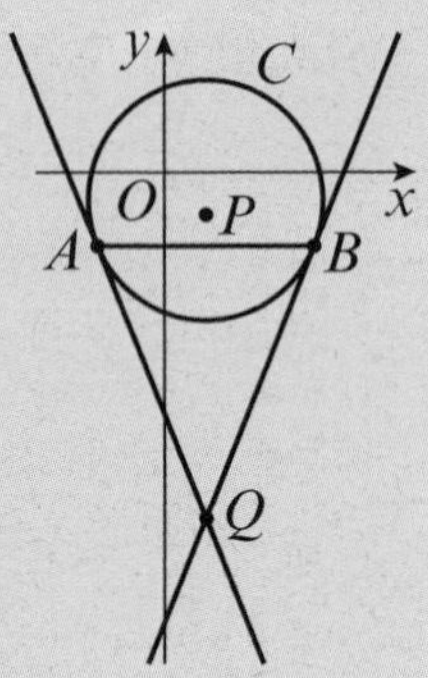

The diagram shows a sketch of the circle C with centre P.

The circle has equation $(x - 4)^2 + (y + 5)^2 = \frac{625}{4}$

a Write down the coordinates of P. **(2 marks)**

b Find the radius of C. **(1 mark)**

The chord AB of C is parallel to the x-axis, lies below the x-axis and has length 24.

c Find the coordinates of A and B. **(5 marks)**

d Show that angle $APB = 147.5°$, to the nearest $0.1°$. **(2 marks)**

e Find the length AQ, giving your answer to 3 significant figures. **(2 marks)**

Gold

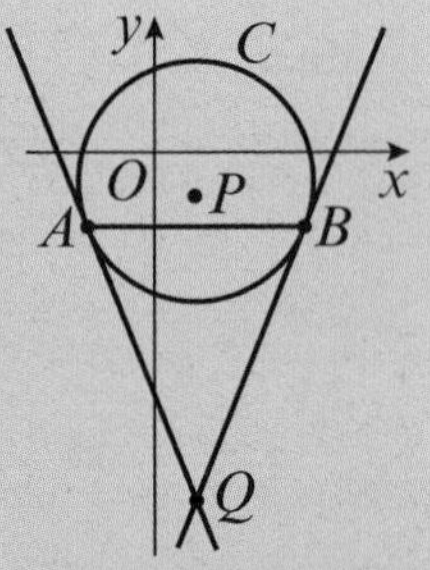

The diagram shows a sketch of the circle C with centre P.

The circle has equation $(x - 3)^2 + (y + 2)^2 = \frac{289}{4}$

The chord AB of C is parallel to the x-axis, lies below the x-axis and has length 15.

a Find the coordinates of A and B. **(8 marks)**

The tangents to C at the points A and B meet at the point Q.

b Find the length of AQ, giving your answer to 3 significant figures. **(4 marks)**

Now try this → **Exam question bank Q28, Q42, Q58, Q70, Q82, Q87, Q92**

7.1 Algebraic fractions

1 Simplify these fractions:

a $\frac{6x^4 + 8x^3 - 4x^2}{2x}$ b $\frac{12x - 24x^2 + 9x^3}{3x}$

c $\frac{8x^6 - 4x^5 + 2x^4}{-2x^2}$

Hint Divide each term in the numerator by the denominator.

2 Write each of these fractions in their simplest form:

a $\frac{x^2 - 9}{x - 3}$ b $\frac{x^2 - 3x + 2}{x - 2}$

c $\frac{2x^2 + 5x - 3}{x + 3}$

Hint To simplify an algebraic fraction, factorise the numerator and denominator where possible and cancel common factors.

3 Write each of these fractions in their simplest form:

a $\frac{2x^2 - 3x - 2}{x^2 + x - 6}$ b $\frac{3x^2 + x - 2}{2x^2 + 5x + 3}$ c $\frac{6x^2 + 11x - 10}{2x^2 - 3x - 20}$

P 4 Show that $\frac{15x^2 - 16x - 15}{3x^2 - 5x}$ can be written in the form $A + \frac{B}{x}$, where A and B are integers to be found. **(3 marks)**

P 5 $\frac{x^2 - 2x + k}{x^2 + 2x - 3} = \frac{x - 5}{x - 1}$ where k is a constant. Work out the value of k. **(3 marks)**

P 6 $\frac{4x^2 + 8x - 5}{6x^2 - 7x + k} = \frac{2x + 5}{3x - 2}$ where k is a constant. Work out the value of k. **(4 marks)**

P 7 $\frac{6x^3 - 2x^2 - 4x}{2x^2 - 7x + 5} = \frac{ax(bx + c)}{px + q}$, where a, b, c, p and q are constants.

Work out the values of a, b, c, p and q. **(5 marks)**

7.2 Dividing polynomials

1 Write each polynomial in the form $(x \pm p)(ax^2 + bx + c)$ by dividing:

a $2x^3 - 5x^2 + 8x - 5$ by $(x - 1)$

b $3x^3 + 8x^2 + 3x - 2$ by $(x + 2)$

c $2x^3 + x^2 - 17x - 12$ by $(x - 3)$

d $4x^3 + 13x^2 - 11x + 4$ by $(x + 4)$

Hint You can use long division to divide a polynomial by $(x \pm p)$, where p is a constant.

For example:

$$\begin{array}{r} 2x^2 \\ x - 1\overline{)2x^3 - 5x^2 + 8x - 5} \\ \underline{2x^3 - 2x^2} \end{array}$$

and so on

2 Divide:

a $3x^4 + 8x^3 - x^2 - 13x - 6$ by $(x + 2)$

b $4x^4 - 8x^3 + x^2 - x - 2$ by $(2x + 1)$

c $9x^4 - 3x^3 - 17x^2 + 13x - 2$ by $(3x - 2)$

d $4x^4 - 12x^3 - 5x^2 + 15x + 9$ by $(2x - 3)$

Hint The answer you obtain following the division is called the **quotient**.

3 Divide:

> **Hint** Write the polynomial in part **a** as $2x^3 + 6x^2 + 0x - 4$ before dividing.

a $2x^3 + 6x^2 - 4$ by $(x + 1)$

b $3x^3 + 7x^2 + 18$ by $(x + 3)$

c $4x^3 - 11x - 10$ by $(x - 2)$

d $2x^3 + 7x^2 + 75$ by $(x + 5)$

4 Find the remainder when:

> **Hint** If there is a remainder, then the linear expression $(x \pm p)$ is not a factor.
> The polynomial can be written as $(x \pm p)(ax^2 + bx + c) + r$ where r is the remainder.

a $x^3 + 3x^2 + 5x - 8$ is divided by $(x + 4)$

b $2x^3 - 5x^2 + 12x - 20$ is divided by $(x - 3)$

c $3x^3 + 2x^2 - 40x + 45$ is divided by $(x + 5)$

(E) 5 Find the remainder when $-15x^3 + 26x^2 - 13x + 5$ is divided by $(5x - 2)$. **(2 marks)**

(E/P) 6 $f(x) = 6x^3 - 13x^2 - 13x + 30$

a Find the remainder when $f(x)$ is divided by $(x + 3)$. **(2 marks)**

b Given that $(x - 2)$ is a factor of $f(x)$, factorise $f(x)$ completely. **(4 marks)**

(E/P) 7 $f(x) = 2x^3 + 3x^2 - 4x + k$ where k is a constant.

Given that $(x + 3)$ is a factor of $f(x)$:

a find the value of k **(2 marks)**

b express $f(x)$ in the form $(x + 3)(ax^2 + bx + c)$ where a, b and c are constants **(2 marks)**

c show that $f(x) = 0$ has exactly one real solution. **(2 marks)**

(E/P) 8 $f(x) = 3x^3 + 10x^2 - 8x - 5$

a Find the remainder, r, when $f(x)$ is divided by $(x - 2)$. **(2 marks)**

b Express $f(x)$ in the form $(x - 2)(ax^2 + bx + c) + r$ where a, b, c and r are constants. **(4 marks)**

(E/P) 9 $f(x) = 10x^3 - 29x^2 + 4x + 15$

a Given that $(x - 1)$ is a factor of $f(x)$, express $f(x)$ in the form $(x - 1)(ax^2 + bx + c)$, where a, b and c are constants. **(2 marks)**

b Hence factorise $f(x)$ completely. **(4 marks)**

c Write down all the solutions to the equation $f(x) = 0$. **(2 marks)**

7.3 The factor theorem

1 Use the factor theorem to show that:

> **Hint** The **factor theorem** states that if $f(x)$ is a polynomial, then:
> - if $f(p) = 0$ then $(x - p)$ is a factor of $f(x)$
> - if $(x - p)$ is a factor of $f(x)$ then $f(p) = 0$

a $(x + 1)$ is a factor of $2x^3 + 7x^2 - 5$

b $(x + 2)$ is a factor of $x^3 + 4x^2 + 3x - 2$

c $(x - 3)$ is a factor of $2x^3 - 3x^2 - 7x - 6$

d $(x - 4)$ is a factor of $x^4 - 3x^3 - 15x - 4$

2 Use the factor theorem to show that the linear expression is a factor of the polynomial f(x) and factorise f(x) completely:

a $(x - 2)$, $2x^3 + x^2 - 13x + 6$

b $(x + 3)$, $2x^3 + 17x^2 + 38x + 15$

c $(x - 1)$, $6x^3 - x^2 - 11x + 6$

d $(x + 4)$, $15x^3 + 61x^2 - 2x - 24$

Hint When you have used the factor theorem to show the linear expression is a factor, you can use long division to find the quadratic factor. Factorise the quadratic factor to write the polynomial as a product of three linear factors.

3 Fully factorise each expression:

a $x^3 + 2x^2 - 21x + 18$

b $2x^3 + 13x^2 + 13x - 10$

c $3x^3 + 2x^2 - 41x - 60$

Hint Try values of p in each expression for f(x), e.g. $p = -1, 1, 2, 3, \ldots$ until you find $f(p) = 0$. Then use the factor theorem to deduce that $(x - p)$ is a factor of f(x).

4 For each of the following polynomials,

i fully factorise each polynomial f(x).

ii Hence sketch the graph of $y = f(x)$.

a $2x^3 - 11x^2 + 5x + 18$

b $2x^3 - 3x^2 - 39x + 20$

c $6x^3 + 37x^2 + 50x - 21$

Hint To sketch the graph, you need to identify the points where the curve crosses the axes.

Set $x = 0$ to find the y-intercept and $y = 0$ to find the x-intercepts.

The general shapes of cubic graphs are:

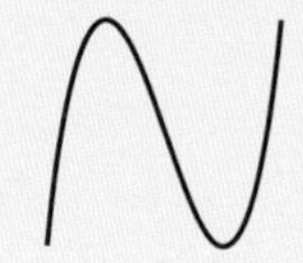

if the coefficient of x^3 is positive

if the coefficient of x^3 is negative

← Section 4.1

P 5 $f(x) = 6x^3 - 17x^2 - 15x + 36$

Given that $(x - 3)$ is a factor of f(x), find all the solutions to $f(x) = 0$. **(5 marks)**

P 6 $f(x) = 9x^3 + 24x^2 - 44x + 16$

a Use the factor theorem to show that $(x + 4)$ is a factor of f(x). **(2 marks)**

b Hence show that f(x) can be written in the form $f(x) = (x + 4)(px + q)^2$, where p and q are integers to be found. **(4 marks)**

E 7 $f(x) = 2x^3 - 3x^2 - 5x + 6$. Factorise f($x$) completely. **(5 marks)**

P 8 $g(x) = x^3 + 2x^2 - 19x + k$

Given that $(x + 1)$ is a factor of g(x),

a show that $k = -20$ **(2 marks)**

b express g(x) as a product of three linear factors. **(3 marks)**

c Sketch the curve with equation $y = x^3 + 2x^2 - 19x - 20$, indicating the values where the curve crosses the x-axis and the y-axis. **(4 marks)**

E/P 9 $p(x) = 25x^3 + 55x^2 - 56x + 12$

a Use the factor theorem to show that $(x + 3)$ is a factor of $p(x)$. **(2 marks)**

b Fully factorise $p(x)$. **(3 marks)**

c Hence show that there are exactly two real roots of the equation $p(x) = 0$. **(5 marks)**

7.4 Mathematical proof

1 Prove that $(x - 4)(x + 6)(2x + 3) \equiv 2x^3 + 7x^2 - 42x - 72$

Hint The identity symbol $\equiv$ means 'is equivalent to'. This means the expressions are equal for all values of x. To prove an identity you should:

- start with the expression on one side of the identity
- algebraically manipulate the expression until it matches the other side
- show every step in your algebraic working.

2 Prove that the triangle with vertices at $P(10, 14)$, $Q(-6, 2)$ and $R(12, 8)$ is right angled.

Hint In a mathematical proof you must:

- state any information or assumptions you are using
- clearly show every step of your working
- make sure each step follows logically from the previous step
- write a statement of proof at the end of your working.

You could find the lengths of the sides in this triangle and use Pythagoras' theorem, or use gradients of line segments. ← Section 5.4

3 Prove that $y = kx$ is a tangent to the curve with equation $y = 4x^2 - 5x + 4$ when $k = 3$ and $k = -13$.

Hint If a line is a tangent to a curve, then there is only one point of intersection. Solve the equations simultaneously and use the discriminant. ← Section 2.5

4 Prove that the points $A(-2, 3)$, $B(2, 1)$ and $C(14, -5)$ are collinear.

Hint If points are collinear, they lie on the same straight line. ← Section 5.1

E/P **5** Prove that $x^2 - 6x + 10 > 0$ for all values of x. **(3 marks)**

E/P **6** Prove that $(x + y)^2 - (x - y)^2 \equiv 4xy$ **(3 marks)**

P 7 Prove that the difference between the squares of two consecutive odd numbers is a multiple of 8. **(4 marks)**

P 8 Prove that the difference between the cube and the square of an odd number is even. **(4 marks)**

P 9 The equation $x^2 + (k - 3)x + (3 - 2k) = 0$ where k is a constant has no real roots.
Prove that k satisfies $-3 < k < 1$. **(6 marks)**

7.5 Methods of proof

1 $f(n) = n^2 + n + 17$ for $n \in \mathbb{N}$.

a Prove that for $5 \leqslant n \leqslant 10$, $f(n)$ is prime.

b Find a counter-example to prove that $f(n)$ is not prime for all positive integers, n.

> **Hint** You can prove a mathematical statement is not true by a **counter-example**. A counter-example is one example that does not work for the statement. One example is sufficient to disprove the statement.

2 Prove that the following statement is **not** true.
'The difference of two prime numbers is always even.'

3 **a** Prove that for positive values of x, $(1 + 2x)^2 > 1 + 4x^2$

b Find a counter-example to prove that the inequality $(1 + 2x)^2 > 1 + 4x^2$ is not true for all values of x.

4 A student is trying to prove that $\frac{x(x+2)^2 + x}{x}$, is a positive numbers of all real values of x.

The student writes:

$$\frac{x(x+2)^2 + x}{x} = \frac{x(x+2)^2}{x} + \frac{x}{x} = (x+2)^2 + 1$$

As $(x + 2)^2 \geqslant 0$ for all values of x then $(x + 2)^2 + 1 > 0$ for all values of x.

a Identify the error made in the proof.

b State the value of x for which the statement is not true.

P 5 Use a counter-example to show that the following statement is false.
'If you add 2 to a number and square it, the result is always greater than the square of the number.' **(2 marks)**

P 6 Prove that $(2n + 1)^2 > 2^{n+1}$ for all positive integers less than 7. **(3 marks)**

P 7 Amir claims that $x^2 \geqslant x$.
Determine whether Amir's claim is always true, sometimes true or never true, justifying your answer. **(2 marks)**

P 8 Prove that $(x + 5)^2 \geqslant 4x + 9$ for all real values of x. **(3 marks)**

P 9 Prove that if n is an integer and $2 \leqslant n \leqslant 7$, then $n^2 + 2$ is not divisible by 4. **(3 marks)**

Problem solving Set A

Bronze

$p(x) = x^3 - 9x^2 + 2x + 48$

a Find the remainder, r, when $p(x)$ is divided by $(x - 4)$. **(2 marks)**

b Hence write $p(x)$ in the form $(x - 4)(x^2 + ax + b) + r$ where a, b and r are constants. **(2 marks)**

c Given that $(x + 2)$ is a factor of $p(x)$, factorise $p(x)$ completely. **(4 marks)**

d Prove that if n is an integer and $3 < n < 8$, then $n^3 - 9n^2 + 2n + 48 < 0$ **(3 marks)**

Silver

$p(x) = 2x^3 - 15x^2 + 32x + 20$

Given that $(2x + 1)$ is a factor of $p(x)$,

a use long division to express $p(x)$ in the form $(2x + 1)(x^2 + ax + b)$, where a and b are constants to be found. **(3 marks)**

b Hence show that $p(x) = 0$ has only one real root. **(4 marks)**

c Prove that $2x^3 - 15x^2 + 32x + 20 > 0$ for all real positive values of x. **(3 marks)**

Gold

$p(x) = 9x^3 - 6x^2 - 20x - 8$

Given that $p(x) = (x + d)(ax^2 + bx + c)$, where a, b, c and d are integers.

a find the values of a, b, c and d. **(6 marks)**

b Prove that $p(x) = 0$ has exactly two real roots. **(3 marks)**

c Sketch the curve with equation $y = 9x^3 - 6x^2 - 20x - 8$, indicating the values where the curve cuts the x-axis and the y-axis. **(4 marks)**

d Hence or otherwise, use set notation to write down the solutions to the inequality

$9x^3 - 6x^2 < 20x + 8$ **(3 marks)**

Problem solving Set B

Bronze

$f(x) = x^3 - x^2 + px + q$ where p and q are integers.

Given that $(x + 1)$ is a factor of $f(x)$,

a show that $q - p = 2$. **(3 marks)**

Given that $(x + 3)$ is also a factor of $f(x)$,

b show that $q - 3p = 36$. **(3 marks)**

c Hence find the value of p and the corresponding value of q. **(2 marks)**

d Factorise $f(x)$ completely. **(2 marks)**

Silver

$f(x) = 2x^3 - x^2 + px + q$ where p and q are integers.

Given that $(x + 2)$ is a factor of $f(x)$,

a show that $q - 2p - 20 = 0$. **(3 marks)**

Given that $(x - 3)$ is also a factor of $f(x)$,

b find the value of p and the corresponding value of q. **(5 marks)**

c Factorise $f(x)$ completely. **(2 marks)**

Gold

$f(x) = x^3 + (p + 4)x^2 + 8x + q$ where p and q are integers.

Given that $(x - 2)$ is a factor of $f(x)$,

a show that $4p + q + 40 = 0$. **(3 marks)**

Given that $(x + p)$ is also a factor of $f(x)$, and that $p > 0$,

b show that $4p^2 - 8p + q = 0$. **(3 marks)**

c Hence find the value of p and the corresponding value of q. **(5 marks)**

d Factorise $f(x)$ completely. **(2 marks)**

Now try this → **Exam question bank Q30, Q34, Q41, Q57, Q69, Q75, Q83**

8 The binomial expansion

8.1 Pascal's triangle

1 Use Pascal's triangle to write down the expansion of:

a $(3 - x)^5$ **b** $(2x + y)^4$

c $(3x - 2y)^5$

Hint The coefficients in the expansion of $(a + b)^n$ are in the $(n + 1)$th row of Pascal's triangle.

2 Find the coefficient of x^3 in the expansion of:

a $(5 + 2x)^4$ **b** $(2 - 3x)^5$

c $\left(3 - \frac{1}{2}x\right)^4$

Hint The term in x^n is the $(n + 1)$th term in the expansion in ascending powers of x. When substituting a term with two or more characters, such as $2x$, into the expansion of $(a + b)^n$ use brackets to make sure you don't make a mistake.

3 Fully expand the expression $(1 - 2x)(1 + 3x)^3$.

Hint Expand $(1 + 3x)^3$, then multiply each term by 1 and by $-2x$.

(E) 4 **a** Find all the terms of the expansion of $(2 + 3x)^4$. Give each term in its simplest form. **(2 marks)**

b Write down the expansion of $(2 - 3x)^4$ in ascending powers of x, giving each term in its simplest form. **(1 mark)**

(E/P) 5 The coefficient of x^2 in the expansion of $(3 + kx)^4$ is 1350. Find two possible values of the constant, k. **(3 marks)**

(E/P) 6 **a** Expand $(4 + k)^3$. **(2 marks)**

b Hence, or otherwise, write down the expansion of $(4 + x^2 - x)^3$ in ascending powers of x. **(3 marks)**

(E/P) 7 The coefficient of x^2 in the expansion of $(5 + x)(2 - kx)^3$ is 18. Find two possible values of the constant, k. **(3 marks)**

8.2 Factorial notation

1 Without using a calculator, work out:

a $0!5!$ **b** $\frac{12!}{9!}$ **c** $\frac{20!}{18!}$

Hint By definition, $0! = 1$. For parts **b** and **c**, you can simplify by cancelling. For example, 12! can be written as $12 \times 11 \times 10 \times 9!$

2 Use a calculator to work out:

a $\binom{16}{9}$ **b** $^{12}C_5$

c $\binom{22}{10}$ **d** $^{18}C_{11}$

Hint $^nC_r = \binom{n}{r} = \frac{n!}{r!(n - r)!}$, $n, r \in \mathbb{N}$.

You can use the ^nC_r and ! functions on your calculator.

3 Write each value, p, q, r and s, from Pascal's triangle using ${}^nC_r$ notation.

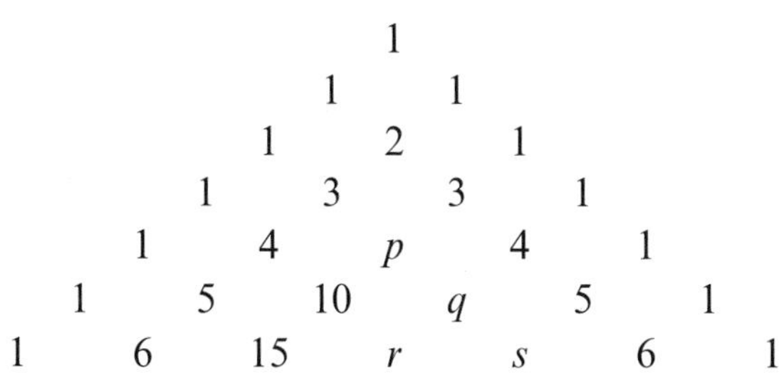

4 Write the 8th number in the 15th row of Pascal's triangle in ${}^nC_r$ form.

Hint The rth entry in the nth row of Pascal's triangle is given by ${}^{n-1}C_{r-1}$ or $\binom{n-1}{r-1}$.

5 The probability of rolling n threes when a fair dice is rolled 10 times is given by

$$\binom{10}{n}\left(\frac{1}{6}\right)^n\left(\frac{5}{6}\right)^{10-n},\ 0 \leqslant n \leqslant 10$$

Find the probability of rolling:

a no threes **b** 5 threes

6 Given that $\binom{25}{11} = \frac{25!}{11!p!}$, write down the value of p. **(1 mark)**

7 Given that $\binom{30}{q} = \frac{30!}{q!8!}$, write down the value of q. **(1 mark)**

P **8** **a** Show that ${}^{k+2}C_k = \frac{(k+2)(k+1)}{2}$ **(3 marks)**

b Given that ${}^{k+2}C_k = 36$, find the value of k. **(3 marks)**

c Explain why there is only one solution for k in part **b**. **(1 mark)**

8.3 The binomial expansion

1 Use the binomial theorem to find the expansion of:

a $(1 + x)^5$ **b** $(2 + x)^6$ **c** $(10 + x)^4$

2 Use the binomial theorem to find the expansion of:

a $(3 - 2x)^4$ **b** $\left(4 + \frac{1}{2}x\right)^5$

c $(x - 2)^6$

Hint The binomial expansion is:

$$(a + b)^n = a^n + \binom{n}{1}a^{n-1}b + \binom{n}{2}a^{n-2}b^2 + \dots + \binom{n}{r}a^{n-r}b^r + \dots + b^n\ (n \in \mathbb{N})$$

where ${}^nC_r = \binom{n}{r} = \frac{n!}{r!(n-r)!}$, $(n, r \in \mathbb{N})$

3 Use the binomial theorem to find the first four terms, in ascending powers of x, in the expansion of:

Hint You need to expand each expression up to the term in x^3.

a $(1 + x)^{10}$ **b** $(1 + x)^{15}$ **c** $(2 + x)^9$

4 Use the binomial theorem to find the first four terms, in ascending powers of x, in the expansion of:

a $\left(1 - \frac{x}{2}\right)^{10}$ b $\left(2 - \frac{1}{3}x\right)^9$

c $(3 + 2x)^8$

Hint Use brackets when you substitute terms into the binomial expansion formula, and include any negative signs.

For example, the 4th term of $\left(2y - \frac{x}{2}\right)^5$ is

$\binom{5}{3}(2y)^2\left(-\frac{x}{2}\right)^3 = 10(4y^2)\left(-\frac{x^3}{8}\right) = -5y^2x^3$

(E) 5 Find the first four terms, in ascending powers of x, of the binomial expansion of $(2 - 3x)^5$, giving each term in its simplest form. **(4 marks)**

(E) 6 Given that $(3 - x)^{12} \equiv A + Bx + Cx^2 + \dots$, find the values of the integers A, B and C. **(4 marks)**

(E) 7 Find the first four terms, in ascending powers of x, of the binomial expansion of $\left(1 - \frac{2}{3}x\right)^9$. **(4 marks)**

(E/P) 8 Find the binomial expansion of $\left(x - \frac{2}{x}\right)^5$, giving each term in its simplest form. **(4 marks)**

8.4 Solving binomial problems

1 Find the coefficient of x^3 in the expansion of:

a $(1 + x)^{14}$ b $(2 + x)^{10}$ c $(5 - x)^7$

2 Find the coefficient of x^4 in the expansion of:

a $(3 - 2x)^9$ b $(5 + 3x)^7$

c $\left(\frac{1}{2} + 8x\right)^{12}$

Hint The **general term** in the expansion of $(a + b)^n$ is $\binom{n}{r}a^{n-r}b^r$

3 Find the coefficient of x^3 in the expansion of $(3 - x)(2 + 3x)^6$.

Hint First expand $(2 + 3x)^6$ up to the term in x^3 then multiply by $(3 - x)$.

There are two ways of making a term in x^3: (constant term × x^3 term) and (x term × x^2 term)

(E) 4 Find the coefficient of x^3 in the expansion of $\left(1 + \frac{2}{3}x\right)^6$ **(3 marks)**

(E/P) 5 The coefficient of x^2 in the expansion of $(1 + 2x)^n$ is 60. Given that $n > 0$, find the value of n. **(3 marks)**

(P) 6 The coefficient of x^2 in the expansion of $(3 + ax)^5$ is 4320. Find two possible values of the constant, a.

(P) 7 The coefficient of x^3 in the expansion of $(2 - bx)^6$ is -20. Find the value of the constant, b.

P 8 **a** Find the first four terms, in ascending powers of x, of the binomial expansion of $(2 - 3x)^8$, giving each term in its simplest form. **(4 marks)**

b Hence, or otherwise, find the coefficient of x^3 in the expansion of $\left(1 + \frac{x}{2}\right)(2 - 3x)^8$. **(3 marks)**

P 9 **a** Find the first three terms, in ascending powers of x, of the binomial expansion of $(3 + px)^6$, where p is a non-zero constant. Give each term in its simplest form. **(4 marks)**

b Given that, in this expansion, the coefficient of x^2 is 5 times the coefficient of x, find the value of p. **(2 marks)**

8.5 Binomial estimation

1 **a** Find the first four terms, in ascending powers of x, of the binomial expansion of $\left(1 - \frac{x}{5}\right)^9$ Give each term in its simplest form.

b Use your expansion to estimate the value of 0.985^9, giving your answer to 4 decimal places.

Hint For part **b**, work out the value of x to substitute into the expansion by letting $1 - \frac{x}{5} = 0.985$.

2 **a** Find the first three terms, in ascending powers of x, of the binomial expansion of $(1 - 2x)^6$. Give each term in its simplest form.

b If x is so small that x^3 and higher powers can be ignored, show that $(3 + x)(1 - 2x)^6 \approx 3 - 35x + 168x^2$

Hint Multiply your answer to part **a** by $(3 + x)$ and collect like terms.

3 **a** Write down the first four terms, in ascending powers of x, of the binomial expansion of $\left(3 + \frac{x}{4}\right)^8$

b By substituting an appropriate value for x, find an approximate value for 3.05^8.

Hint For part **b**, work out the value of x to substitute into the expansion by letting $\frac{x}{4} = 0.05$.

4 **a** Find the first four terms, in ascending powers of x, of the binomial expansion of $\left(1 + \frac{x}{4}\right)^9$ Give each term in its simplest form. **(4 marks)**

b Use your expansion to estimate the value of 1.015^9, giving your answer to 3 decimal places. **(3 marks)**

P 5 **a** Find the first four terms, in ascending powers of x, of the binomial expansion of $(1 - 3x)^6$. Give each term in its simplest form. **(4 marks)**

b If x is small, so that x^3 and higher powers can be ignored, show that

$$\left(1 + \frac{x}{2}\right)(1 - 3x)^6 \approx 1 - \frac{35x}{2} + 126x^2$$ **(2 marks)**

E/P 6 **a** Find the first three terms, in ascending powers of x, of the binomial expansion of $\left(2-\frac{x}{2}\right)^7$, giving each term in its simplest form. **(4 marks)**

b Explain how you would use your expansion to give an estimate for the value of 1.985^7. **(1 mark)**

E/P 7 **a** Find the first four terms, in ascending powers of x, of the binomial expansion of $\left(3-\frac{x}{5}\right)^{10}$
Give each term in its simplest form. **(4 marks)**

b Use your answer to part **a** to estimate the value of 2.995^{10}, giving your answer to 6 decimal places. **(3 marks)**

c Use your calculator to evaluate 2.995^{10} and calculate the percentage error in your answer to part **b**. **(2 marks)**

Problem solving Set A

Bronze

In the binomial expansion of $(3-kx)^6$, where k is a constant, the coefficient of x^3 is -4320.
Calculate:

a the value of k **(4 marks)**

b the value of the coefficient of x^4 in the expansion. **(2 marks)**

Silver

a Write down the first three terms, in ascending powers of x, of the binomial expansion of $(2-px)^{11}$, where p is a non-zero constant, giving each term in its simplest form. **(4 marks)**

b Given that, in the expansion of $(2-px)^{11}$, the coefficient of x is q and the coefficient of x^2 is $10q$, find the value of p and the value of q. **(3 marks)**

Gold

In the binomial expansion of $\left(3+\frac{x}{k}\right)^8$, the coefficient of x^2 is 3 times the coefficient of x^3.
Find the value of the constant, k. **(7 marks)**

Problem solving Set B

Bronze

$f(x) = (2 + kx)(1 - 5x)^4$, where k is a constant.
The expansion, in ascending powers of x, of $f(x)$ up to and including the term in x^2 is
$A - 37x + Bx^2$, where A and B are constants.

a Find the first three terms, in ascending powers of x, of the binomial expansion of $(1 - 5x)^4$, giving each term in its simplest form. **(4 marks)**

b Write down the value of A. **(1 mark)**

c Find the value of k. **(2 marks)**

d Hence find the value of B. **(2 marks)**

Silver

$f(x) = (1 + kx)(2 - 5x)^4$, where k is a constant.
The expansion, in ascending powers of x, of $f(x)$ up to and including the term in x^2 is
$A - 64x + Bx^2$, where A and B are constants.

Find the values of A, B and k. **(7 marks)**

Gold

$f(x) = (3 + px)(1 + qx)^5$, where p and q are positive constants.
The expansion, in ascending powers of x, of $f(x)$ up to and including the term in x^3 is
$3 + 17x + \frac{70}{3}x^2 + kx^3$ where k is a constant.

a Find the value of p and the value of q. **(7 marks)**

b Hence find the value of k. **(1 mark)**

Now try this → **Exam question bank Q9, Q20, Q29, Q40, Q51, Q61**

9 Trigonometric ratios

9.1 The cosine rule

1 In triangle ABC, $AB = 7.2$ cm, $BC = 9.3$ cm and $\angle ABC = 110°$. Find the length of side AC.

Hint Use this version of the cosine rule if you know two sides and the angle between them and want to find the missing side:

$$a^2 = b^2 + c^2 - 2bc\cos A$$

2 Find the size of the largest angle in a triangle with sides of length 5 cm, 6.5 cm and 10 cm.

Hint Use this version of the cosine rule to find an angle if you know all three sides:

$$\cos A = \frac{b^2 + c^2 - a^2}{2bc}$$

In any triangle, the largest angle is opposite the longest side.

3 In each of these triangles calculate the value of x. Give your answers to 3 significant figures.

a A, B, C; 3.6 cm, x cm, 30°, 5.2 cm

b A, C, B; 1.8 cm, 110°, 3.1 cm, x cm

c A, B, C; 5.8 cm, 50°, 6.2 cm, x cm

d A, B, C; 5 cm, x, 9 cm, 12 cm

e A, B, C; 4.9 cm, x, 7.1 cm, 7.1 cm

f A, B, C; 9.3 cm, 18.2 cm, x, 10.4 cm

4 Three towns A, B and C are positioned such that:

B is 40 km from A on a bearing of 105°

C is 48 km from A on a bearing of 135°

Calculate the distance between B and C.

Hint Draw your own clearly labelled diagram in questions where one is not provided.

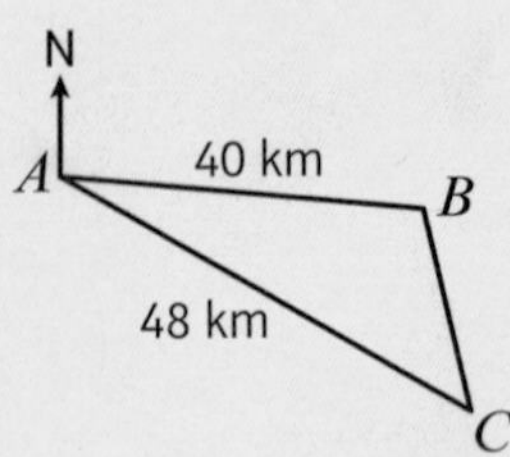

Bearings are measured from north in a clockwise direction.

(E) **5** In triangle ABC, $AB = 5$ cm, $BC = 8$ cm, $AC = 6$ cm and $\angle BAC = \theta$.
Show that $\cos\theta = -\frac{1}{20}$ **(2 marks)**

6 The diagram shows a lighthouse L and two boats P and Q which are assumed to be in the same horizontal plane.

Boat P is 600 m due north of L and boat Q is 800 m from L on a bearing of 020°.

Calculate the distance between boat P and boat Q. **(3 marks)**

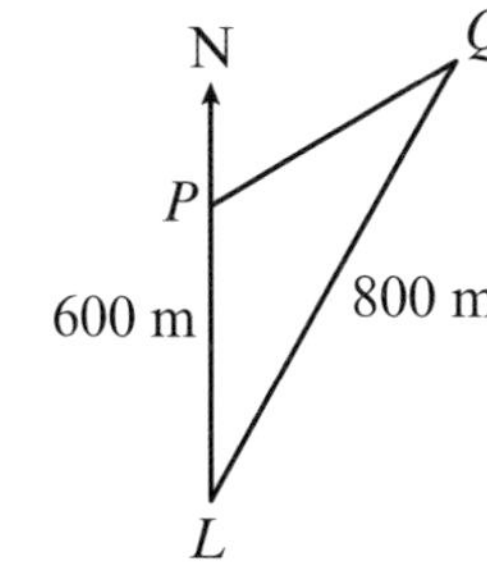

P 7 In triangle ABC, $AB = 7$ cm, $BC = 12$ cm and $AC = 10$ cm.
Use the cosine rule to show that triangle ABC does **not** contain an obtuse angle. **(3 marks)**

P 8 $ABCD$ is a quadrilateral. $AB = 8$ cm, $BC = 10$ cm and $AD = 9$ cm.
$\angle ABC = 80°$ and $\angle ADC = 90°$.

Calculate the perimeter of $ABCD$. **(5 marks)**

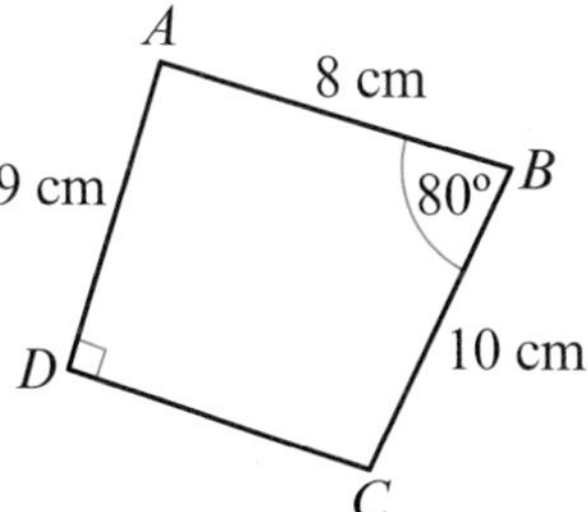

P 9 In triangle ABC, $AB = (x + 2)$ cm, $AC = (x + 1)$ cm and $BC = x$ cm.
Given that $\cos \angle BAC = \frac{3}{4}$, find the value of x. **(6 marks)**

9.2 The sine rule

1 Work out the length of the side marked x in each triangle:

a

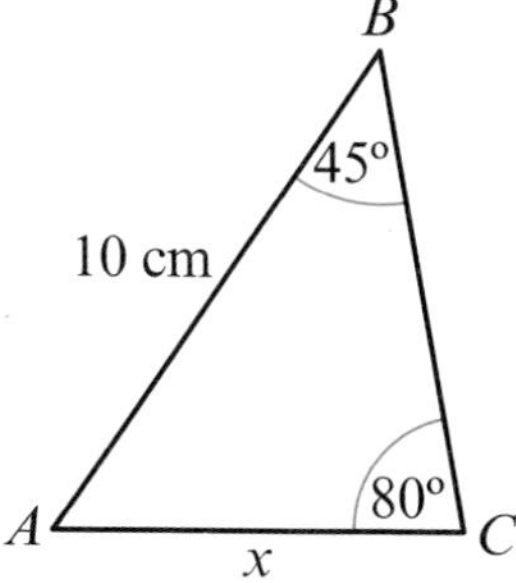

b

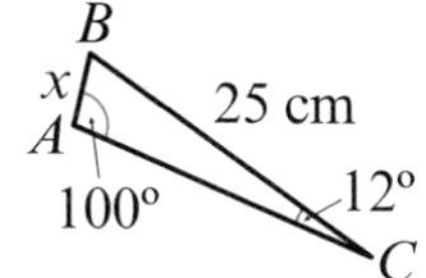

c

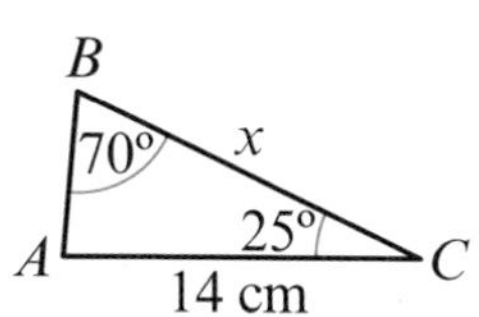

Hint Use this version of the sine rule to find the length of a missing side:

$$\frac{a}{\sin A} = \frac{b}{\sin B} = \frac{c}{\sin C}$$

If you are given two angles in a triangle you can find the third angle using

Sum of angles in a triangle = 180°

2 Work out the size of the angle marked θ in each triangle:

a

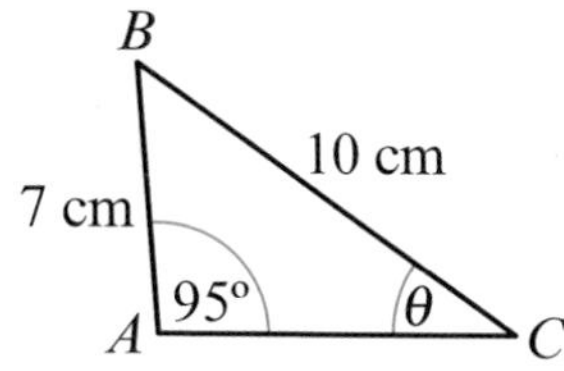

b

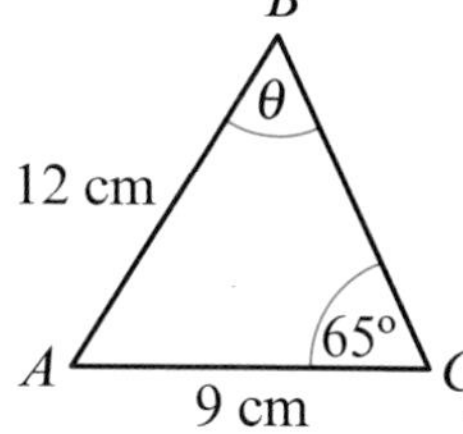

c

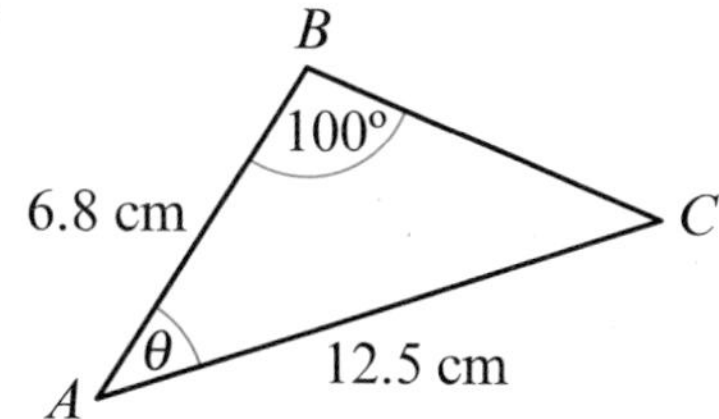

Hint Use this version of the sine rule to find a missing angle:

$$\frac{\sin A}{a} = \frac{\sin B}{b} = \frac{\sin C}{c}$$

3 In the diagram find:

a $\angle ADC$

b the length of CD.

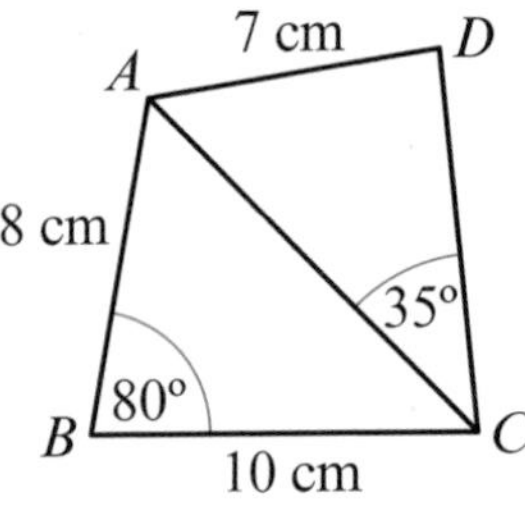

Hint Start by using trigonometry on triangle ABC to find the length of AC.

4 In triangle ABC, $AB = 6$ cm, $AC = 4$ cm and $\angle ABC = 40°$.

a Calculate the two possible values of $\angle ACB$.

b Draw a diagram to illustrate your answers.

Hint As $\sin\theta = \sin(180° - \theta)$, the sine rule sometimes produces two possible solutions for a missing angle. You need to check the sum of the angles of the triangle to see if there are two possible solutions.

(E) 5 In triangle ABC, $AC = 18$ cm, $BC = 20$ cm and $\angle BAC = 50°$. Calculate the size of $\angle ACB$. **(3 marks)**

(E/P) 6 In triangle ABC, $AB = 9$ cm, $AC = 8$ cm, $\angle ABC = 25°$ and $\angle BCA = x$.

a Find the value of $\sin x$, giving your answer to 3 decimal places. **(2 marks)**

b Given that there are two possible values of x, find these values of x, giving your answers to 2 decimal places. **(3 marks)**

(E/P) 7 Three supermarkets A, B and C lie in the same horizontal plane and are positioned such that:

B is 6 km from A on a bearing of 015°

C is on a bearing of 045° from A

The distance between B and C is 14 km. Calculate:

a the distance of C from A **(3 marks)**

b the bearing of C from B to the nearest degree. **(4 marks)**

(E/P) 8 In the diagram, $AC = 7$ cm, $BC = 4$ cm. $\angle OBA = 65°$, $\angle CAD = 15°$ and $\angle BAC = \theta$.

Find:

a the size of angle θ **(4 marks)**

b the length of AD **(4 marks)**

c the length of CD. **(3 marks)**

(E/P) 9 In the diagram, $AC = 2x + 1$, $BC = 4x - 3$, $\sin A = \frac{\sqrt{3}}{2}$ and $\sin B = \frac{3}{4}$

Show that the exact value of x is $\frac{11 + 5\sqrt{3}}{8}$ **(7 marks)**

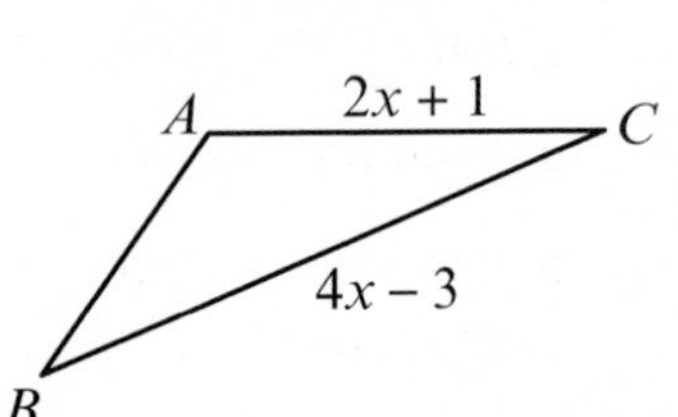

9.3 Areas of triangles

1 Calculate the area of each triangle.

a

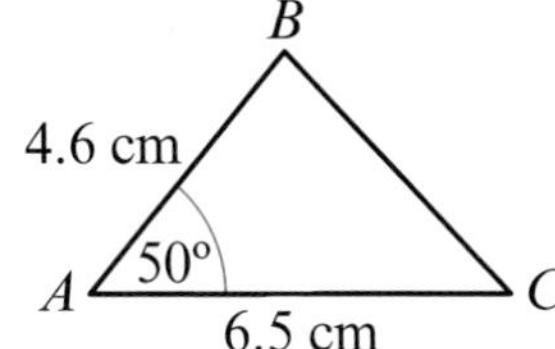

b

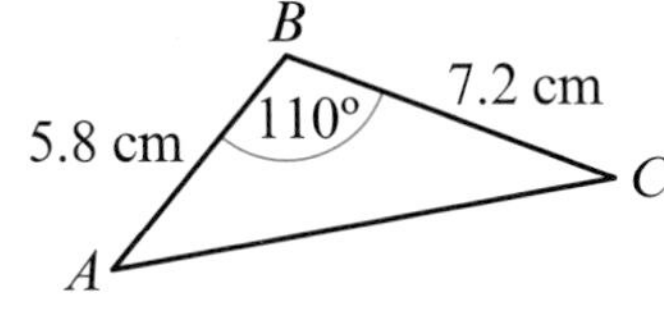

c

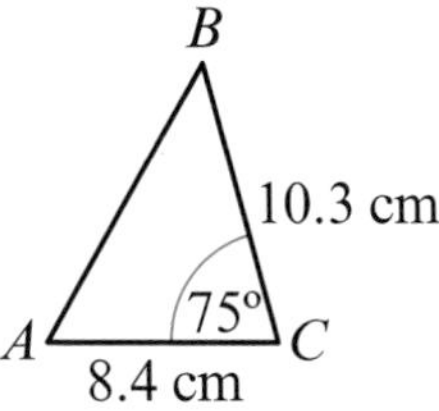

> **Hint** You can find the area of a triangle when you know two sides and the angle between them.
>
> Area $= \frac{1}{2}ab\sin C$

2 Work out the possible sizes of angle x in each triangle.

a

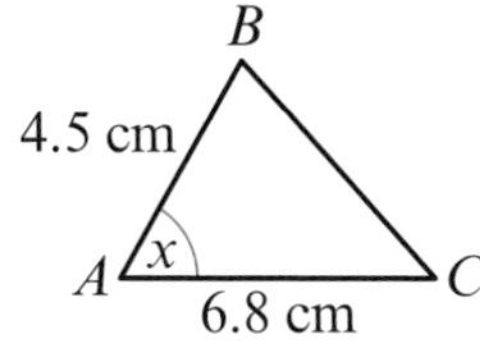

Area = 8.2 cm²

b

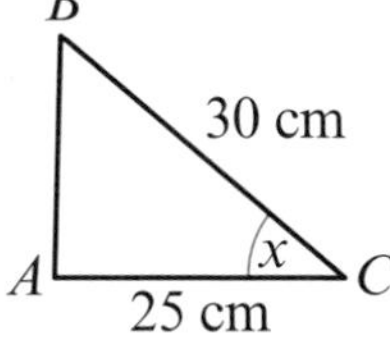

Area = 300 cm²

c

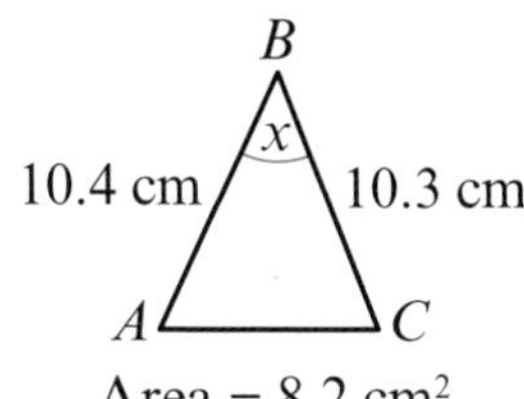

Area = 8.2 cm²

> **Hint** There are two possible solutions for each triangle.
>
> Use $\sin\theta = \sin(180° - \theta)$ to find one acute angle and one obtuse angle.

3 The area of a triangular plot of land is 1500 m². The lengths of two sides of the triangle are 75 m and 80 m, and the angle between them is θ.

a Show that a possible size of angle θ is 150°.

b Work out the total perimeter of the plot of land.

> **Hint** For part **b**, use the cosine rule to work out the length of the longest side, as you know two sides and the angle between them.

E **4** The diagram shows a triangle ABC. The length of AC is 16.4 cm.

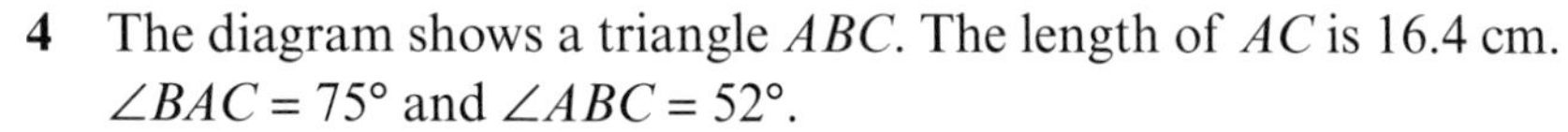

$\angle BAC = 75°$ and $\angle ABC = 52°$.

a Find the length of BC to the nearest 0.1 cm. **(3 marks)**

b Calculate the area of triangle ABC, giving your answer to the nearest cm². **(3 marks)**

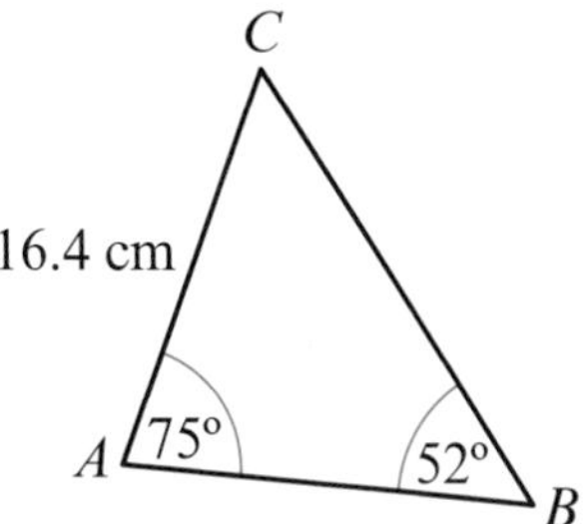

(E/P) **5** In triangle PQR, $PR = x + 3$, $QR = 6 - x$ and $\sin \angle PRQ = \frac{2}{5}$
The area of the triangle is A cm².

a Find an expression for A in the form $\frac{ax^2 + bx + c}{d}$ where a, b, c and d are integers. **(3 marks)**

b By completing the square or otherwise, find the maximum value of A and state the corresponding value of x. **(4 marks)**

(E/P) **6** In triangle ABC, $AB = 3x$ and $AC = 2x + 1$.

Given that $\sin \angle BAC = \frac{3}{8}$ and the area of the triangle is 43 cm², calculate the value of x, giving your answer to 3 significant figures. **(5 marks)**

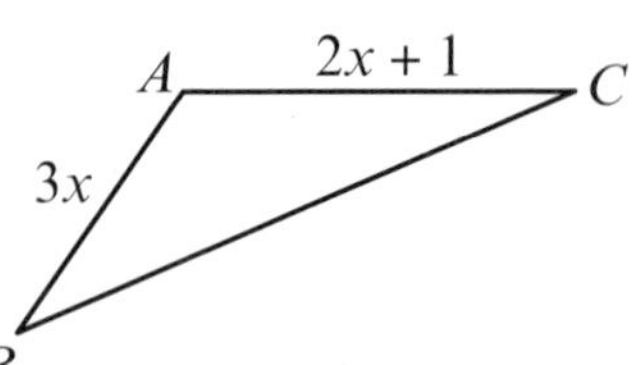

(E/P) **7** A triangular floor space is modelled by the triangle ABC shown in the diagram. The length of AB is 20 m, $\angle BAC = 80°$ and $\angle ABC = 70°$.

a Calculate the area of the floor space to 3 significant figures. **(4 marks)**

b Why is your answer unlikely to be accurate to the nearest square metre? **(1 mark)**

9.4 Solving triangle problems

1 In each triangle find the values of x, y and z. Give your answers to 3 significant figures.

a

b

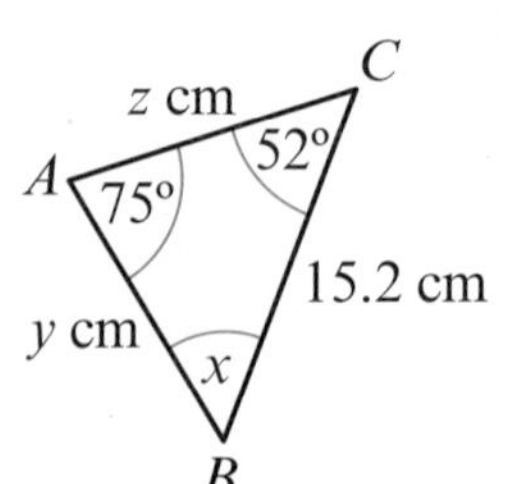

c

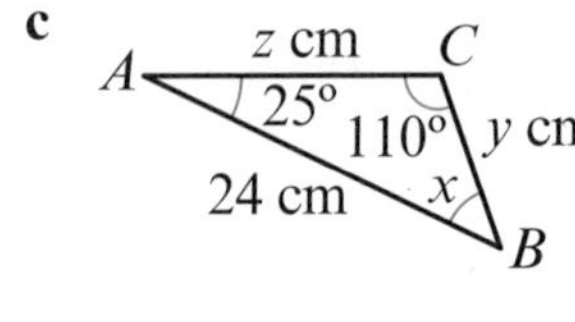

Hint If you have a choice of method, the sine rule is often easier to use than the cosine rule.

2 In each triangle find the value of x.
Give your answers to 3 significant figures.

a

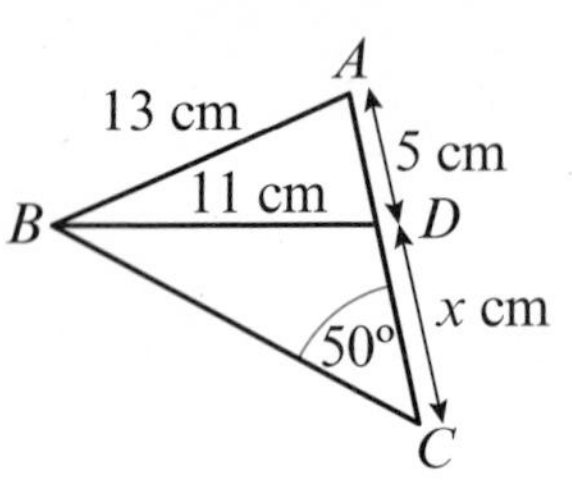

b

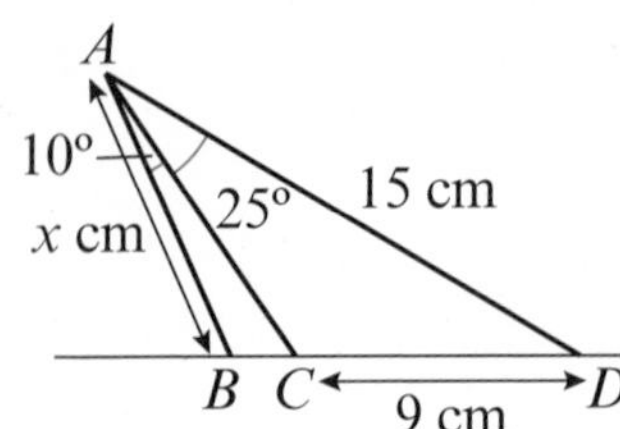

c

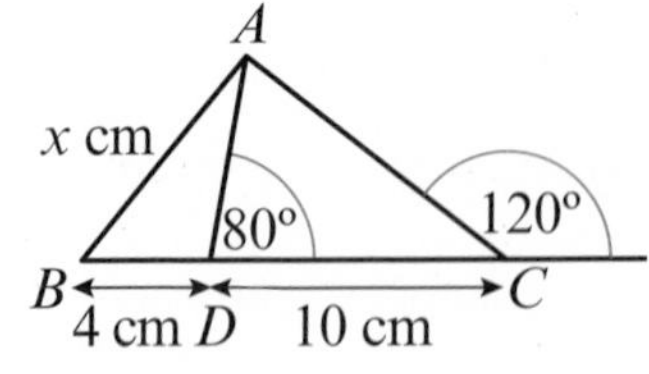

3 The diagram shows the dimensions of a field. $AB = 50$ m, $BC = 65$ m and $CD = 80$ m. $\angle ABC = 130°$ and $\angle BCD = 52°$.

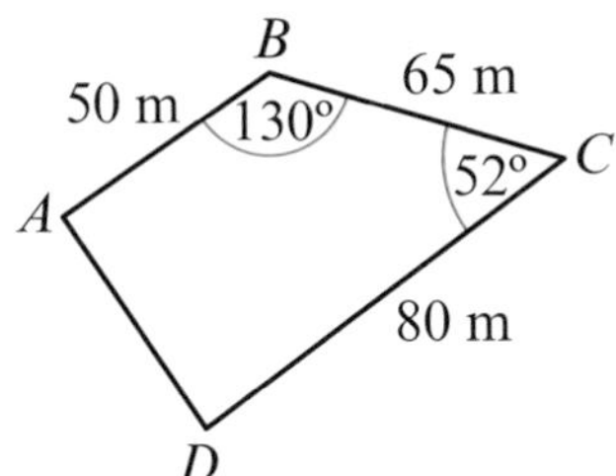

Hint Split the quadrilateral into two triangles, ABC and ACD.

Calculate the area of the field.

P 4 $ABCD$ is a quadrilateral. $AB = 8$ cm, $AD = 7$ cm and $BC = 10$ cm. $\angle ABC = 80°$ and $\angle ADC = 90°$.
Calculate the area of quadrilateral $ABCD$. **(7 marks)**

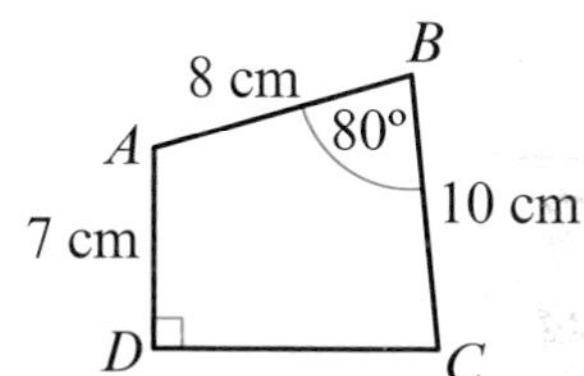

P 5 In triangle ABC, $AB = 15$ cm, $BC = 8$ cm and $\angle ABC = \theta$.
The area of triangle ABC is 48 cm^2.

a Find the two possible values of $\cos\theta$. **(4 marks)**

b Given that AC is the longest side of the triangle, find the exact length of AC. **(2 marks)**

P 6 The diagram shows a trapezium $ABCD$.
$AB = 8$ cm, $AD = 9$ cm and $DE = 5$ cm.
$\angle DAE = 30°$ and $\angle BEC = 40°$.

Find the area of triangle ABE. **(7 marks)**

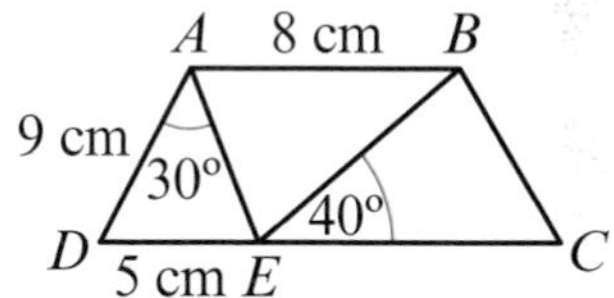

P 7 A ship A is 10 km from a lighthouse L on a bearing of 080°.
A ship B is 6.5 km from L on a bearing of 200°.

a Find the distance between A and B. **(5 marks)**

b Find the bearing of A from B to the nearest degree. **(5 marks)**

9.5 Graphs of sine, cosine and tangent

1 **a** Sketch the graph of $y = \sin\theta$ for $-360° \leqslant \theta \leqslant 360°$.

b State the period of the graph of $y = \sin\theta$.

c State the maximum and minimum values of y.

Hint The **period** of a graph is the interval covered on the horizontal axis before the graph starts to repeat itself.

2 **a** Sketch the graph of $y = \cos\theta$ for $-360° \leqslant \theta \leqslant 360°$.

b State the period of the graph of $y = \cos\theta$.

c State the maximum and minimum values of y.

Hint The graph of $y = \cos\theta$ has the same shape as the graph of $y = \sin\theta$, but translated horizontally.

3 **a** Sketch the graph of $y = \tan\theta$ for $-360° \leqslant \theta \leqslant 360°$.

b State the period of the graph of $y = \tan\theta$.

Hint The graph of $y = \tan\theta$ has no maximum or minimum values.

It has vertical asymptotes at intervals at $\theta = \ldots, -270°, -90°, 90°, 270°, \ldots$

4 Use your answers to questions **1–3** to identify, for $-360° \leqslant \theta \leqslant 360°$, the values of θ for which:

a $\sin\theta = 1$ **b** $\cos\theta = -1$ **c** $\tan\theta = 0$

(E/P) **5** **a** Use a calculator to find a solution to $\sin\theta = 0.5$. **(1 mark)**

b Sketch the graphs of $y = \sin\theta$ and $y = 0.5$ for $-360° \leqslant \theta \leqslant 360°$ on the same set of axes. **(3 marks)**

c Use your graphs to solve the equation $\sin\theta = 0.5$ for $-360° \leqslant \theta \leqslant 360°$. **(2 marks)**

(E/P) **6** **a** Use a calculator to find a solution to $\tan\theta = -1$. **(1 mark)**

b Sketch the graphs of $y = \tan\theta$ and $y = -1$ for $-360° \leqslant \theta \leqslant 360°$ on the same set of axes. **(3 marks)**

c Use your graphs to solve the equation $\tan\theta = -1$ for $-360° \leqslant \theta \leqslant 360°$. **(2 marks)**

(E/P) **7** **a** Sketch the graph of $y = \cos\theta$ for $-360° \leqslant \theta \leqslant 360°$. **(2 marks)**

b Given $\cos 45° = \frac{1}{\sqrt{2}}$, use your graph to find all solutions for $-360° \leqslant \theta \leqslant 360°$ to:

i $\cos\theta = \frac{1}{\sqrt{2}}$ **ii** $\cos\theta = -\frac{1}{\sqrt{2}}$ **(4 marks)**

9.6 Transforming trigonometric graphs

1 **a** On the same set of axes, sketch the graphs of $y = \cos x$, $y = 2\cos x$ and $y = -\cos x$ for $0 \leqslant x \leqslant 360°$.

b Describe the transformation which maps the graph of $y = \cos x$ onto the graph of:

i $y = 2\cos x$ **ii** $y = -\cos x$

Hint A transformation from the graph of $y = \mathrm{f}(x)$ onto the graph of $y = a\mathrm{f}(x)$ is a **stretch** by scale factor a in the vertical direction.

The graph of $y = -\mathrm{f}(x)$ is a **reflection** of the graph of $y = \mathrm{f}(x)$ in the x-axis.

← Section 4.6

2 **a** On the same set of axes, sketch the graphs of $y = \sin x$ and $y = \sin 3x$ for $0 \leqslant x \leqslant 360°$.

b On a second set of axes, sketch the graphs of $y = \tan x$ and $y = \tan(-x)$ for $-180° \leqslant x \leqslant 180°$.

c Describe the transformation which maps the graph of:

i $y = \sin x$ onto the graph of $y = \sin 3x$

ii $y = \tan x$ onto the graph of $y = \tan(-x)$

Hint A transformation from the graph of $y = \mathrm{f}(x)$ onto the graph of $y = \mathrm{f}(ax)$ is a **stretch** by scale factor $\frac{1}{a}$ in the horizontal direction.

The graph of $y = \mathrm{f}(-x)$ is a **reflection** of the graph of $y = \mathrm{f}(x)$ in the y-axis.

← Section 4.6

3 a On the same set of axes, sketch the graphs of $y = \sin x$ and $y = \sin x + 1$ for $0 \leqslant x \leqslant 360°$.

b Describe the transformation which maps the graph of $y = \sin x$ onto the graph of $y = \sin x + 1$.

Hint A transformation from the graph of $y = \text{f}(x)$ onto the graph of $y = \text{f}(x) + a$ is a **translation** by vector $\binom{0}{a}$.

← Section 4.5

4 a On the same set of axes, sketch the graphs of $y = \tan x$ and $y = \tan(x + 30°)$ for $0 \leqslant x \leqslant 360°$.

b Describe the transformation which maps the graph of $y = \tan x$ onto the graph of $y = \tan(x + 30°)$.

Hint A transformation from the graph of $y = \text{f}(x)$ onto the graph of $y = \text{f}(x + a)$ is a **translation** in the x-axis by vector $\binom{-a}{0}$

← Section 4.5

P 5 a On the same set of axes, sketch the graphs of $y = \cos x$ and $y = \cos 2x$ for $0 \leqslant x \leqslant 360°$. **(3 marks)**

b Describe the transformation which maps the graph of $y = \cos x$ onto the graph of $y = \cos 2x$. **(2 marks)**

c State the period of the graph of $y = \cos 2x$. **(1 mark)**

d Write down the values of x for which $\cos 2x = 0$ in the interval $0 \leqslant x \leqslant 360°$. **(2 marks)**

P 6 a On the same set of axes, sketch the graphs of $y = \sin x$ and $y = \sin(x - 20°)$ for $0 \leqslant x \leqslant 360°$. **(3 marks)**

b Describe the transformation which maps the graph of $y = \sin x$ onto the graph of $y = \sin(x - 20°)$. **(2 marks)**

c Identify the value of x in the interval $0 \leqslant x \leqslant 360°$ for which:

i $\sin(x - 20°) = 1$ ii $\sin(x - 20°) = -1$ **(2 marks)**

P 7 a On the same set of axes, sketch the graphs of $y = \cos x$ and $y = 5\cos x$ for $-180° \leqslant x \leqslant 180°$. **(3 marks)**

b Describe the transformation which maps the graph of $y = \cos x$ onto the graph of $y = 5\cos x$. **(2 marks)**

c Identify the maximum and minimum values of $5\cos x$. **(1 mark)**

d Given that the equation $5\cos x = k$, for $-180° \leqslant x \leqslant 180°$, where k is constant, has one solution $x = \theta$, state, in terms of θ, the second solution. **(1 mark)**

P 8 a Sketch the graphs of $y = \sin x$ and $y = \cos x$ on the same set of axes for $0 \leqslant x \leqslant 360°$. **(3 marks)**

b Use your graphs to identify the number of solutions to the equation $\sin x = \cos x$ for $0 \leqslant x \leqslant 360°$. **(1 mark)**

c State the solutions to the equation $\sin x = \cos x$, for $0 \leqslant x \leqslant 360°$. **(2 marks)**

d Find one possible value for k, where k is a constant, for which:

i $\sin x = \cos(x + k)$ ii $\cos x = \sin(x + k)$ **(2 marks)**

Problem solving Set A

Bronze

The diagram shows a triangle ABC.

The length of AC is 20.3 cm, $\angle BAC = 82°$ and $\angle ABC = 40°$.

Calculate:

a the length of BC correct to the nearest 0.1 cm **(3 marks)**

b the area of triangle ABC, giving your answer to the nearest cm^2 **(3 marks)**

c the perimeter of triangle ABC, giving your answer to the nearest 0.1 cm. **(3 marks)**

Silver

A hiker walking due north on a straight path sees a wind turbine, W, on a bearing of 040°.

After walking 500 m, the bearing of W from the hiker's new position is 075°. Find:

a the distance between W and the hiker's new position **(5 marks)**

b the distance of W from the path. **(3 marks)**

Gold

A man walks 750 m from a point A to a point C on a bearing of 042°.

He then walks along a straight path from C on a bearing of 162° towards a point B.

When the man reaches a point X on the path, his distance from point A is the same as the distance between point B and point A.

Given that the distance $AB = 680$ m, calculate:

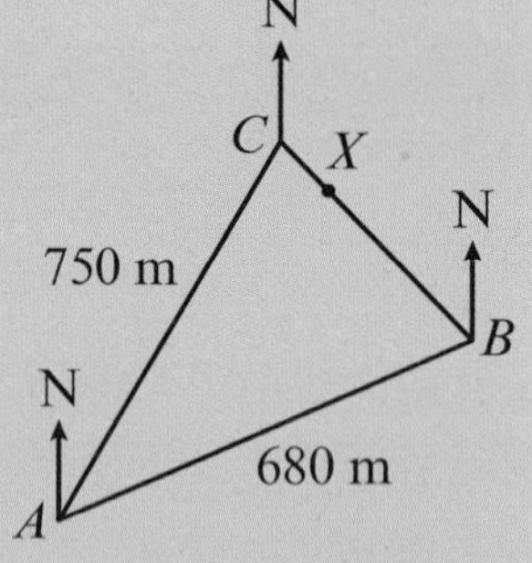

a the distance XB **(7 marks)**

b the bearing of A from B. **(3 marks)**

Problem solving Set B

Bronze

The graph shows the curve with equation $y = \cos x + k$, where k is a constant.

The graph passes through the point with coordinates (0, 4).

a Find the value of k. **(2 marks)**

b Use your value of k to find the coordinates of the minimum points for the graph of $y = \cos x + k$ for $-180° \leqslant x \leqslant 360°$. **(2 marks)**

c Find, in terms of A, the coordinates of the point where the graph with equation $y = \cos x + A$ crosses the y-axis. **(2 marks)**

Silver

The graph shows the curve with equation $y = \cos(x + k)$ where k is a positive constant.

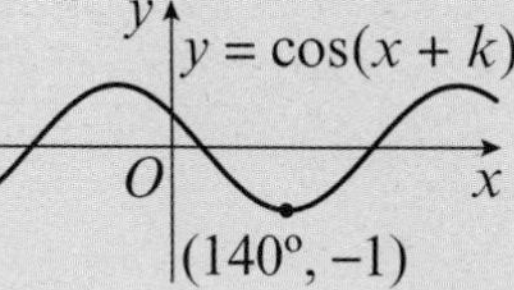

The graph passes through the point with coordinates (140°, −1).

a State the smallest possible value of k. **(2 marks)**

b Use your value of k to find the solutions to the equation $\cos(x + k) = 0$ for $-180° \leqslant x \leqslant 360°$. **(2 marks)**

c Identify another possible value for k for the curve. Explain your answer. **(2 marks)**

Gold

$f(x) = \cos kx$, $0 \leqslant k \leqslant 360°$, where k is a positive constant.

a In the case when k is a positive integer, write, in terms of k, the number of solutions to the equation $f(x) = 0.5$. **(2 marks)**

b In the case when $k = 1.5$, sketch the graph of $y = f(x)$, showing clearly any points where the curve cuts the coordinate axes. **(2 marks)**

c Given that the equation $f(x) = 0.5$ has exactly three real roots, state the range of possible values of k. **(3 marks)**

Now try this → **Exam question bank Q17, Q31, Q37, Q39, Q45, Q63, Q73, Q78**

10 Trigonometric identities and equations

10.1 Angles in all four quadrants

1 Use the unit circle to determine the quadrants where the following trigonometric functions are positive.

a $\sin\theta$

b $\cos\theta$

c $\tan\theta$

d all of $\sin\theta$, $\cos\theta$ and $\tan\theta$

Hint $P(x, y)$ moves on a unit circle such that OP makes an angle θ with the positive x axis.

$\cos\theta = x = x$-coordinate of P

$\sin\theta = y = y$-coordinate of P

$\tan\theta = \frac{y}{x}$ = gradient of OP

This means that the sign, (+ or –), of the trigonometric functions depends on the sign of x and y.

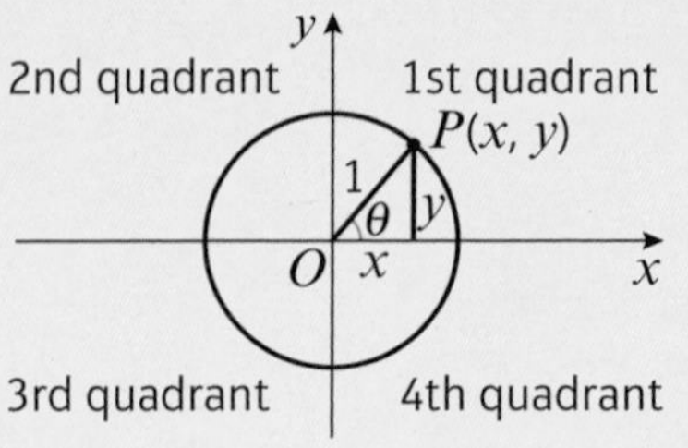

2 Identify the quadrant that θ lies in as OP moves around the unit circle and hence state whether the trigonometric function of θ is positive or negative.

a $\theta = 116°$, $\cos\theta$

b $\theta = 240°$, $\tan\theta$

c $\theta = 420°$, $\sin\theta$

d $\theta = -100°$, $\tan\theta$

e $\theta = 350°$, $\sin\theta$

f $\theta = -80°$, $\cos\theta$

3 Without using a calculator, write down the values of:

a $\cos 90°$

b $\cos(-180°)$

c $\sin 540°$

d $\sin(-90°)$

e $\tan 135°$

f $\tan(-45°)$

g $\sin 225°$

h $\cos(-45°)$

i $\tan(-90°)$

Hint When OP makes an angle of 45° with the positive x-axis the triangle formed is isosceles.

$\cos 45° = x = \frac{1}{\sqrt{2}}$

$\sin 45° = y = \frac{1}{\sqrt{2}}$

$\tan 45° = \frac{y}{x} = 1$

4 Given that θ is an acute angle, express in terms of $\sin\theta$, $\cos\theta$ or $\tan\theta$:

a $\sin(180° + \theta)$

b $\cos(360° - \theta)$

c $\tan(180° - \theta)$

d $\sin(-\theta)$

e $\tan(180° + \theta)$

f $\cos(\theta - 180°)$

g $\sin(\theta - 360°)$

h $\cos(540° - \theta)$

i $\tan(\theta - 180°)$

Hint

y
180° − θ
θ
θ θ
θ θ
O
x
180° + θ
360° − θ

5 Express in terms of trigonometric ratios of acute angles:

a $\sin 220°$ **b** $\cos 160°$ **c** $\tan 210°$
d $\sin(-100°)$ **e** $\tan 130°$ **f** $\cos 280°$
g $\sin 95°$ **h** $\cos(-200°)$ **i** $\tan(-115°)$

10.2 Exact values of trigonometric ratios

1 Express the following as trigonometric ratios of either 30°, 45° or 60° and hence state the exact value.

a $\sin 150°$ **b** $\sin(-45°)$
c $\sin 315°$ **d** $\sin 240°$
e $\cos 150°$ **f** $\cos(-60°)$
g $\cos(-135°)$ **h** $\cos 120°$
i $\tan 225°$ **j** $\tan(-150°)$
k $\tan 405°$ **l** $\tan(-120°)$

Hint The exact values of the trigonometric ratios of 30° and 60° can be worked out using Pythagoras' theorem by considering an equilateral triangle of side 2 units.

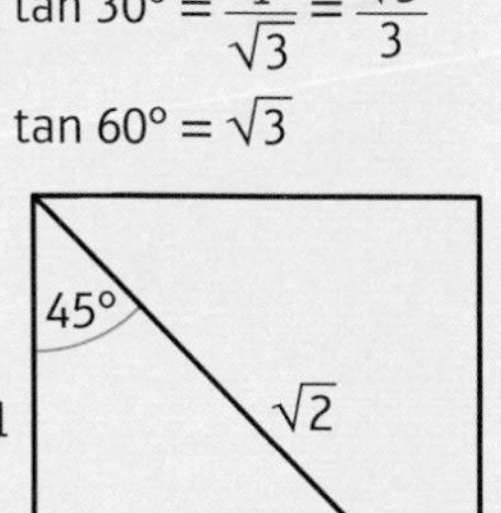

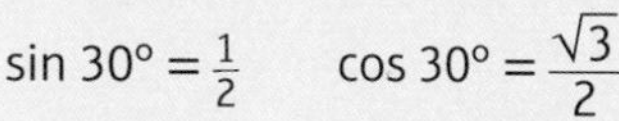

$\sin 30° = \frac{1}{2}$ $\cos 30° = \frac{\sqrt{3}}{2}$ $\tan 30° = \frac{1}{\sqrt{3}} = \frac{\sqrt{3}}{3}$

$\sin 60° = \frac{\sqrt{3}}{2}$ $\cos 60° = \frac{1}{2}$ $\tan 60° = \sqrt{3}$

The exact values of the trigonometric ratios of 45° can be worked out by considering a square of side 1 unit.

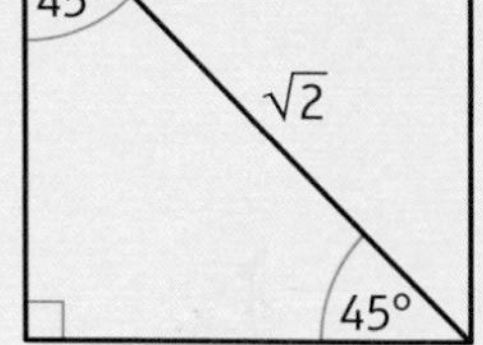

$\sin 45° = \frac{1}{\sqrt{2}}$

$\cos 45° = \frac{1}{\sqrt{2}}$ $\tan 45° = 1$

2 The table shows some of the values of x and y that satisfy the equation $y = a \sin x + b$.

x	0°	30°	60°	90°	120°	150°	180°
y	3	4	$3+\sqrt{3}$	5	$3+\sqrt{3}$	4	3

Find the value of y when $x = 45°$.

3 The diagram shows parallelogram $ABCD$, where $AB = DC = x$ cm and $AD = BC = 10$ cm.

Given that $AC = 2x$ cm and $\angle BAC = 30°$, show that

$$x^2 = \frac{100(5 + 2\sqrt{3})}{13}$$

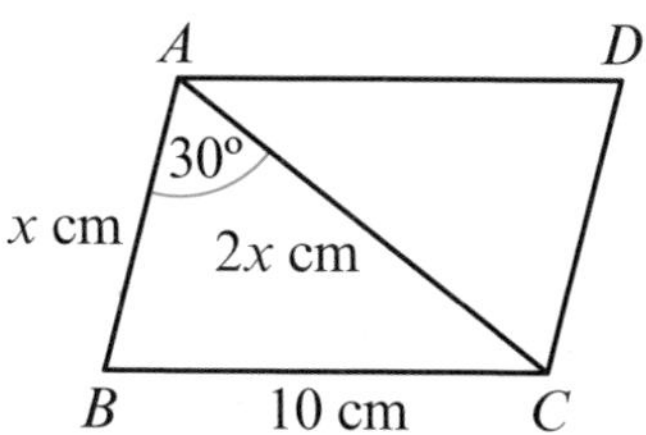

10.3 Trigonometric identities

1 Write down the value of:

Hint Use the trigonometric identity $\sin^2\theta + \cos^2\theta \equiv 1$.

a $2\sin^2\theta - (1 - \cos^2\theta) + \cos^2\theta$

b $6\sin^2\frac{\theta}{2} + 6\cos^2\frac{\theta}{2}$ **c** $(1 + \cos x)^2 + (1 - \cos x)^2 + 2\sin^2 x$

2 Write in terms of $\tan x$:

a $\dfrac{2\sin x}{\cos x}$ **b** $\dfrac{3\sin x}{\sqrt{1-\sin^2 x}}$ **c** $\dfrac{1-\cos^2 x}{1-\sin^2 x}$

Hint Use the trigonometric identity $\tan x \equiv \dfrac{\sin x}{\cos x}$ for all values of x such that $\cos x \neq 0$

3 Prove that $\dfrac{\sin^2 x - \cos^2 x}{1-\sin^2 x} \equiv \tan^2 x - 1$

Hint To prove an identity, start from the LHS and manipulate the expression until it matches the RHS. You can quote the standard identities in your proof. ← **Sections 7.4, 7.5**

4 Given that $\sin x = \frac{3}{8}$ and that x is obtuse, find the exact value of:

a $\cos x$ **b** $\tan x$

Hint You can use $\cos^2 x \equiv 1 - \sin^2 x$ to work out the exact value of $\cos x$, then $\tan x \equiv \dfrac{\sin x}{\cos x}$. Remember you need to check the sign of cosine and tangent in the required quadrant. You can check your answer using basic trigonometry and Pythagoras' theorem in a right-angled triangle.

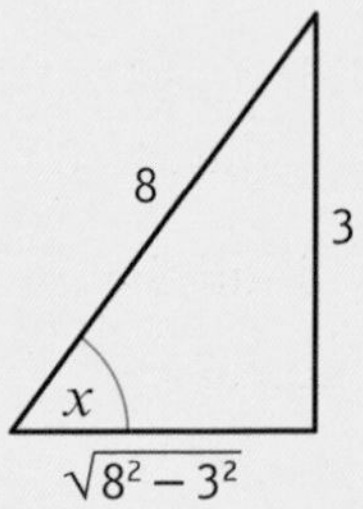

(Right-angled triangle with hypotenuse 8, opposite side 3, angle x, adjacent side $\sqrt{8^2-3^2}$)

(E) **5** **a** Given that $\sin\theta = 6\cos\theta$, find the value of $\tan\theta$. **(1 mark)**

b Express $4\cos^2\theta - 5\sin^2\theta$ in terms of $\sin\theta$. **(1 mark)**

(E/P) **6** Show that the equation $4\cos^2 x - 3\sin^2 x = 2$ can be written as $7\cos^2 x = 5$. **(2 marks)**

(E/P) **7** Prove that $\dfrac{\cos^2 x - \sin^2 x}{\cos^2 x - \sin x\cos x} \equiv 1 + \tan x$ **(3 marks)**

(E/P) **8** The diagram shows the triangle ABC with $AB = 7$ cm, $BC = 12$ cm and $AC = 9$ cm.

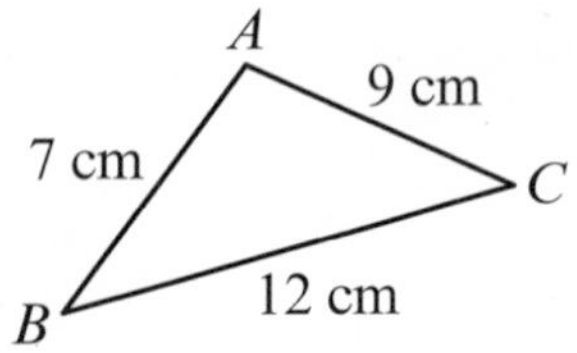

(Triangle ABC: AB = 7 cm, AC = 9 cm, BC = 12 cm)

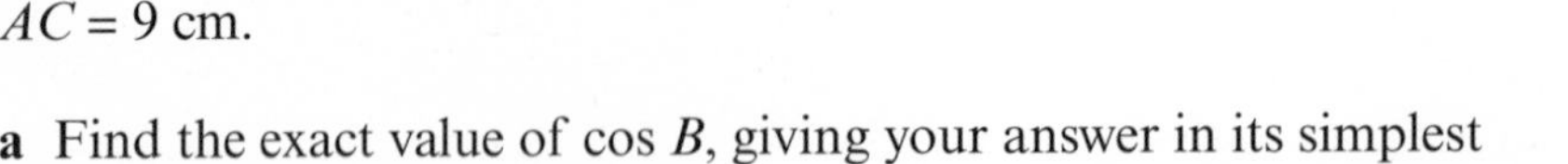

a Find the exact value of $\cos B$, giving your answer in its simplest form. **(3 marks)**

b Hence find the exact value of $\sin B$. **(2 marks)**

(E/P) **9** The diagram shows triangle PQR with $PR = 9$ cm, $QR = 10$ cm and angle $PQR = 45°$.

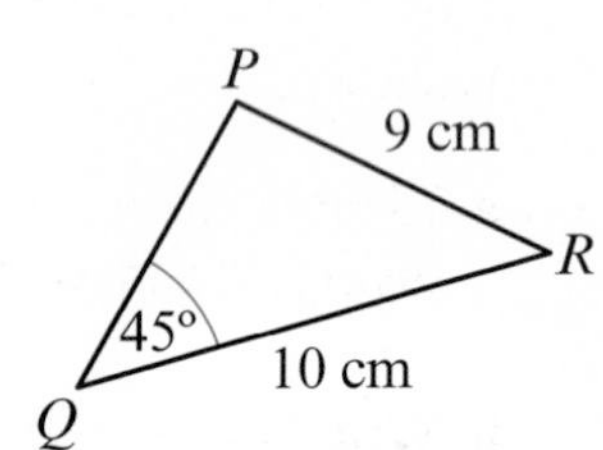

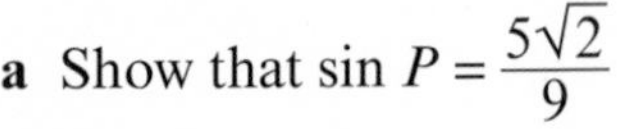

a Show that $\sin P = \dfrac{5\sqrt{2}}{9}$ **(3 marks)**

b Given that angle P is obtuse, find the exact value of $\cos P$. **(2 marks)**

10.4 Simple trigonometric equations

1 The diagram shows a sketch of $y = \sin x$.

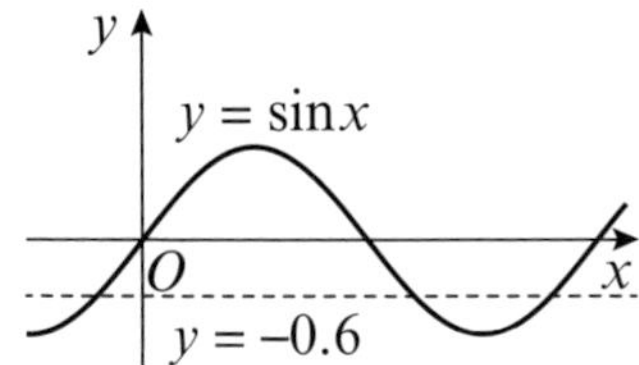

a Use your calculator to find the principal solution to the equation $\sin x = -0.6$.

b Use the graph and your answer to part **a** to find solutions to the equation $\sin x = -0.6$ in the interval $0 \leqslant x \leqslant 360°$.

Hint Use the inverse trigonometric function on your calculator to get the **principal value**.

Your calculator will give the principal values in these ranges:

$\cos^{-1}$: $0 \leqslant x \leqslant 180°$

$\sin^{-1}$: $-90° \leqslant x \leqslant 90°$

$\tan^{-1}$: $-90° < x < 90°$

2 Find the solutions to the equation $\cos\theta = \frac{1}{\sqrt{2}}$ in the interval $0 \leqslant \theta \leqslant 360°$.

Hint Find the first solution using your calculator or your knowledge of exact values of trigonometric functions. Draw the graph of $y = \cos\theta$ for the given interval and use the properties of the graph to find the second solution. ← **Section 9.5**

3 Solve the equation $4\sin x = 3$ in the interval $0 \leqslant x \leqslant 360°$.

Hint First rewrite the equation in the form $\sin x = \ldots$

4 Solve the equation $\sin\theta = -10\cos\theta$ in the interval $0 < \theta \leqslant 360°$.

Hint You can use the identity $\tan\theta \equiv \frac{\sin\theta}{\cos\theta}$ to solve trigonometric equations.

The principal value of $\tan\theta$ is not in the given interval. As the period for the graph of $y = \tan\theta$ is 180°, you can find successive solutions by adding or subtracting multiples of 180°.

E 5 A student's attempt to solve the equation $5\tan x = 9\sin x$ in the interval $-90° < x < 90°$ is shown.

$$5\tan x = 9\sin x$$
$$5\frac{\sin x}{\cos x} = 9\sin x$$
$$5\sin x = 9\sin x\cos x$$
$$5 = 9\cos x$$
$$\cos x = \frac{5}{9}$$
$$x = 56.3° \text{ (to 3 s.f.)}$$

a Identify two mistakes made by this student, giving a brief explanation of each mistake. **(2 marks)**

b Write down the correct solutions to the original equation. **(1 mark)**

E/P **6** **a** Given that $3 \sin \theta = 5 \cos \theta$, find the value of $\tan \theta$. **(1 mark)**

b Hence, or otherwise, find the values of θ in the interval $0 \leqslant x \leqslant 360°$ for which $3 \sin \theta = 5 \cos \theta$, giving your answers to 1 decimal place. **(3 marks)**

E/P **7** Find all the values of x for which $\cos^2 x = \frac{3}{4}$ in the interval $-180° \leqslant x < 180°$. **(5 marks)**

E/P **8** **a** Show that the equation $4 \cos^2 x - 3 \sin^2 x = 1$ can be written as $7 \cos^2 x = 4$. **(2 marks)**

b Hence solve the equation $4 \cos^2 x - 3 \sin^2 x = 1$ for $0 \leqslant x \leqslant 360°$, giving your answers to 1 decimal place. **(7 marks)**

E/P **9** **a** Solve $(2 + \tan \theta)(4 \sin \theta - 1) = 0$ for $-180° \leqslant \theta < 180°$. **(4 marks)**

b Solve $\dfrac{6 \sin x}{\tan x} = 1$ for $0 \leqslant x < 360°$. **(6 marks)**

10.5 Harder trigonometric equations

1 Find the values of θ in the interval $0 \leqslant \theta \leqslant 360°$ for which:

a $\cos 3\theta = \dfrac{\sqrt{3}}{2}$ **b** $\sin \dfrac{\theta}{2} = \dfrac{1}{\sqrt{2}}$

c $\tan 2\theta = -\sqrt{3}$

Hint In part **a**, let $X = 3\theta$. If $0 \leqslant \theta \leqslant 360°$, then $0 \leqslant 3\theta \leqslant 1080°$.

Use the graph of $y = \cos X$ to solve $\cos X = \frac{\sqrt{3}}{2}$ for $0 \leqslant X \leqslant 1080°$.

There will be six solutions. Divide the solutions for X by 3 to find the solutions for θ.

2 Solve the equations in the given intervals:

a $\sin(\theta + 30°) = 0.2$, $0 \leqslant \theta \leqslant 360°$

b $\cos(60° - \theta) = -0.4$, $-180° \leqslant \theta \leqslant 180°$

c $\tan(\theta - 20°) = 9$, $0 \leqslant \theta \leqslant 360°$

Hint In part **a**, let $X = \theta + 30°$ and solve $\sin X = 0.2$ for $30° \leqslant X \leqslant 390°$.

Subtract 30° from each of the solutions for X to find solutions for θ.

3 Solve the equations in the given intervals:

a $\cos 3\theta = \sqrt{3} \sin 3\theta$, $-180° \leqslant \theta \leqslant 180°$ **b** $2 \sin 4\theta - 5 \cos 4\theta = 0$, $0 \leqslant \theta \leqslant 360°$

c $5 \sin(\theta + 40°) + 8 \cos(\theta + 40°) = 0$, $-180° \leqslant \theta \leqslant 180°$

4 Solve the equations in the given intervals:

a $\sin(2x - 30°) = 0.75$, $0 \leqslant x \leqslant 360°$

b $\cos(3x + 20°) = -0.4$, $0 \leqslant x \leqslant 180°$

c $\tan(4x - 45°) = 2$, $0 \leqslant x \leqslant 180°$

d $2 \sin(2x + 10°) = 3 \cos(2x + 10°)$, $-180° \leqslant x \leqslant 180°$

Hint In part **a**, let $X = 2x - 30°$. The end points in the interval for X are $2 \times 0 - 30° = -30°$ and $2 \times 360° - 30° = 690°$. Find the solutions for X, then use $2x - 30° = X$ for each solution to find the solutions for x.

E 5 Solve, for $0 \leqslant x < 360°$,

a $\cos 3x = -\frac{1}{2}$ **(6 marks)**

b $\sin(x - 20°) = -\frac{1}{\sqrt{2}}$ **(4 marks)**

E 6 a Solve $5 \sin 2x = 3 \cos 2x$ for $0 \leqslant x \leqslant 360°$, giving your answers to 1 decimal place. **(5 marks)**

b Find the solutions of the equation $\sin(3x - 15°) = \frac{1}{\sqrt{2}}$ for which $0 \leqslant x \leqslant 180°$. **(6 marks)**

E 7 a Find all values of θ, to 1 decimal place, in the interval $0 \leqslant \theta \leqslant 360°$ for which $5 \sin(\theta + 30°) = 4$. **(4 marks)**

b Find all values of θ, to the nearest degree, in the interval $0 \leqslant \theta \leqslant 360°$ for which $\tan^2 \frac{\theta}{2} = 16$. **(5 marks)**

E 8 Solve $\tan(5x - 20°) = \frac{5}{2}$ for $-90° < x < 90°$, giving answers to 1 decimal place. **(6 marks)**

E 9 a Show that the equation $\tan 2x = 4 \sin 2x$ can be written in the form $(1 - 4 \cos 2x) \sin 2x = 0$ **(2 marks)**

b Hence solve $\tan 2x = 4 \sin 2x$ for $0 \leqslant x \leqslant 180°$, giving your answers to 1 decimal place where appropriate. You must show clearly how you obtained your answers. **(5 marks)**

10.6 Equations and identities

1 Solve the equations for $0 \leqslant \theta \leqslant 360°$.
Give your answers to 3 significant figures where they are not exact.

Hint You need to be able to solve quadratic equations in trigonometric functions, which can lead to two sets of solutions.

← Section 2.1

a $9 \sin^2 \theta = 1$ b $4 \cos^2 \theta - 1 = 0$

c $5 \sin^2 \theta - \sin \theta = 0$ d $\tan^2 \theta + \tan \theta - 12 = 0$

e $(3 \cos \theta - 2)^2 = 1$ f $\tan^2 3\theta = 4$

2 Solve the equations for $0 \leqslant \theta \leqslant 360°$.
Give your answers to 3 significant figures where they are not exact.

Hint There are no valid solutions to $\sin x = k$ or $\cos x = k$ when $k < -1$ or $k > 1$.

a $\cos^2 3\theta = 1$ b $\tan^2 \theta = 5\tan \theta$

c $\sin^2 \theta - 3 \sin \theta - 4 = 0$ d $4 \sin(\theta + 30°) = \tan(\theta + 30°)$

3 a Show that the equation $2 \cos^2 x - \sin^2 x - 3 \cos x + 2 = 0$ can be written as $3 \cos^2 x - 4 \cos x + 1 = 0$

Hint You can use $\sin^2 x \equiv 1 - \cos^2 x$ to convert to a quadratic equation in $\cos x$.

b Hence solve the equation $2 \cos^2 x - \sin^2 x - 3 \cos x + 2 = 0$ for $0 \leqslant x \leqslant 360°$.

4 Solve the equations for $0 \leqslant \theta \leqslant 360°$.
Give your answers to 1 decimal place where they are not exact.

a $5\cos^2\theta = 6(1 - \sin\theta)$

b $8 - 7\cos\theta = 6\sin^2\theta$ c $3\tan\theta = 2\cos\theta$

Hint You can use $\cos^2\theta \equiv 1 - \sin^2\theta$ and $\tan\theta \equiv \frac{\sin\theta}{\cos\theta}$ to convert to a quadratic equation.

E/P 5 a Show that the equation $2\sin^2 x - 3\cos^2 x = 1$ can be written as $5\sin^2 x = 4$. **(2 marks)**

b Hence solve the equation $2\sin^2 x - 3\cos^2 x = 1$ for $0 \leqslant x < 360°$, giving your answers to one decimal place. **(7 marks)**

E/P 6 Solve $2\cos^2 x + \sin^2 x = \frac{10}{9}$ for $0 < x < 180°$, giving your answers to one decimal place. **(4 marks)**

E/P 7 Find all the solutions of $2\cos^2 x + 2 = 7\sin x$ for $0 \leqslant x \leqslant 360°$. **(6 marks)**

8 a Show that the equation $\sin\theta\tan\theta = 2\cos\theta + 3$ can be written in the form $3\cos^2\theta + 3\cos\theta - 1 = 0$ **(3 marks)**

b Hence solve $\sin\theta\tan\theta = 2\cos\theta + 3$ for $0 \leqslant \theta < 360°$, showing each stage of your working. **(5 marks)**

E/P 9 Solve $15\sin^2 x + 8\cos x - 16 = 0$ for $360° \leqslant x < 540°$. Give your answers to one decimal place. **(5 marks)**

Problem solving Set A

Bronze

a Given that $2\sin x = 5\cos x$, write down the value of $\tan x$. **(1 mark)**

b Hence solve $2\sin(\theta + 20°) = 5\cos(\theta + 20°)$ for $0 \leqslant \theta < 360°$, giving your answers to one decimal place. **(5 marks)**

Silver

Solve the equation $\frac{\sin 3x + 1}{1 - \sin 3x} = 2$, for $0 \leqslant x \leqslant 180°$.

Give your answers to one decimal place. **(6 marks)**

Gold

Solve the equation $4\cos(2x - 15°) + 3 = 0$, for $-180° \leqslant x \leqslant 180°$.

Give your answers to one decimal place. **(7 marks)**

Problem solving Set B

Bronze

a Show that the equation $\dfrac{3\sin^2 x + \cos^2 x}{\cos^2 x} = 5$ can be written as $\tan^2 x = \frac{4}{3}$ **(2 marks)**

b Hence solve the equation $\tan^2 x = \frac{4}{3}$ for $0 \leqslant x < 360°$, giving your answers to one decimal place. **(6 marks)**

Silver

a Show that the equation $4\cos^2 x + 9\sin x - 6 = 0$ can be written as $4\sin^2 x - 9\sin x + 2 = 0$. **(2 marks)**

b Hence solve $4\cos^2 x + 9\sin x - 6 = 0$ for $0 \leqslant x < 720°$, giving your answers to one decimal place **(6 marks)**

Gold

Solve $5(\cos 3x - 1) = 3\sin 3x \tan 3x$, for $0 \leqslant x < 90°$, giving your answers, where appropriate, to one decimal place. **(9 marks)**

Now try this → **Exam question bank Q7, Q47, Q60, Q66, Q76, Q84, Q85**

11 Vectors

11.1 Vectors

1 $ABCDEF$ is a hexagon. $\overrightarrow{AF} = \mathbf{a}$ and $\overrightarrow{AO} = \mathbf{b}$.

Find in terms of **a** and **b**:

a $\overrightarrow{AD}$ **b** $\overrightarrow{EB}$

c $\overrightarrow{CB}$ **d** $\overrightarrow{AE}$

e $\overrightarrow{OF}$ **f** $\overrightarrow{FC}$

g $\overrightarrow{CE}$ **h** $\overrightarrow{BF}$

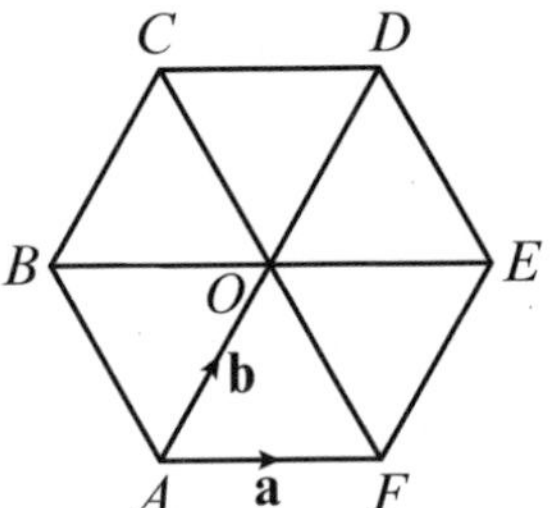

Hint The vector $\overrightarrow{FA} = -\mathbf{a}$ and the vector $\overrightarrow{OA} = -\mathbf{b}$.

2 In the diagram, $\overrightarrow{WX} = \mathbf{a}$, $\overrightarrow{WP} = \mathbf{b}$, $\overrightarrow{WZ} = \mathbf{c}$ and $\overrightarrow{XY} = \mathbf{d}$.

Find in terms of **a**, **b**, **c** and **d**:

a $\overrightarrow{XP}$ **b** $\overrightarrow{PY}$

c $\overrightarrow{PZ}$ **d** $\overrightarrow{ZY}$

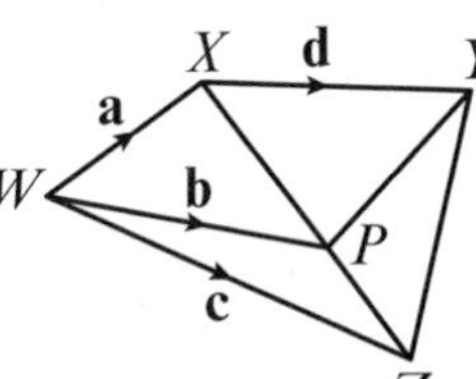

Hint You can use answers to previous parts if it helps. For example, $\overrightarrow{PY} = \overrightarrow{PX} + \overrightarrow{XY}$.

3 In the parallelogram $PQRS$, $PQ = \mathbf{a}$ and $QR = \mathbf{b}$. The midpoint of PR is M.

Find in terms of **a** and **b**:

a $\overrightarrow{PR}$ **b** $\overrightarrow{PM}$

c $\overrightarrow{SM}$

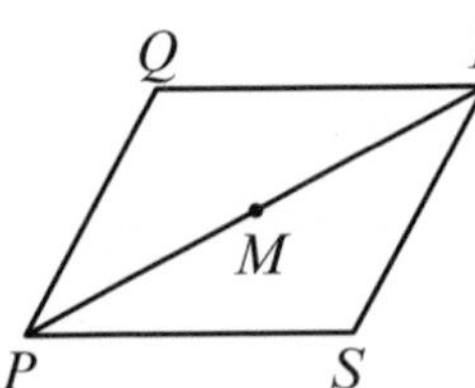

Hint The midpoint M is halfway between P and R, so $\overrightarrow{PM} = \frac{1}{2}\overrightarrow{PR}$.

← Section 6.1

(E) **4** The vector $8\mathbf{a} - 10\mathbf{b}$ is parallel to the vector $p\mathbf{a} + q\mathbf{b}$.

State, in simplest terms, the value of $\frac{p}{q}$ **(1 mark)**

(E/P) **5** In the triangle PQR, $\overrightarrow{PQ} = \mathbf{a}$ and $\overrightarrow{PR} = \mathbf{b}$.

X divides PQ in the ratio $1:3$ and Y divides QR in the ratio $2:3$.

Find $\overrightarrow{YX}$ in terms of **a** and **b**. **(4 marks)**

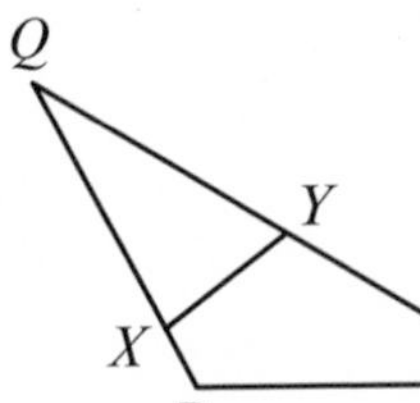

Hint If X divides PQ in the ratio $1:3$ then it is $\frac{1}{4}$ of the way along PQ from P.

(E/P) **6** In the triangle ABC, $\overrightarrow{AB} = \mathbf{a}$ and $\overrightarrow{BC} = \mathbf{b}$.

D and E are the midpoints of AB and BC respectively.

Use vectors to prove that $\overrightarrow{AC}$ is parallel to $\overrightarrow{DE}$. **(3 marks)**

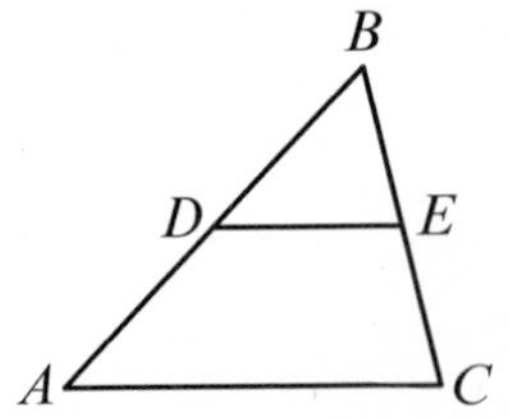

E/P 7 $EFGH$ is a trapezium with FG parallel to EH and $\overrightarrow{FG} = k\,\overrightarrow{EH}$.

M is the midpoint of EF and N is the midpoint of GH.

$\overrightarrow{EH} = \mathbf{a}$, $\overrightarrow{EF} = \mathbf{b}$ and $\overrightarrow{HG} = \mathbf{c}$.

a Prove that $\overrightarrow{MN} = \left(\frac{1+k}{2}\right)\mathbf{a}$. **(3 marks)**

b Hence, or otherwise, explain why MN is parallel to EH and FG. **(1 mark)**

11.2 Representing vectors

1 $\mathbf{a} = -7\mathbf{i} + 3\mathbf{j}$ and $\mathbf{b} = 2\mathbf{i} - 5\mathbf{j}$.

Find, in terms of $\mathbf{i}$ and $\mathbf{j}$:

a $3\mathbf{a}$ **b** $\mathbf{a} - 2\mathbf{b}$ **c** $3\mathbf{b} - \mathbf{a}$

Hint: Add the **i** components and add the **j** components.

2 Given that $\mathbf{a} = \begin{pmatrix} -2 \\ 4 \end{pmatrix}$, $\mathbf{b} = \begin{pmatrix} 6 \\ -1 \end{pmatrix}$ and $\mathbf{c} = \begin{pmatrix} -3 \\ -5 \end{pmatrix}$,

find, as a column vector:

a $\frac{1}{2}\mathbf{a}$ **b** $\mathbf{a} + \mathbf{b} + \mathbf{c}$ **c** $\mathbf{c} + 2\mathbf{b} - \frac{3}{2}\mathbf{a}$

Hint: $\begin{pmatrix} -2 \\ 4 \end{pmatrix}$ is another way of writing $-2\mathbf{i} + 4\mathbf{j}$.

3 Find the resultant of the vectors $\mathbf{a} = -6\mathbf{i} + 7\mathbf{j}$ and $\mathbf{b} = 3\mathbf{i} - 5\mathbf{j}$.

Hint: The resultant is the vector sum of two or more vectors.

E **4** Given that $\overrightarrow{AB} = -8\mathbf{i} + 7\mathbf{j}$ and $\overrightarrow{AC} = 11\mathbf{i} - \mathbf{j}$, find $\overrightarrow{BC}$. **(2 marks)**

E **5** $\mathbf{a} = \begin{pmatrix} 2 \\ p \end{pmatrix}$, $\mathbf{b} = \begin{pmatrix} -3 \\ -4 \end{pmatrix}$ and $\mathbf{c} = \begin{pmatrix} q \\ 11 \end{pmatrix}$

Given that $2\mathbf{b} - \mathbf{c} = \mathbf{a}$, find the value of p and the value of q. **(3 marks)**

E/P **6** $\mathbf{a} = -2\mathbf{i} + 7\mathbf{j}$, $\mathbf{b} = 3\mathbf{i} - 2\mathbf{j}$, $\mathbf{c} = -\mathbf{i} - 6\mathbf{j}$ and $\mathbf{d} = 2\mathbf{i} - 2\mathbf{j}$.

Show that the vector $2\mathbf{a} + 5\mathbf{b}$ is not parallel to the vector $2\mathbf{d} - \mathbf{c}$. **(3 marks)**

E/P **7** $\mathbf{a} = 2p\mathbf{i} + 3q\mathbf{j}$ and $\mathbf{b} = -4q\mathbf{i} + p\mathbf{j}$.

Given that $\mathbf{b} - \mathbf{a} = -10\mathbf{i} + 10\mathbf{j}$, find the value of p and the value of q. **(3 marks)**

11.3 Magnitude and direction

1 Find the exact magnitude of the following vectors.

a $6\mathbf{i} - 8\mathbf{j}$ **b** $5\mathbf{i} + 4\mathbf{j}$

c $-3\mathbf{i} + 6\mathbf{j}$ **d** $-8\mathbf{i} - 8\mathbf{j}$

Hint: The magnitude of the vector $\mathbf{a} = x\mathbf{i} + y\mathbf{j}$ is $|\mathbf{a}| = \sqrt{x^2 + y^2}$. If you are asked to find the exact magnitude, leave your answer in simplified surd form.

← Section 1.5

2 Find the angle that each of these vectors makes with the positive x-axis.

a $2\mathbf{i} + 4\mathbf{j}$ b $-7\mathbf{i} - \mathbf{j}$

c $6\mathbf{j}$ d $5\mathbf{i} - 5\mathbf{j}$

Hint Draw a diagram and calculate the angle θ.

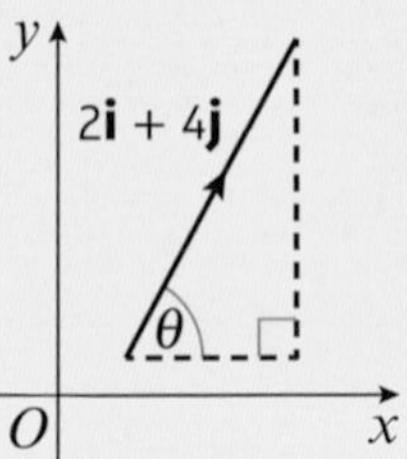

3 Find the unit vector in the direction $-5\mathbf{i} + 12\mathbf{j}$.

Hint The unit vector in the direction of **a** is $\frac{\mathbf{a}}{|\mathbf{a}|}$

4 Write the vector shown in the diagram in the form $p\mathbf{i} + q\mathbf{j}$, where p and q are exact constants to be found.

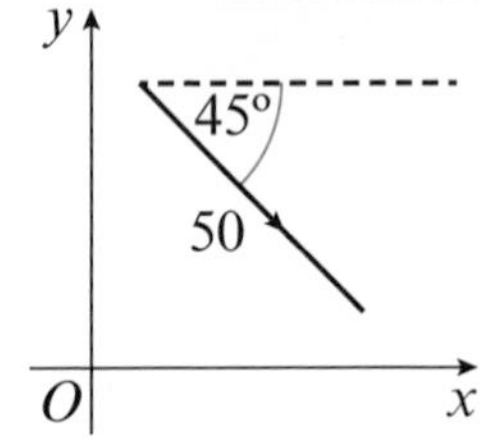

Hint Use trigonometry to find the lengths of the x- and y-components of the vector.

E/P 5 The vector $\mathbf{a} = p\mathbf{i} + q\mathbf{j}$, where p and q are positive constants, is such that $|\mathbf{a}| = 11$. Given that **a** makes an angle of 20° with the positive y-axis, find the values of p and q. **(2 marks)**

E/P 6 Given that $|2p\mathbf{i} + 3p\mathbf{j}| = 7\sqrt{13}$, find the possible values of p. **(3 marks)**

E 7 Given that $\overrightarrow{OA} = 5\mathbf{i} + 4\mathbf{j}$, $\overrightarrow{AB} = 7\mathbf{i} - 4\mathbf{j}$ and $\overrightarrow{CB} = 10\mathbf{i} + 7\mathbf{j}$, find:

a $\angle AOB$ **(2 marks)**

b $\angle BOC$ **(3 marks)**

c $\angle AOC$ **(1 mark)**

E/P 8 Vector $\mathbf{a} = p\mathbf{i} + q\mathbf{j}$ has magnitude 6.5 and makes an angle θ with the positive x-axis where $\cos\theta = -\frac{5}{12}$

Find the possible values of p and q. **(4 marks)**

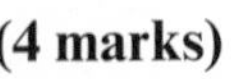

11.4 Position vectors

1 The points P and Q have coordinates $(-2, 3)$ and $(7, 0)$ respectively.

Find, in terms of **i** and **j**:

a the position vector of P

b the position vector of Q c the vector $\overrightarrow{PQ}$.

Hint $\overrightarrow{PQ} = \overrightarrow{OQ} - \overrightarrow{OP}$ where $\overrightarrow{OP}$ and $\overrightarrow{OQ}$ are the position vectors of P and Q respectively.

2 The coordinates of the triangle ABC, relative to the origin O, are $A(6, -2)$, $B(-1, -1)$ and $C(1, 3)$. Find, in terms of **i** and **j**:

a $\overrightarrow{OA}$ b $\overrightarrow{BC}$ c $\overrightarrow{CA}$

Hint $\overrightarrow{BC} = \overrightarrow{OC} - \overrightarrow{OB}$ where $\overrightarrow{OB}$ and $\overrightarrow{OC}$ are the position vectors of B and C respectively.

3 $\overrightarrow{OD} = -6\mathbf{i} + 8\mathbf{j}$ and $\overrightarrow{ED} = -9\mathbf{i} + 3\mathbf{j}$. Find, in surd form:

a $|\overrightarrow{OD}|$ **b** $|\overrightarrow{DE}|$ **c** $|\overrightarrow{OE}|$

Hint Use the given vectors to calculate $\overrightarrow{OE}$.

E 4 $\overrightarrow{OA} = \begin{pmatrix} -7 \\ -6 \end{pmatrix}$ and $\overrightarrow{AB} = \begin{pmatrix} 3 \\ 11 \end{pmatrix}$

Find:

a the position vector of B **(1 mark)**

b the exact value of $|\overrightarrow{OB}|$, giving your answer in surd form. **(2 marks)**

E/P 5 The point P lies on the line with equation $y = 2x - 3$. Given that $|OP| = \sqrt{53}$, find possible expressions for $\overrightarrow{OP}$ in the form $p\mathbf{i} + q\mathbf{j}$. **(4 marks)**

11.5 Solving geometric problems

1 Relative to a fixed origin O, the points P and Q have position vectors $\mathbf{p}$ and $\mathbf{q}$ respectively. The point N divides PQ in the ratio 1 : 3.

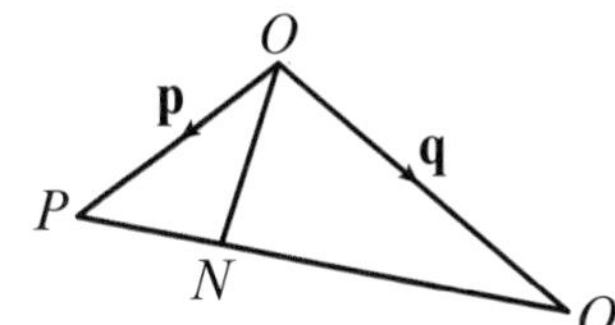

a Write down $\overrightarrow{PQ}$ in terms of $\mathbf{p}$ and $\mathbf{q}$.

b Copy and complete this expression by filling the blank space:

$\overrightarrow{PN} = \_\_\overrightarrow{PQ}$

Hint Use the ratio to work out what fraction of $\overrightarrow{PQ}$ is represented by $\overrightarrow{PN}$.

c Copy and complete this expression by filling the blank space:

$\overrightarrow{ON} = \overrightarrow{OP} + \_\_\_$

d Find the position vector of N.

2 Relative to a fixed origin O, the points A and B have position vectors $\mathbf{a}$ and $\mathbf{b}$ respectively. The point M divides AB in the ratio 2 : 3.

Find the position vector of M in terms of $\mathbf{a}$ and $\mathbf{b}$.

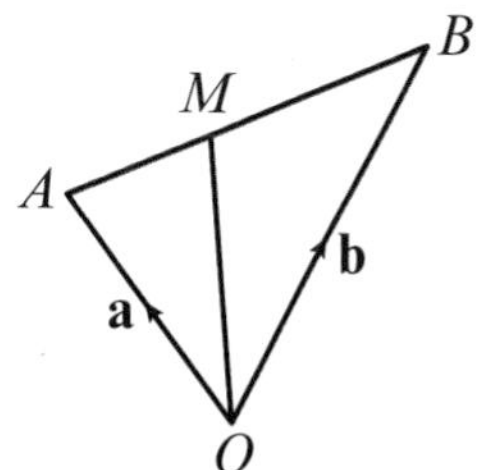

Hint If M divides AB in the ratio 2 : 3 then it is $\frac{2}{5}$ of the way along AB from A.

3 $ABCD$ is a parallelogram. $\overrightarrow{AB} = \mathbf{a}$ and $\overrightarrow{AD} = \mathbf{b}$. X divides AB in the ratio 3 : 1 and Y divides XD in the ratio 2 : 3.

Find $\overrightarrow{XY}$.

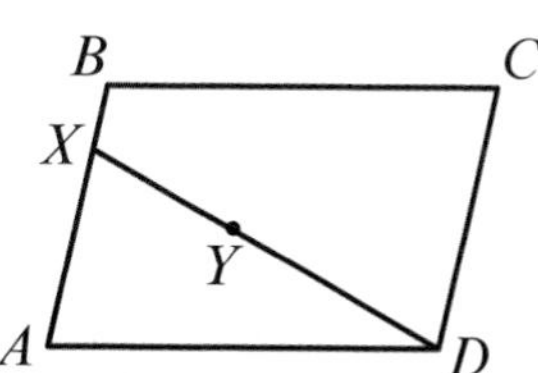

Hint First find $\overrightarrow{XD}$.

E 4 In triangle PQR, $\overrightarrow{PQ} = 8\mathbf{i} + \mathbf{j}$ and $\overrightarrow{PR} = 2\mathbf{i} - 6\mathbf{j}$.

Find the size of $\angle QPR$ in degrees, correct to one decimal place. **(5 marks)**

E/P **5** Points P, Q and R are the midpoints of JK, KL and JL respectively.

Let $\overrightarrow{JK} = \mathbf{a}$ and $\overrightarrow{JL} = \mathbf{b}$.

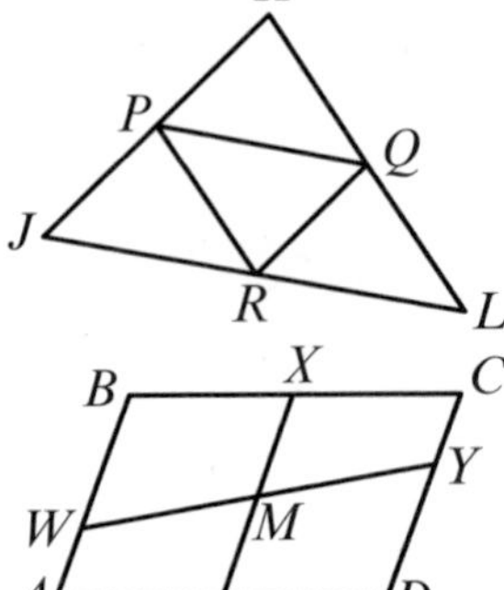

a Prove that $\overrightarrow{PR} = \frac{1}{2}\overrightarrow{KL}$. **(3 marks)**

b Explain how you know that $\overrightarrow{PR}$ and $\overrightarrow{KL}$ must be parallel. **(1 mark)**

E/P **6** $ABCD$ is a parallelogram. Let $\overrightarrow{AB} = \mathbf{a}$ and $\overrightarrow{AD} = \mathbf{b}$. Points X and Z are the midpoints of BC and AD respectively. Points W and Y divide AB and CD in the ratio $1:2$ respectively.

a Find:

i $\overrightarrow{WY}$ **ii** $\overrightarrow{XZ}$ **(2 marks)**

b Given that $\overrightarrow{WM} = \lambda\overrightarrow{WY}$, where λ is a constant, show that $\overrightarrow{AM} = \frac{1}{3}(1 + \lambda)\mathbf{a} + \lambda\mathbf{b}$. **(2 marks)**

c Given also that $\overrightarrow{XM} = \mu\overrightarrow{XZ}$, where μ is a constant, find an expression for $\overrightarrow{AM}$ in terms of a, b, and μ. **(2 marks)**

d Hence, or otherwise, prove that M is the midpoint of WY and XZ. **(4 marks)**

E/P **7** OAB is a triangle where P divides OA in the ratio $1 : k$ and Q divides BA in the ratio $1 : k$.

a Prove that $\overrightarrow{PQ}$ is parallel to $\overrightarrow{OB}$. **(4 marks)**

b Given that $|\overrightarrow{PQ}| = \frac{1}{3}|\overrightarrow{OB}|$, find the value of k. **(2 marks)**

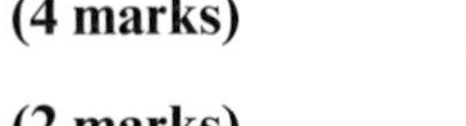

11.6 Modelling with vectors

1 Find the speed of a particle moving with these velocities:

Hint The speed of a particle travelling with velocity $(x\mathbf{i} + y\mathbf{j})$ m s^{-1} is $\sqrt{x^2 + y^2}$ m s^{-1}.

a $(5\mathbf{i} - 12\mathbf{j})$ m s^{-1} **b** $(-6\mathbf{i} - 8\mathbf{j})$ km s^{-1}

c $(6\mathbf{i} + \mathbf{j})$ mm s^{-1} **d** $14\mathbf{j}$ cm s^{-1}

2 A particle moves with velocity $(-4\mathbf{i} + 3\mathbf{j})$ m s^{-1}. Find:

a the speed of the particle

b the distance travelled by the particle in 4 seconds

Hint Distance travelled = speed × time

c the time taken for the particle to travel 27.5 m.

3 At time $t = 0$, a particle P moves with velocity $(-7\mathbf{i} - 24\mathbf{j})$ km s^{-1}. The particle accelerates constantly such that, at time $t = 6$, it is moving with velocity $(2\mathbf{i} - 51\mathbf{j})$ km s^{-1}.

The acceleration vector of the particle is given by the formula $\mathbf{a} = \frac{\mathbf{v} - \mathbf{u}}{t}$

Find the acceleration of P in terms of $\mathbf{i}$ and $\mathbf{j}$.

4 Initially a particle is moving with velocity $(\mathbf{i} - 4\mathbf{j})$ m s^{-1}. The particle then accelerates with acceleration $(2\mathbf{i} - 3\mathbf{j})$ m s^{-2} for 4 seconds.

The velocity, $\mathbf{v}$, of the particle after t seconds is given by $\mathbf{v} = \mathbf{u} + \mathbf{a}t$, where $\mathbf{u}$ is the initial velocity and $\mathbf{a}$ is the acceleration. Find:

a the velocity of the particle after 4 seconds

b the speed of the particle after 4 seconds.

Hint Substitute the given values of $\mathbf{u}$ and $\mathbf{a}$ into the formula $\mathbf{v} = \mathbf{u} + \mathbf{a}t$ and then simplify your answer.

(E) **5** Town A is 3 km west and 6 km north a fixed point O.

Town B is 8 km east and 2 km north of O.

a Write down the position vector of A relative to O. **(1 mark)**

Find:

b $|\overrightarrow{AB}|$ **(3 marks)**

c the bearing of B from A. **(3 marks)**

(E) **6** The forces $\mathbf{F}_1$ and $\mathbf{F}_2$ are given by the vectors $\mathbf{F}_1 = (p\mathbf{i} + q\mathbf{j})$ N and $\mathbf{F}_2 = (3\mathbf{i} - 7\mathbf{j})$ N. The resultant force $\mathbf{R} = \mathbf{F}_1 + \mathbf{F}_2$ acts in a direction parallel to the vector $(2\mathbf{i} - 3\mathbf{j})$.

a Find the angle between $\mathbf{R}$ and the positive x-axis. **(2 marks)**

b Show that $3p + 2q = 5$. **(3 marks)**

c Given that $p = 3$, find the magnitude of $\mathbf{R}$. **(2 marks)**

(E/P) **7** A forest fire starts at point A. The fire spreads and covers the area shown by the triangle in the diagram.

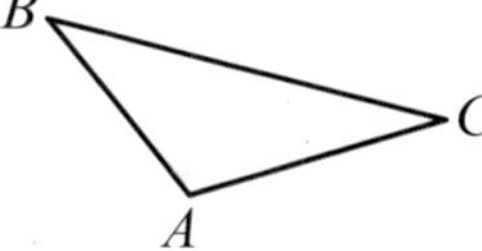

The vector $\overrightarrow{AB} = (-3\mathbf{i} + 5\mathbf{j})$ km and the vector $\overrightarrow{AC} = (6\mathbf{i} + \mathbf{j})$ km. Find the area of forest which is on fire. **(7 marks)**

(E) **8** A particle P moves under the action of two forces, $\mathbf{F}_1$ and $\mathbf{F}_2$, where $\mathbf{F}_1 = (a\mathbf{i} + b\mathbf{j})$ N and $\mathbf{F}_2 = (-3\mathbf{i} + 7\mathbf{j})$ N. The resultant force acting on the particle is $\mathbf{R} = (8\mathbf{i} + 32\mathbf{j})$ N.

a Find, to 1 decimal place, the angle between the resultant force and $\mathbf{i}$. **(2 marks)**

b Find the value of a and the value of b. **(3 marks)**

A third force $\mathbf{F}_3 = (c\mathbf{i} + 10\mathbf{j})$ N is also applied to P, such that the resultant of all three forces acts parallel to $\mathbf{j}$.

c Find the value of c. **(3 marks)**

Problem solving Set A

Bronze

In triangle ABC, $\overrightarrow{AB} = 5\mathbf{i} + 12\mathbf{j}$ and $\overrightarrow{AC} = 12\mathbf{i} + 5\mathbf{j}$. Find:

a $\overrightarrow{BC}$ **(1 mark)**

b the lengths AB, AC and BC **(3 marks)**

c $\angle BAC$ **(2 marks)**

Silver

The position vectors of the vertices in triangle ABC are $\overrightarrow{OA} = -\mathbf{i} + 4\mathbf{j}$, $\overrightarrow{OB} = 8\mathbf{j}$ and $\overrightarrow{OC} = 7\mathbf{i} + 9\mathbf{j}$. Find:

a $\overrightarrow{BC}$ **(1 mark)**

b $\angle BAC$ **(4 marks)**

c the area of triangle ABC. **(2 marks)**

Gold

In triangle ABC, $\overrightarrow{AB} = 8\mathbf{i} - 2\mathbf{j}$ and $\overrightarrow{AC} = 2\mathbf{i} - 7\mathbf{j}$. Find the area of triangle ABC. **(7 marks)**

Problem solving Set B

Bronze

In triangle OAB, $\overrightarrow{OA} = \mathbf{a}$ and $\overrightarrow{OB} = \mathbf{b}$.

R divides AB in the ratio 3 : 2.

$\overrightarrow{OP} = 5\overrightarrow{OR}$.

a Show that $\overrightarrow{OP} = 2\mathbf{a} + 3\mathbf{b}$. **(3 marks)**

Point Q is such that $\overrightarrow{OQ} = -\mathbf{a}$.

b Prove that the points Q, B and P are collinear. **(3 marks)**

Silver

In the diagram, M is the midpoint of OA and N lies on OC.

The point B lies on MC such that $MB : BC = 1 : 4$.

$\overrightarrow{OM} = \mathbf{a}$ and $\overrightarrow{ON} = \mathbf{b}$. Given that $\overrightarrow{AB} = k\overrightarrow{AN}$,

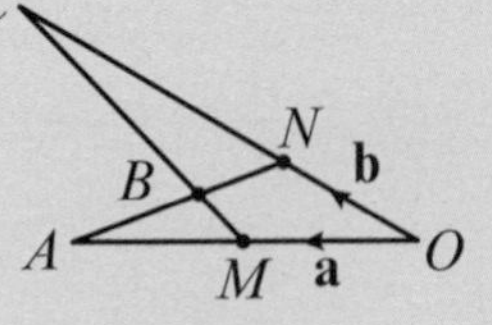

a find the value of k **(4 marks)**

b write $\overrightarrow{OC}$ as a scalar multiple of $\mathbf{b}$. **(1 marks)**

Gold

$ABCD$ is a square. P divides AB in the ratio 4 : 1 and Q divides CD in the ratio 3 : 2. M is the midpoint of AD.

Prove that PQ bisects MC.

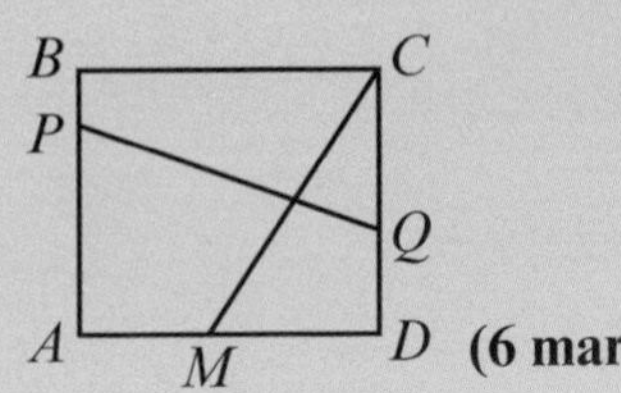

(6 marks)

Now try this → Exam question bank Q5, Q14, Q16, Q19, Q56, Q80

12.1 Gradients of curves

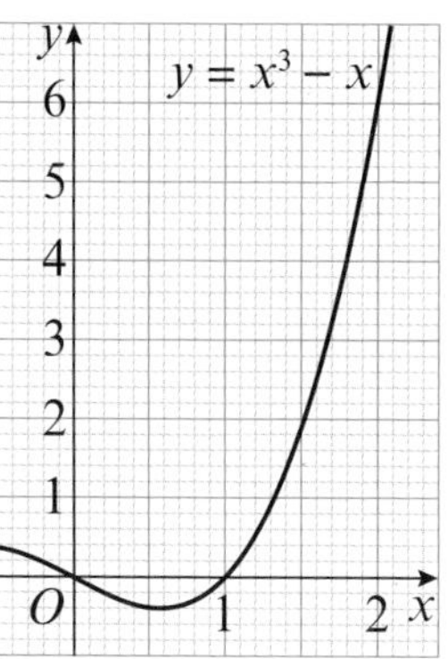

1 The diagram shows the curve with equation $y = x^3 - x$.

a Copy and complete this table showing estimates for the gradient of the curve at different values of x.

x-coordinate	0	1	1.5	2
Estimate for gradient of curve				

b Write a hypothesis about the gradient of the curve at the point where $x = p$.

c Test your hypothesis by estimating the gradient of the curve at the point (0.5, −0.375).

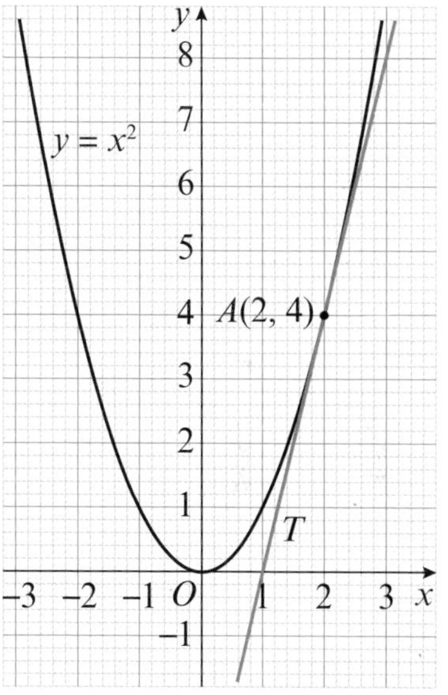

2 The diagram shows the curve with equation $y = x^2$.
The tangent, T, to the curve at the point $A(2, 4)$ is shown.

a Calculate the gradient of the tangent, T.

b Calculate the gradient of the chord AP when P has coordinates:

i (3, 9) **ii** (2.5, 6.25)

iii (2.1, 4.41) **iv** (2.01, 4.0401)

v $(2 + h, (2 + h)^2)$

12.2 Finding the derivative

1 For the function $f(x) = 12x$, use the definition of the derivative to show that:

a $f'(4) = 12$ **b** $f'(-2) = 12$

c $f'(0) = 12$

2 For the function $f(x) = 2x^2$, use the definition of the derivative to show that:

a $f'(3) = 12$ **b** $f'(-1) = -4$

c $f'(0) = 0$

Hint The derivative is defined by

$$f'(x) = \lim_{h \to 0} \frac{f(x + h) - f(x)}{h}$$

Substitute the value given for x into the definition of the derivative. Then use the function to expand and simplify. In each case you should be able to cancel h from each term in the numerator.

/P **3** Prove, from first principles, that the derivative of $8x$ is 8. **(3 marks)**

/P **4** Prove, from first principles, that the derivative of $3x^2$ is $6x$. **(4 marks)**

(E/P) **5** Prove, from first principles, that the derivative of x^3 is $3x^2$. **(4 marks)**

(E/P) **6** The curve C has equation $6 - 3x^2$. The curve C passes through the point Q with x-coordinate 1. Use differentiation from first principles to find the value of the gradient of the tangent to C at Q. **(5 marks)**

12.3 Differentiating x^n

1 Find $f'(x)$ given that $f(x)$ equals:

a x^5 **b** x^{-6} **c** $\sqrt{x}$ **d** $\frac{1}{\sqrt[4]{x}}$

Hint First write each function in the form x^n.

2 Find $\frac{dy}{dx}$ given that y equals:

a $5x^4$ **b** $5x^{-2}$ **c** $\frac{3x^2}{12x^5}$ **d** $\frac{9}{\sqrt[3]{x^2}}$

Hint If $y = ax^n$, then $\frac{dy}{dx} = anx^{n-1}$.

3 Find the gradient of the curve $y = 2x^3$ at the point where:

a $x = 3$ **b** $x = -5$

c $x = \frac{1}{2}$ **d** $x = \frac{3}{4}$

Hint First differentiate $2x^3$ and then substitute each value of x.

(E) **4** Find the gradient of the curve $y = \frac{5}{x^3}$ at the point where $x = 2$. **(2 marks)**

(E) **5** Find the gradient of the curve $y = -\frac{8}{\sqrt{x}}$ at the point where $x = 16$. **(2 marks)**

(E/P) **6** The gradient of the curve $y = 2x^4$ at the point $x = a$ is 27. Find the value of a. **(3 marks)**

(E/P) **7** A curve C has equation $y^3 - 64x^2 = 0$, $y \geqslant 0$. Find the gradient of C at the point where $x = 27$. **(4 marks)**

12.4 Differentiating quadratics

1 Find $f'(x)$ given that $f(x)$ equals:

a $3x^2 - 7x + 12$ **b** $\frac{3}{4}x^2 + \frac{1}{2}x - 1$

c $12 - 5x^2$

Hint Differentiate the terms one at a time. If $f(x) = ax^2 + bx + c$, then $f'(x) = 2ax + b$.

2 Find the gradient of the curve with the equation:

a $y = -4x^2 + 11x - 1$ at the point (2, 5)

b $y = 3x^2 - 20$ at the point (4, 28)

c $y = 18 + \frac{7}{2}x - \frac{1}{4}x^2$ at the point (8, 30).

Hint Substitute the x-value into the expression for $\frac{dy}{dx}$ to find the gradient at that point.

(P) **3** Find the coordinates of the point on the curve with equation $y = 2x^2 - 3x + 4$ where the gradient is 9.

Hint Set $\frac{dy}{dx} = 9$ and then solve to find the value of x. Then find the corresponding value of y.

E/P 4 Find the gradient of the curve with the equation $y = (2 - x)(2 + x)$ at the point where $x = 5$. **(3 marks)**

E/P 5 The curve with equation $y = 6x^2 - 5x + 3$ has gradient -29 when $x = p$. Find the value of p. **(3 marks)**

E 6 The curve with equation $y = x^2 - 3x + 1$ intersects the line with equation $y = 2x - 5$ at the points P and Q. Find:

a the coordinates of the points P and Q **(3 marks)**

b the gradient of the curve at the points P and Q. **(3 marks)**

E 7 The curve with equation $y = 5x^2 - 6x - 1$ intersects the line with equation $y = 7$ at the points A and B. Find:

a the coordinates of the points A and B **(3 marks)**

b the gradient of the curve at the points A and B. **(3 marks)**

12.5 Differentiating functions with two or more terms

1 Differentiate:

a $5x - 7x^2$ **b** $8x^{\frac{5}{2}} - 6x^4$ **c** $8x^{\frac{5}{2}} - 4x^2 - 6x^{\frac{1}{3}}$

Hint Differentiate the terms one at a time.

2 Find $f'(x)$ given that $f(x)$ equals:

a $\sqrt{x} + \frac{1}{\sqrt{x}}$ **b** $\frac{1}{7x^2} + \sqrt[4]{x}$

Hint First write each term in the form ax^n.

3 Find $\frac{dy}{dx}$ when y equals:

a $x(2x^2 - 6x + 5)$ **b** $(x^2 - 2)\left(x + \frac{2}{x}\right)$

c $\frac{1}{\sqrt{x}}\left(6x^{\frac{5}{2}} - 8x^{\frac{3}{2}}\right)$

Hint Multiply out the brackets and write each term in the form ax^n.

4 In each case, find the gradient of the curve with equation $y = f(x)$ at the point A:

a $f(x) = \frac{1}{2}x^4 - 8x^2 + 2$; A is $(-2, -22)$ **b** $f(x) = \frac{6}{\sqrt{x}} - 3x - 2x^2$; A is $(1, 1)$

E 5 Find the gradient of the curve $f(x) = \frac{4x^2 - 8x}{x^3}$ at the point $P(-2, -4)$. **(3 marks)**

E/P 6 Given that $f(x) = px^2 - 8p\sqrt{x}$ and that $f'(4) = 15$, find the value of p. **(4 marks)**

E/P 7 $h(x) = px^3 + 4px + \frac{q}{x}$, where p and q are constants.

Given that $h(-1) = -41$ and $h'(-1) = 67$, find the values of p and q. **(5 marks)**

E/P 8 $f(x) = (1 + 2x)^6$, $g(x) = (1 - 2x)^6$

a Find the first three terms, in ascending powers of x, of the binomial expansion of $f(x)$, giving each term in its simplest form. **(2 marks)**

b If x is small, so that x^2 and higher powers can be ignored, find:

i $f'(x)$ **ii** $g'(x)$ **(4 marks)**

12.6 Gradients, tangents and normals

Hint The tangent to the curve $y = \mathrm{f}(x)$ at the point where $x = a$ has equation
$$y - \mathrm{f}(a) = \mathrm{f}'(a)(x - a)$$

1 Find the equation of the tangent to the curve:

a $y = x^3 - 4x^2$ at the point $(2, -8)$

b $y = \frac{4}{x} - 3x^2$ at the point $(-2, -14)$

c $y = 6x^{\frac{3}{2}} - 4x^{\frac{1}{2}}$ at the point where $x = 4$

d $y = \dfrac{12 - x^2}{2x}$ at the point where $x = -3$.

Hint The normal to the curve $y = \mathrm{f}(x)$ at the point where $x = a$ has equation
$$y - \mathrm{f}(a) = -\frac{1}{\mathrm{f}'(a)}(x - a)$$

2 Find the equation of the normal to the curve:

a $y = \frac{1}{2}x^2 + \frac{6}{x}$ at the point $\left(3, \frac{13}{2}\right)$

b $y = \frac{4}{x} - 3x^2$ at the point where $x = 4$.

3 $\mathrm{g}(x) = 10 - 2x - x^2$

a Find the equation of the tangent to the curve $y = \mathrm{g}(x)$ at the point where $x = 1$.

b Find the equation of the normal to the curve $y = \mathrm{g}(x)$ at the point where $x = 1$.

(E) 4 $\mathrm{f}(x) = x^2 - 5x + 4$

a Find the equation of the tangent to the curve $y = \mathrm{f}(x)$ at the point where $x = 5$. **(3 marks)**

b Find the equation of the normal to the curve $y = \mathrm{f}(x)$ at the point where $x = 3$. **(3 marks)**

The tangent found in part **a** intersects the normal found in part **b** at the point P.

c Find the coordinates of point P. **(4 marks)**

(E) 5 $\mathrm{g}(x) = x^3 - 4x$

a Find the equation of the tangent to the curve $y = \mathrm{g}(x)$ at the point where $x = -1$. **(3 marks)**

This tangent meets the x-axis at the point A.

b Find the coordinates of A. **(1 mark)**

c Show that A lies on the curve $y = \mathrm{g}(x)$. **(1 mark)**

(E/P) 6 The point P with x-coordinate 6 lies on the curve with equation $y = \frac{1}{2}x^2 - 8x + 6$.

The normal to the curve at P intersects the curve at the points P and Q.

Find the coordinates of Q. **(6 marks)**

(E/P) 7 The point $P(4, 2.5)$ lies on the curve C with equation $y = \sqrt{x} + \dfrac{1}{\sqrt{x}}$

Giving your answer in the form $ax + by + c = 0$, find:

a the equation of the tangent to C at P **(3 marks)**

b the equation of the normal to C at P. **(3 marks)**

The tangent to C at P intersects the y-axis at A and the normal to C at P intersects the y-axis at B.

c Find the area of triangle ABP. **(4 marks)**

12.7 Increasing and decreasing functions

1 The graph shows the function $y = f(x)$ where $f(x) = \frac{1}{2}x^4 - x^2$.

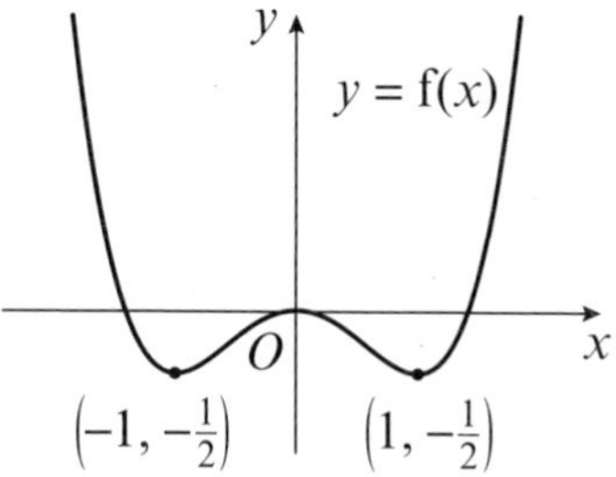

a State the coordinates of the three points where the gradient of $y = f(x)$ is zero.

b State whether the function is increasing, decreasing or neither over the interval:

i $(-\infty, -1)$

ii $[-1, 0]$

iii $[0, 1]$

iv $(-1, 1)$

v $[1, \infty)$

vi $[0, 2]$

Hint The function $f(x)$ is increasing on $[a, b]$ if $f'(x) \geqslant 0$ for all values of x such that $a < x < b$.

The function $f(x)$ is decreasing on $[a, b]$ if $f'(x) \leqslant 0$ for all values of x such that $a < x < b$.

If neither of these conditions holds for all values on a given interval, then the function is neither increasing nor decreasing on that interval.

2 Find the range of values of x for which $f(x)$ is increasing, given that $f(x)$ equals:

a $5x^2 - 8x + 7$ **b** $14 - 9x - 3x^2$ **c** $x^3 - 12x + 8$ **d** $18x - 2x^3$

3 Find the range of values of x for which $f(x)$ is decreasing, given that $f(x)$ equals:

a $3x^2 + 7x - 2$ **b** $15 - 6x - 5x^2$ **c** $-2x^5 - 8x^3$ **d** $5x(3 - x)$

/P 4 Show that the function $f(x) = x^5 + 2x^3 + 8x - 18$ is increasing for all real values of x. **(3 marks)**

/P 5 Show that the function $g(x) = -2x^3 - 18x^2 - 54x + 10$ is decreasing for all real values of x. **(3 marks)**

/P 6 $h(x) = x^3 + qx$

Find the range of values of q for which $h(x)$ is increasing for all real values of x. **(3 marks)**

/P 7 A curve has equation $y = 3x^3 + 2x^2 - 5x - 6$.

a Find $\frac{dy}{dx}$ **(2 marks)**

b Hence find the range of values of x for which y is increasing. Write your answer using set notation. **(4 marks)**

12.8 Second order derivatives

1 Find $f'(x)$ and $f''(x)$ when $f(x)$ equals:

a $9x^2 - 6x + 7$ **b** $-x^3 + 4x^{-1}$ **c** $4x^{-\frac{1}{2}} + 8x^{\frac{3}{2}}$

Hint First differentiate $f(x)$ to obtain $f'(x)$. Then differentiate $f'(x)$ to obtain $f''(x)$.

2 Find $\frac{dy}{dx}$ and $\frac{d^2y}{dx^2}$ when y equals:

a $\frac{6}{x^3} - \frac{1}{2}x^4 - 9x$ **b** $\frac{8}{\sqrt[4]{x}} - \sqrt[6]{x} - 7$ **c** $(x^2 - 3)(5 - x^3)$

Hint Write each term in the form ax^n and then differentiate twice to obtain $\frac{d^2y}{dx^2}$

3 $h(x) = 4x^4 - \sqrt[3]{x} + \frac{1 - x^5}{x^3}$. Find:

a $h'(x)$ b $h''(x)$

E 4 Given that $f(x) = \frac{1}{x} - 6x - 2x^3$, find $f''(-2)$. **(3 marks)**

E 5 $p(x) = (1 - 4x)^3$

a Fully expand $p(x)$. **(2 marks)**

b Find $p''(x)$. **(1 mark)**

E/P 6 $f(x) = ax^3 + bx^2 - 8x + 6$, where a and b are constants.

Given that $f'(2) = -24$ and $f''(-5) = 100$, find the values of a and b. **(4 marks)**

E/P 7 $g(x) = px^3 + qx^2 + 6x - 1$, where p and q are constants.

Given that $g'(-2) = g'\left(\frac{5}{2}\right) = 0$, find:

a the value of p and the value of q **(4 marks)**

b $g''(x)$. **(2 marks)**

12.9 Stationary points

1 $f(x) = (x - 4)(x - 8)$

a Sketch the graph of $y = f(x)$.

b Explain how the graph in part **a** shows that $y = f(x)$ has a local minimum between $x = 4$ and $x = 8$.

c Find $f'(x)$.

d Solve $f'(x) = 0$.

e State the coordinates of the stationary point.

Hint The stationary point occurs where $f'(x) = 0$. The shape of the graph of $y = f(x)$ can be used to determine whether this stationary point is a local minimum or a local maximum.

2 Find the coordinates of the minimum point on the curve with equation:

a $y = x^2 - 6x + 8$ b $y = x^2 - 3x - 11$

c $y = 3x^2 + 2x + 1$

Hint Find $f'(x)$ and then solve $f'(x) = 0$ to obtain the value of x at the stationary point. Then find the corresponding value of $y = f(x)$.

3 Find the coordinates of the maximum point on the curve with equation:

a $y = 8 - 16x - x^2$ b $y = 12x - 4x^2$ c $y = 18 - 30x - 7x^2$

4 a Find the coordinates of the stationary points on the curve with equation $y = -4x^3 + 6x^2 + 24x + 3$

b Find $\frac{d^2y}{dx^2}$

c Use $\frac{d^2y}{dx^2}$ to determine the nature of the stationary points on the curve of y.

Hint If $\frac{dy}{dx} = 0$ and $\frac{d^2y}{dx^2} > 0$, the stationary point is a local minimum.

If $\frac{dy}{dx} = 0$ and $\frac{d^2y}{dx^2} < 0$, the stationary point is a local maximum.

5 $q(x) = -x^3 - 6x^2 - 12x - 6$

a Find the coordinates of the stationary point on the curve with equation $y = q(x)$.

b Show that this stationary point is a point of inflection.

c Sketch the graph of $y = q(x)$.

Hint If $\frac{dy}{dx} = 0$ and $\frac{d^2y}{dx^2} = 0$, the stationary point could be a local minimum, local maximum or a point of inflection.

Look at the gradient either side of the stationary point to determine its nature.

6 $f(x) = 2x^3 - \frac{1}{2}x^2 - x + 2$

Find the coordinates of the stationary points on the curve with equation $y = f(x)$. **(4 marks)**

7 $g(x) = \frac{54}{x} + 4\sqrt{x}$

a Find the coordinates of the stationary point on the curve $y = g(x)$. **(5 marks)**

b Find $g''(x)$. **(1 mark)**

c Show that the stationary point from part **a** is a local minimum. **(1 mark)**

8 $f(x) = -2x^3 + 4x^2 + 8x - 10$

Find the coordinates of the local maximum on the curve with equation $y = f(x)$. **(5 marks)**

9 $h(x) = x^3 - 3x^2 + 3x + 4$

a Find the coordinates of the stationary point on the curve $y = h(x)$ and determine the nature of this stationary point. **(4 marks)**

b Sketch the graph of $y = h(x)$. **(2 marks)**

12.10 Sketching gradient functions

1 The diagram shows the curve with equation $y = f(x)$.

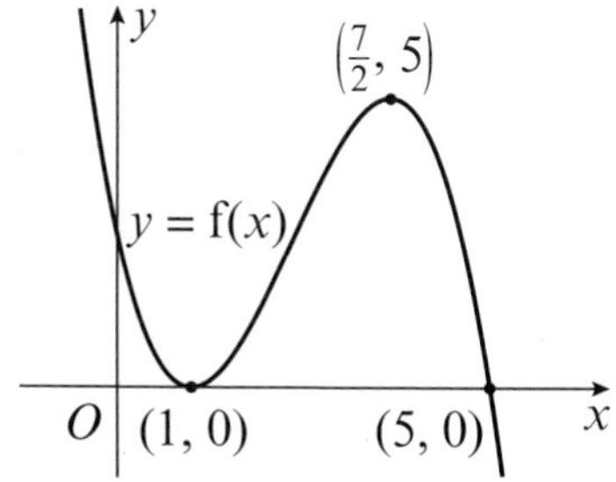

a Write down the coordinates of the stationary points on the curve $y = f(x)$.

b Describe the gradient of $y = f(x)$ between the two stationary points.

c Describe the gradient of $y = f(x)$ for $x < 1$.

d Describe the gradient of $y = f(x)$ for $x > \frac{7}{2}$

e Use your answers to parts **a** to **d** to sketch the graph of $y = f'(x)$.

Hint At the point where $y = f(x)$ has a stationary point, $y = f'(x)$ will intersect the x-axis.

At the point where $y = f(x)$ has positive gradient, $y = f'(x)$ will be above the x-axis.

At the point where $y = f(x)$ has negative gradient, $y = f'(x)$ will be below the x-axis.

2 The diagram shows the curve with equation $y = f(x)$.

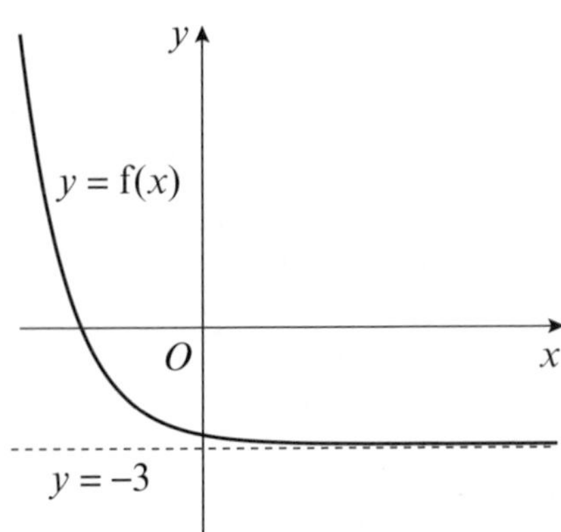

a Describe the gradient of $y = f(x)$ for all values of x.

b Sketch the graph of $y = f'(x)$, giving the equation of any asymptotes.

3 For each graph, sketch the graph of the corresponding gradient function on a separate set of axes. Label the coordinates of any points where the curve cuts or meets the x-axis, and give the equations of any asymptotes.

a

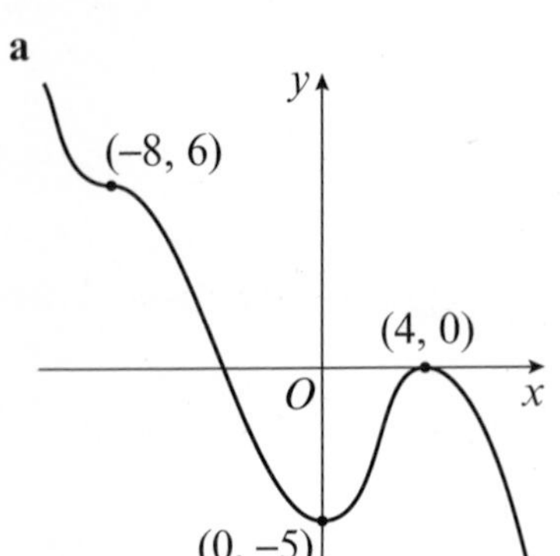

b

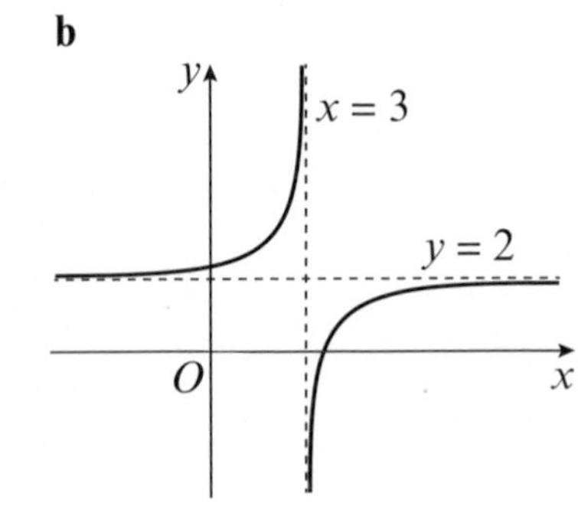

c

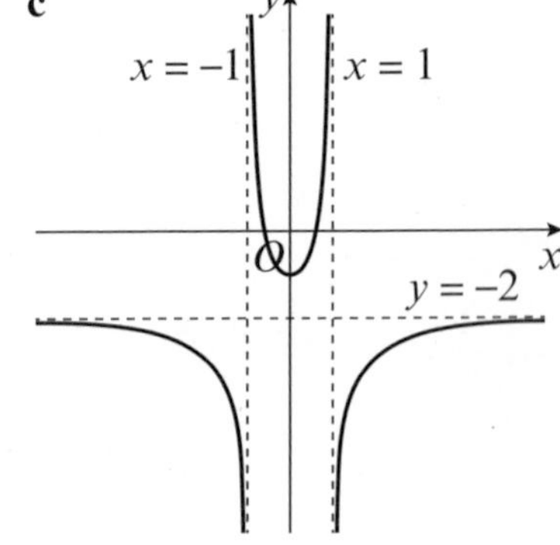

(E) 4 The graph shows the average wave height at a beach during each month.

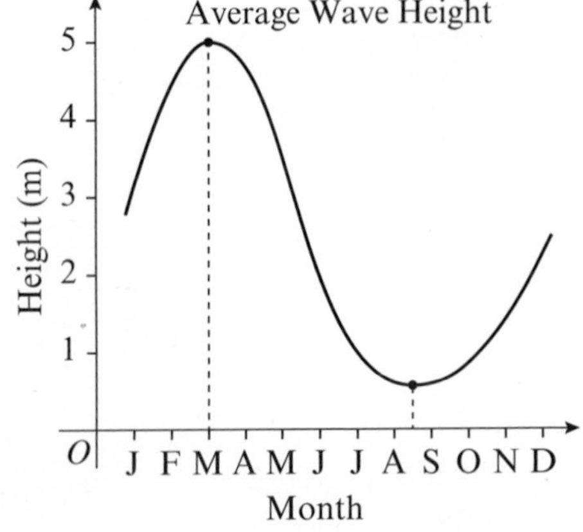

On a separate graph, sketch the gradient function for the wave height. Indicate the coordinates of any points where the gradient function crosses the horizontal axis. **(3 marks)**

12.11 Modelling with differentiation

1 Find $\frac{dA}{dx}$ where $A = 6x^2$.

Hint Finding $\frac{dA}{dx}$ where $A = 6x^2$ is the same as finding $\frac{dy}{dx}$ where $y = 6x^2$.

2 Given that $V = \frac{20\,000}{t^2}$, find:

a $\frac{dV}{dt}$

b the value of $\frac{dV}{dt}$ when $t = 10$

Hint Substitute $t = 10$ into your expression for $\frac{dV}{dt}$

3 When a cyclist reaches the bottom of a large hill, he is travelling with a speed, v, of 8 m s^{-1}. The cyclist then free-wheels along a straight flat road with speed $v = (8 - \sqrt{t})$ m s^{-1}, where t is the time elapsed since he reached the bottom of the hill.

Find:

a the time it takes for the cyclist to come to a stop

b $\frac{dv}{dt}$

c the cyclist's acceleration at time $t = 9$ seconds.

Hint Acceleration is the rate of change of speed with respect to time:

$a = \frac{dv}{dt}$

← **Statistics and Mechanics Year 1, Chapter 11**

4 The volume, V cm^3, of an expanding sphere of radius r cm is given by $V = \frac{4}{3}\pi r^3$. Find:

a the rate of change of the volume with respect to the radius

b the rate of change of volume with respect to radius when the radius is 4 cm.

Hint $\frac{dV}{dr}$ is the rate of change of volume with respect to the radius.

P 5 A fence is 80 m long and is used to enclose an area with five sides, as shown in the diagram. Some of the dimensions of the fence are also shown. The area enclosed by the fence is A m^2.

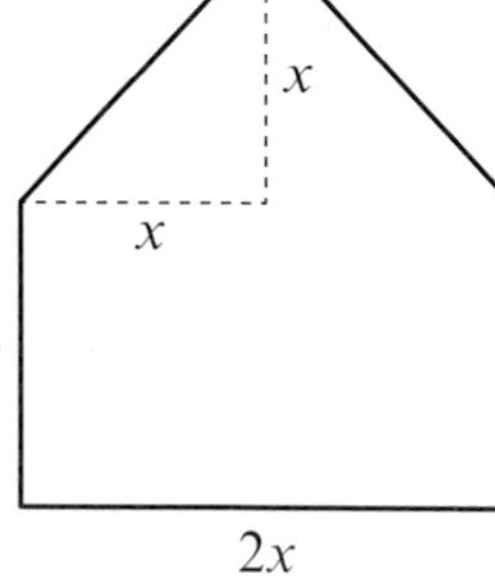

a Find an expression for y in terms of x. **(3 marks)**

b Prove that $A = 80x - x^2(1 + 2\sqrt{2})$. **(3 marks)**

c Given that x can vary, use calculus to find the exact maximum value of A. **(5 marks)**

P 6 A solid cylinder has volume 1024π cm^3 and variable height h and radius r.

a Show that the surface area of the cylinder is given by $S = \frac{2048\pi}{h} + 64\pi\sqrt{h}$. **(3 marks)**

b Hence find the minimum value of the surface area, leaving your answer in terms of π. **(5 marks)**

E/P **7** A rectangular box has a horizontal base but no top.
The cross-section of the box is a square of side length x cm.
The total area of cardboard used to make the box is 608 cm^2.

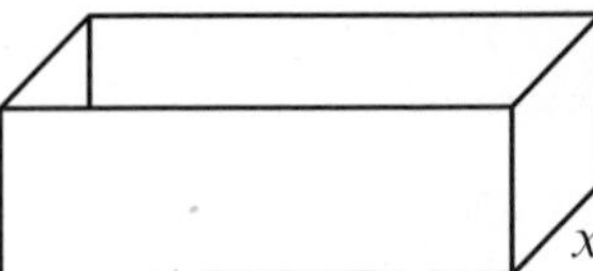

a Show that the volume, V cm^3, of the box is given by

$$V = \frac{608x - 2x^3}{3}$$ **(3 marks)**

b Find the maximum value of V, giving your answer to 1 decimal place. **(3 marks)**

c Explain how you know your answer is a maximum. **(2 marks)**

E/P **8** The diagram shows a design for a storage bin. The bin is made from a cylinder of radius r cm and height h cm attached to a hemisphere of radius r cm.

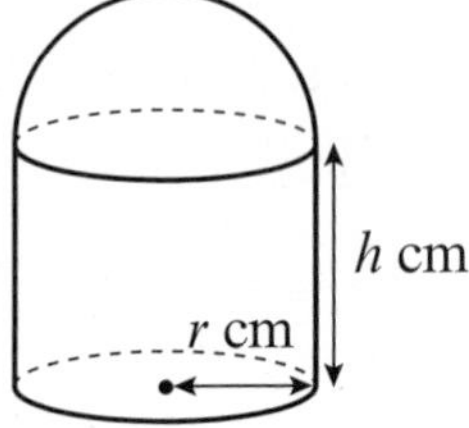

The total surface area of the bin is 800π cm^2, and the volume of the bin is V cm^3.

a Show that $V = \dfrac{\pi r(2400 - 5r^2)}{6}$ **(5 marks)**

b Find the maximum volume of the bin. Give your answer correct to 3 significant figures. **(5 marks)**

Problem solving Set A

Bronze

$h(x) = 14 + 6x - x^2$

a Find the coordinates of the stationary point on the curve with equation $y = h(x)$. **(3 marks)**

b By sketching the graph, or otherwise, determine the nature of the stationary point. **(2 marks)**

Silver

$g(x) = -x^3 - 3x^2 + 4$.

a Find the coordinates of the stationary points on the curve with equation $y = g(x)$. **(4 marks)**

b Hence, find all the solutions to the equation $g(x) = 0$. **(3 marks)**

c By sketching the graph of $y = g(x)$, or otherwise, determine the nature of each stationary point. **(2 marks)**

Gold

$f(x) = \frac{2}{3}\sqrt{x^3} - 10\sqrt{x} - \frac{8}{\sqrt{x}}, x > 0$.

a Find the coordinates of the stationary points on the curve with equation $y = f(x)$. **(5 marks)**

b By considering $f''(x)$, determine the nature of each stationary point. **(4 marks)**

Problem solving Set B

Bronze

$h(x) = (x - 2)(x - 1)(x + 2)$

The curve with equation $y = h(x)$ cuts the x-axis at the points D, E and F as shown in the diagram.

a Write down the coordinates of the points D, E and F. **(1 mark)**

b Find $h'(x)$. **(2 marks)**

c Find the equation of the tangent to the curve at:

i E **ii** F **(3 marks)**

The tangents in part **c** intersect at a point G.

d Find the coordinates of G. 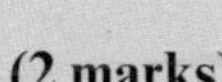 **(2 marks)**

Silver

$g(x) = x(x^2 - x - 2)$

The curve C with equation $y = g(x)$ cuts the x-axis at the origin and at the points P and Q as shown.

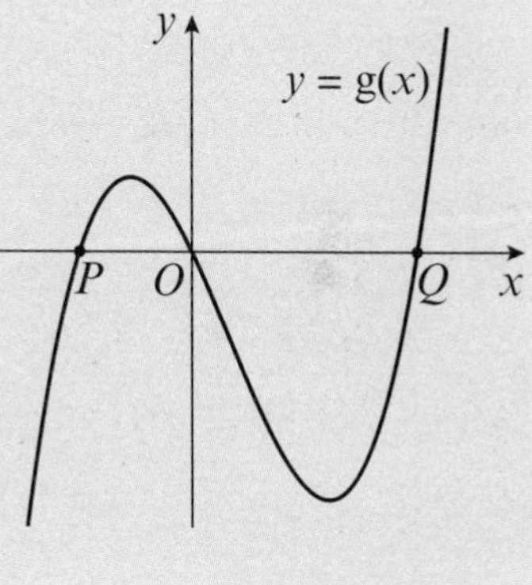

a Find $g'(x)$. **(2 marks)**

b Find the equation of the normal to C at:

i P **ii** Q **(3 marks)**

The two normals found in part **b** intersect at a point R.

c Find the coordinates of R. **(2 marks)**

Gold

$f(x) = \frac{1}{6}x^3 + \frac{1}{3}x^2 - \frac{4}{3}x$

The curve with equation $y = f(x)$ cuts the x-axis at the origin and at the points A and B as shown.

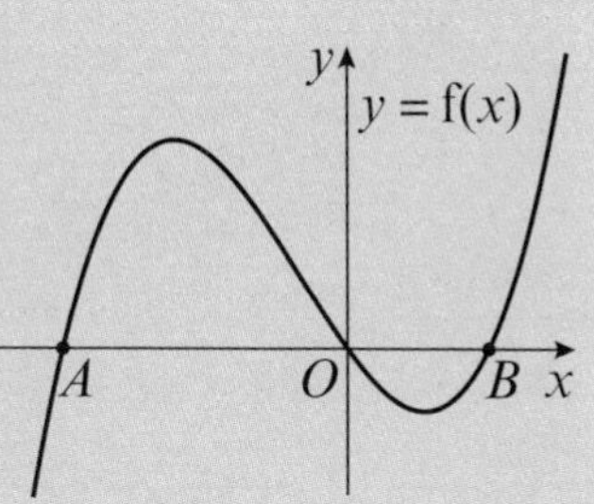

The normal to the curve at A intersects the normal to the curve at B at a point P. Find the area of triangle APB. **(8 marks)**

Now try this → Exam question bank Q11, Q15, Q27, Q50, Q74, Q90

13 Integration

13.1 Integrating x^n

1 Find $f(x)$ when $f'(x)$ equals:

a $9x^2$ **b** $12x^{-3}$

c $6x^{\frac{5}{2}}$ **d** 25

Hint If $f'(x) = x^n$, then $f(x) = \frac{1}{n+1}x^{n+1} + c,\ n \neq -1$

2 Find y when $\frac{dy}{dx}$ equals:

a $\sqrt{x}$ **b** $\frac{3}{\sqrt[3]{x}}$

c $5\sqrt[4]{x}$ **d** $-\frac{1}{6\sqrt{x}}$

Hint First write each term in the form kx^n.
If $\frac{dy}{dx} = kx^n$, then $y = \frac{k}{n+1}x^{n+1} + c,\ n \neq -1$

3 Find $f(x)$ when $f'(x)$ is given by each of the following expressions, giving your coefficients as simplified fractions.

a $x^4 + x^3 - 5x + 6$ **b** $4x^2 + \frac{1}{3}x^{-2} - 5x^{-3}$

c $3x^{-\frac{1}{4}} - 2x^{-\frac{3}{4}}$

Hint Apply the rule of integration to each term in the expression and add c.

4 $f'(x) = (4x + 3)^2$

a Write $f'(x)$ in the form $ax^2 + bx + c$.

b Hence find $f(x)$.

Hint First expand the brackets, then apply the rule of integration to each term in the expression and add the unknown constant.

E/P 5 Find $f(x)$ given $f'(x) = (3x - 5)^2$. **(4 marks)**

E/P 6 Find y given $\frac{dy}{dx} = (1 - 3x)^2$. **(4 marks)**

E/P 7 $h'(x) = (p - 5x)^4$

a The coefficient of the x^3 term in the expansion of $h'(x)$ is -1000.

Find the value of the constant p. **(4 marks)**

b Hence, or otherwise, find $h(x)$. **(2 marks)**

E/P 8 $f'(x) = (4 - 2\sqrt{x})^2$

a Write $f'(x)$ in the form $a + b\sqrt{x} + cx$, where a, b and c are constants to be found. **(2 marks)**

b Hence find $f(x)$. **(3 marks)**

E/P 9 $g'(x) = (3 - 2x)^8$

a Find the first three terms, in ascending powers of x, if the binomial expansion of $g'(x)$ giving each term in its simplest form. **(4 marks)**

b Assuming x is small and higher powers can be ignored, use your answer to part **a** to find an approximation to $g(x)$ in the form $px + qx^2 + rx^3 + c$. **(1 mark)**

13.2 Indefinite integrals

1 Find the following integrals:

a $\int x^4 \,\mathrm{d}x$ **b** $\int -7x^6 \,\mathrm{d}x$

c $\int 4x^{-2} \,\mathrm{d}x$

Hint $\int kx^n \,\mathrm{d}x = \frac{k}{n+1}x^{n+1} + c, n \neq -1$

2 Find the following integrals:

a $\int (5x^2 + 9x^{-3})\,\mathrm{d}x$ **b** $\int \left(\frac{3}{4}x^5 - 4x^{-5} + 8x\right)\mathrm{d}x$

c $\int \left(4x^{\frac{3}{2}} - 10x^{-\frac{3}{2}}\right)\mathrm{d}x$

Hint Integrate each term separately:
$\int (\mathrm{f}(x) + \mathrm{g}(x))\,\mathrm{d}x = \int \mathrm{f}(x)\,\mathrm{d}x + \int \mathrm{g}(x)\,\mathrm{d}x$

3 Find the following integrals:

a $\int (6 - 5\sqrt{x})^2 \,\mathrm{d}x$ **b** $\int \frac{(3x-2)^2}{\sqrt{x}}\,\mathrm{d}x$

c $\int \left(\frac{5 - 7x^3 + 4x^{-\frac{3}{2}}}{x^2}\right)\mathrm{d}x$

Hint First expand the brackets or divide so each term is in the form kx^n, then integrate.

4 Find the following integrals:

a $\int \left(\frac{P}{\sqrt{x}} - Rx^{\frac{3}{2}}\right)\mathrm{d}x$ **b** $\int \left(\frac{A}{x^3} + \frac{B}{x^5}\right)\mathrm{d}x$

c $\int \left(M\sqrt[3]{x} - Nx^{\frac{3}{4}}\right)\mathrm{d}x$

Hint Treat letters other than x as constants.

5 Given that $y^{\frac{1}{2}} = 6x^{\frac{1}{5}} - 5$, find $\int y \,\mathrm{d}x$ **(3 marks)**

6 Given that $\mathrm{g}(x) = 3x^3 - 7\sqrt{x} + \frac{4}{\sqrt[3]{x}} - \frac{2}{x^2}$, find $\int \mathrm{g}(x)\,\mathrm{d}x$ **(4 marks)**

7 $\int \left(abx^2 + \frac{4b}{x^5}\right)\mathrm{d}x = 2x^3 - \frac{2}{x^4} + c$, where a and b are real constants.

Find the values of a and b. **(4 marks)**

8 $\mathrm{g}(x) = (1 + 2x)(4 - 3x)^8$

Given that x is small, and so terms in x^3 and higher powers of x can be ignored,

a find an approximation for $\mathrm{g}(x)$ in the form $P + Qx + Rx^2$ **(4 marks)**

b find an approximation for $\int \mathrm{g}(x)\,\mathrm{d}x$. **(3 marks)**

13.3 Finding functions

1 Find the equation of the curve with the given derivative of y with respect to x that passes through the given point:

a $\frac{\mathrm{d}y}{\mathrm{d}x} = 4x^3 - 10x^2$; point (3, 6)

b $\frac{\mathrm{d}y}{\mathrm{d}x} = \frac{5}{\sqrt{x}} + \frac{1}{2}x^2$; point (16, 65)

Hint Apply the rule of integration and then substitute the given values of x and y to find the constant c.

2 Find the equation of the curve with the given derivative of y with respect to x that passes through the given point:

a $\frac{dy}{dx} = \frac{3 - x^2}{x^4}$; point $(-1, 4)$

b $\frac{dy}{dx} = (x - 5)^2$; point $(2, 18)$

Hint Begin by writing each term in the form ax^n, then integrate and finally substitute to find the value of the constant c.

3 The gradient of a curve, with equation $y = f(x)$, is given by $f'(x) = \frac{8}{\sqrt{x}} - 4\sqrt{x}$
The curve passes through the point $(9, -30)$.
Find the equation of the curve in the form $y = f(x)$.

E 4 $\frac{dy}{dx} = 10 - 6x - 4x^2$. Given that $y = -15$ when $x = 3$, find y in terms of x. **(4 marks)**

E 5 $f'(x) = \frac{5\sqrt{x} - 3x^{\frac{3}{2}}}{x^2}$
The curve with equation $y = f(x)$ passes through the point $(4, 7)$. Find $f(x)$. **(5 marks)**

E 6 The gradient of a curve, with equation $y = f(x)$, is given by $f'(x) = 6x - \frac{4}{\sqrt[3]{x}}$
The curve passes through the point $(8, 112)$.
Find the equation of the curve in the form $y = f(x)$. **(4 marks)**

E/P 7 $p'(x) = \frac{(1 - 5x)^3}{\sqrt{x^3}}$

a Show that $p'(x) = ax^{-\frac{3}{2}} + bx^{-\frac{1}{2}} + cx^{\frac{1}{2}} + dx^{\frac{3}{2}}$ where a, b, c and d are constants to be found. **(4 marks)**

b Given that the point $(1, -24)$ lies on the curve with equation $y = p(x)$, find $p(x)$. **(3 marks)**

E/P 8 Given that $h'(x) = \frac{3}{2}x^2 - \frac{1}{4\sqrt{x}}$ and $h(4) = -10$, find the exact value of $h(2)$. **(4 marks)**

13.4 Definite integrals

1 Evaluate the following definite integrals.

a $\int_2^{10} x^2 dx$ **b** $\int_1^8 6\sqrt[3]{x} dx$

c $\int_2^4 \frac{4}{x^3} dx$ **d** $\int_4^9 \frac{5}{\sqrt{x}} dx$

Hint If $f'(x)$ is the derivative of $f(x)$ for all values of x in the interval $[a, b]$ then the definite integral is defined as $\int_a^b f'(x)dx = [f(x)]_a^b = f(b) - f(a)$

2 Evaluate the following definite integrals.

a $\int_1^2 \left(18x - \frac{4}{x^3}\right) dx$ **b** $\int_1^4 \frac{1}{\sqrt{x}}(14x^3 - 5x^2) dx$ **c** $\int_1^{36} \left(2\sqrt{x} - \frac{6}{x}\right)^2 dx$

3 Use calculus to find the value of $\int_1^2 \left(\frac{5}{x^2} - \frac{3}{x^4}\right) dx$

Hint This question says 'use calculus' so you need to use integration and show the integrated function. Do not use a numerical integration function on your calculator.

4 Use calculus to find the value of $\int_1^{16}\left(\frac{2}{x^2}+4\sqrt{x}\right)dx$ **(4 marks)**

5 Use calculus to find the value of $\int_1^{4}\left(\frac{8x^3}{3}-\frac{6}{x^2}-\frac{2}{\sqrt{x}}\right)dx$ **(4 marks)**

P 6 Given that $\int_4^k 9\sqrt{x}\,dx = 702$, find the value of k. **(4 marks)**

7 $y^{\frac{1}{2}} = 2x^{\frac{1}{3}} - \frac{1}{2}x^{-\frac{1}{3}}$

a Show that $y = ax^{\frac{2}{3}} + bx^{-\frac{2}{3}} + c$, where a, b and c are constants to be found. **(2 marks)**

b Find $\int_1^8 y\,dx$ **(3 marks)**

P 8 Given that $\int_4^9 a\sqrt{x}\,dx = -114$, find the value of a. **(3 marks)**

13.5 Areas under curves

1 The diagram shows part of the curve C with equation

$y = -x^3 - x^2 + x + 1$

Find the area of the shaded region.

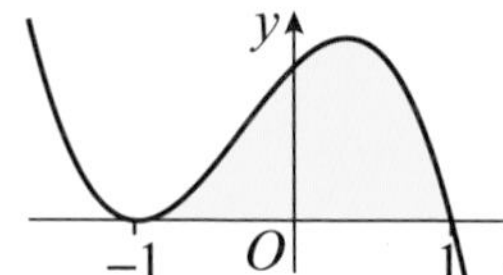

Hint The area between a positive curve, the x-axis and the lines $x = a$ and $x = b$ is given by

Area $= \int_a^b y\,dx$

where $y = f(x)$ is the equation of the curve.

2 The diagram shows a sketch of the curve with equation $y = x^2 - 7x + 10$.

Find:

a the coordinates of point A

b the area of the shaded region R.

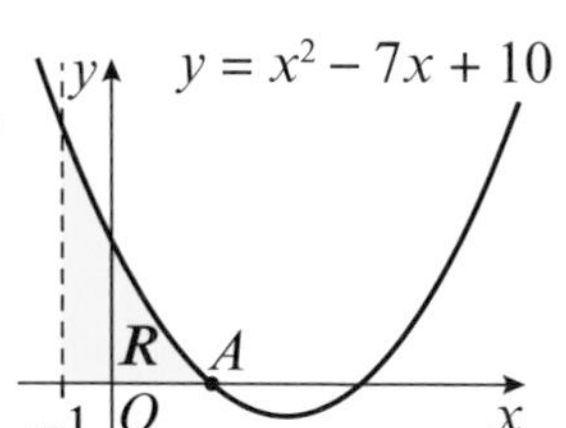

Hint First solve $x^2 - 7x + 10 = 0$. The x-coordinate of the point A will be the smaller of the two answers.

3 $f(x) = 5x^2 + \frac{2}{x^3} - 5$, $x > 0$.

The diagram shows the curve with equation $y = f(x)$.

The region R is bounded by the curve, the x-axis and the lines $x = \frac{1}{2}$ and $x = 2$. Use calculus to find the exact area of R.

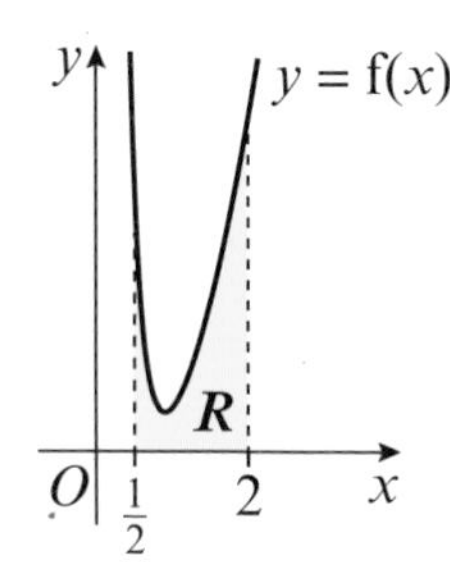

Hint The question says 'use calculus', so you need to show the integrated function in your working.

4 The diagram shows the curve with equation $y = g(x)$, where

$g(x) = 12 + 4x - x^2$, $x \geqslant 0$

The curve meets the x-axis at the points A and B. Find:

a the coordinates of points A and B **(2 marks)**

b the exact area of region R. **(4 marks)**

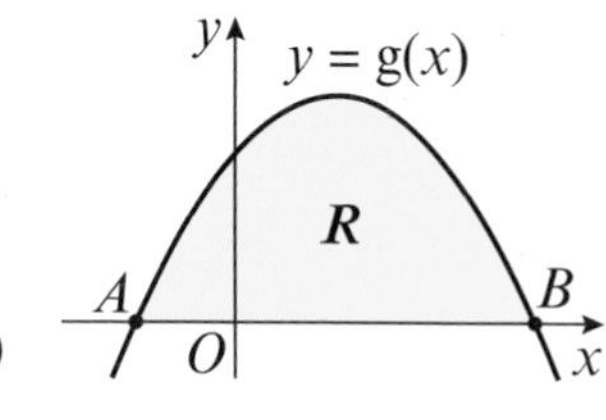

E/P **5** $h(x) = -(x - 1)^2(x - 4)$. The diagram shows the curve with equation $y = h(x)$.

The curve meets the x-axis at the points P and Q.

Use calculus to find the exact area of region R. **(6 marks)**

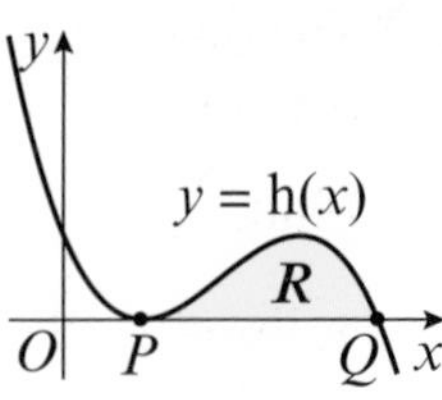

E **6** $p(x) = 8\sqrt{x} - x^2$. The diagram shows the curve with equation $y = p(x)$.

The curve meets the x-axis at the origin and at the point A.

Find:

a the coordinates of point A **(2 marks)**

b the exact area of region R. **(4 marks)**

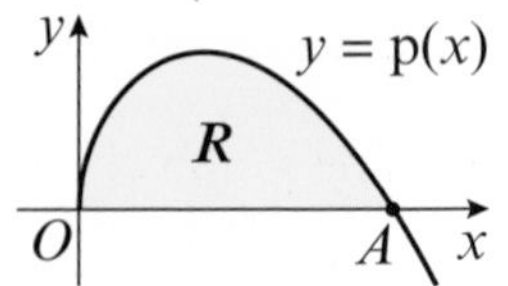

E/P **7** The shaded region is bounded by the curve with equation $y = 6x^2 - 12x + 9$, the x- and y-axes and the line $x = k$. Given that the area of the shaded region is 27, find the value of k. **(6 marks)**

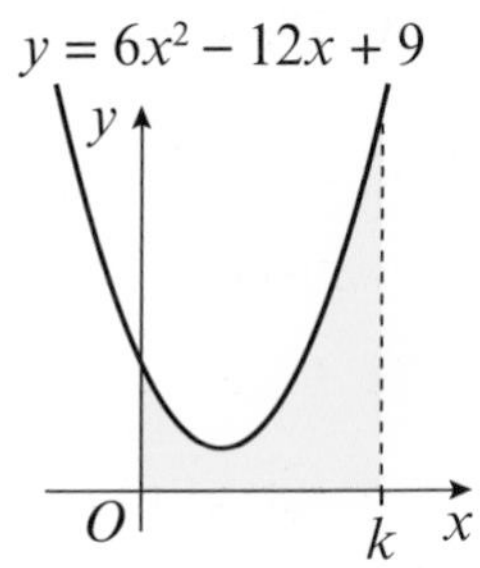

13.6 Areas under the x-axis

1 Find the area of the finite region bounded by the curve $y = x^2 - x - 6$ and the x-axis.

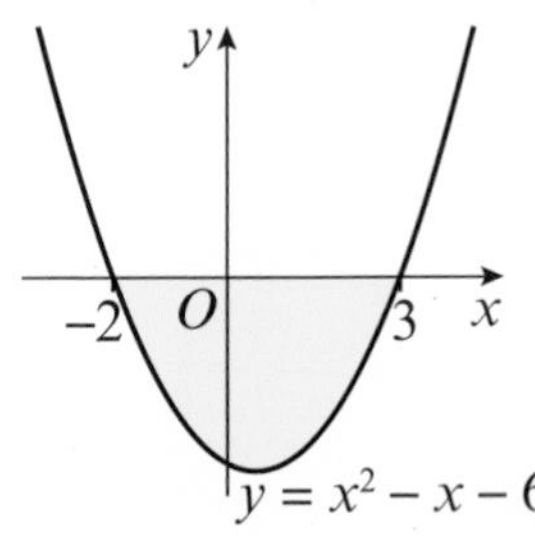

Hint Find $\int_{-2}^{3} (x^2 - x - 6)\,dx$. The value of the definite integral will be negative because the area is below the x-axis, but you should state the positive quantity as the final answer.

2 The graph shows a sketch of part of the curve C with equation $y = (x + 1)(x - 2)(x - 4)$. The curve cuts the x-axis at $(-1, 0)$ and at the points A and B.

a Write down the x-coordinates of the points A and B.

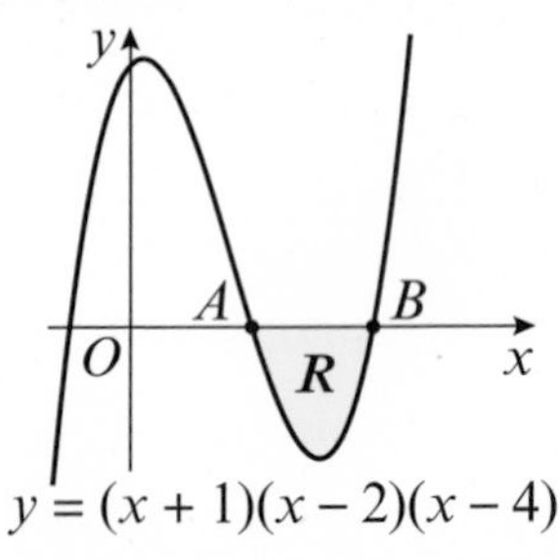

b Find the area of the shaded region R.

Hint The limits of the integral will be the x-coordinates of points A and B.

3 The diagram shows the region bounded by the curve $y = -x^3 + 5x^2 - 6x$ and the x-axis.

Find:

a the area of region R_1

b the area of region R_2

c the total area of the shaded region.

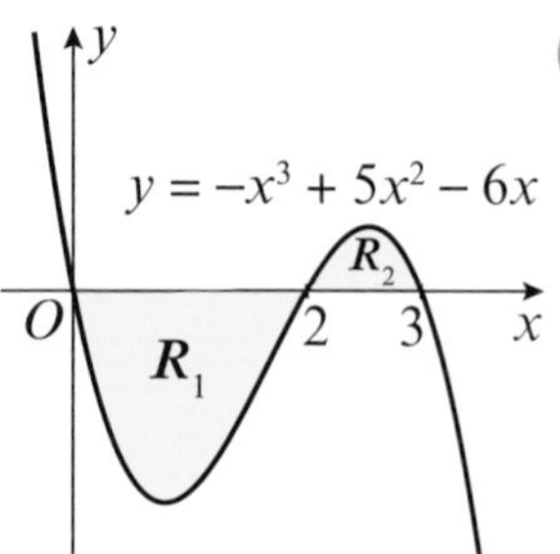

Hint To find the total shaded area, find the areas of the regions above the x-axis and below the x-axis separately.

4 The graph shows a sketch of part of the curve C with equation

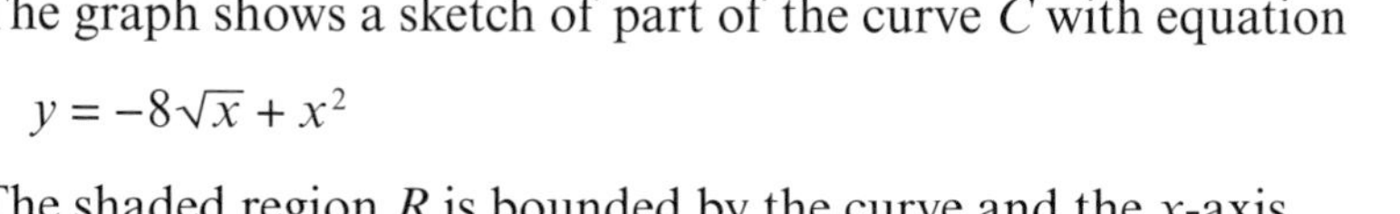

$$y = -8\sqrt{x} + x^2$$

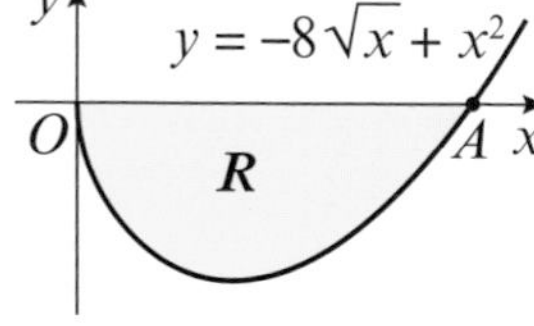

The shaded region R is bounded by the curve and the x-axis.

a Find the coordinates of the point A. **(1 mark)**

b Use calculus to find the exact area of the shaded region R. **(5 marks)**

5 $\mathrm{f}(x) = \frac{1}{2}x^2 - \frac{13}{2}x + 15$. The graph shows a sketch of part of the curve with equation $y = \mathrm{f}(x)$.

a Find the coordinates of points A and B. **(2 marks)**

b Find the exact area of the shaded region R. **(4 marks)**

6 The graph shows a sketch of part of the curve C with equation

$$y = x(x + 4)(x - 1)$$

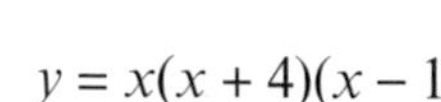

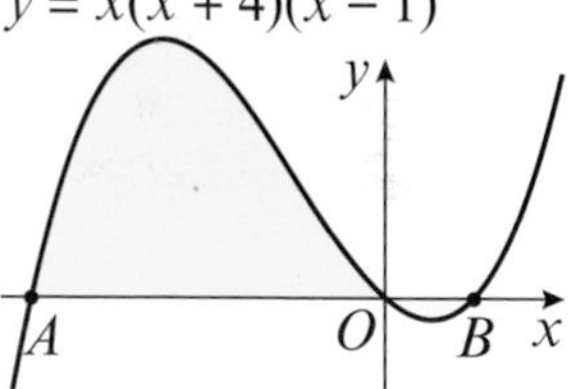

The curve cuts the x-axis at the origin O and at the points A and B.

a Write down the coordinates of points A and B. **(1 mark)**

b Find the total area of the shaded region. **(7 marks)**

7 $\mathrm{f}(x) = x^3 + 6x^2 - 4x - 24$. The graph shows a sketch of part of the curve with equation $y = \mathrm{f}(x)$.

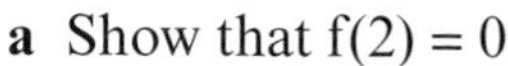

a Show that $\mathrm{f}(2) = 0$. **(1 mark)**

b Hence, or otherwise, factorise $\mathrm{f}(x)$ completely. **(2 marks)**

c Find the total shaded area shown in the sketch. **(5 marks)**

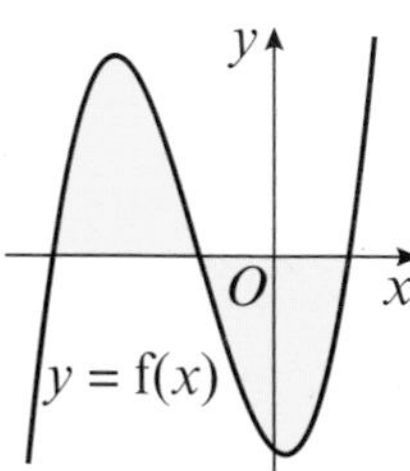

13.7 Areas between curves and lines

1 The diagram shows part of the curve with equation $y = \frac{1}{4}x^2 + 2$ and the line with equation $y = 6$. The line cuts the curve at the points A and B.

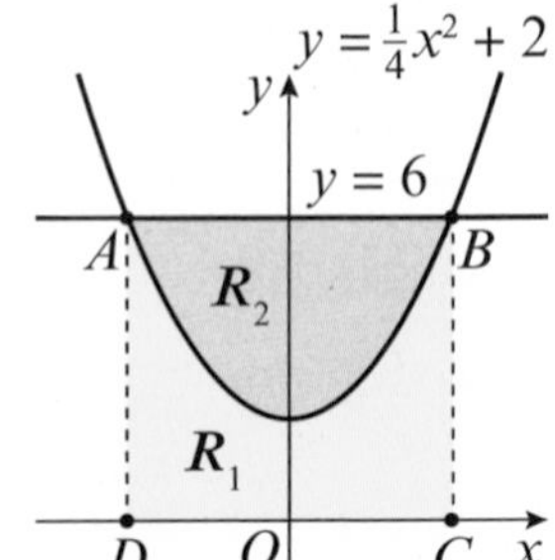

a Find the coordinates of points A and B.

b Use integration to find the shaded region R_1.

c Find the area of the rectangle $ABCD$.

d Find the area of the shaded region R_2.

Hint Area of R_2 = area of rectangle – area of R_1

2 The diagram shows part of the curve with equation $y = -x^2 + 4x$ and the line with equation $y = 4 - x$. The line cuts the curve at the points P and Q.

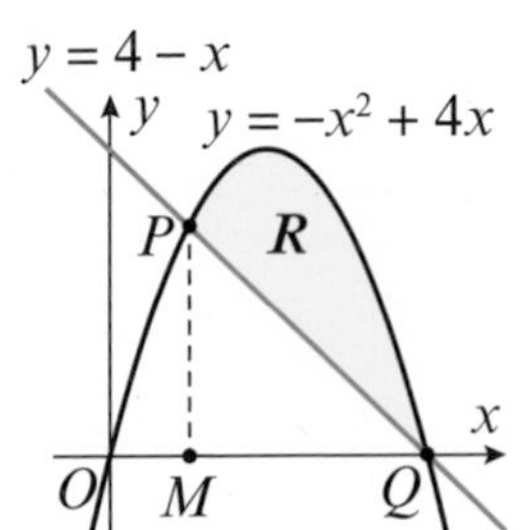

a Find the coordinates of the points P and Q.

b Use integration to find the shaded region bounded by the curve, the line PM and the x-axis.

c Find the area of the triangle PQM.

d Find the area of the shaded region R.

Hint

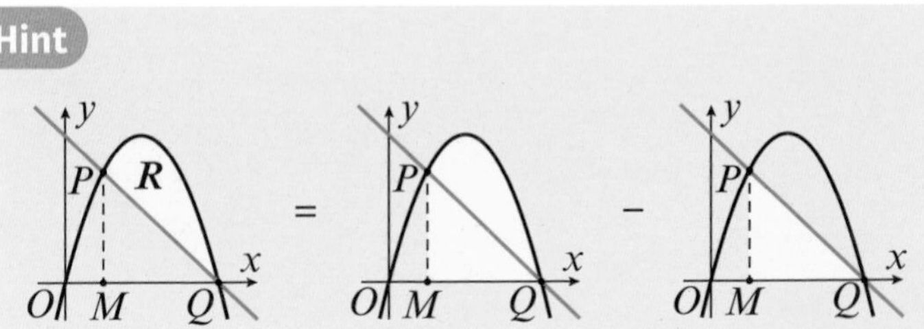

3 The diagram shows part of the curve with equation $y = 4\sqrt{x} - \frac{1}{2}\sqrt{x^3} + 2$ and the line with equation $5x + 8y = 49$.

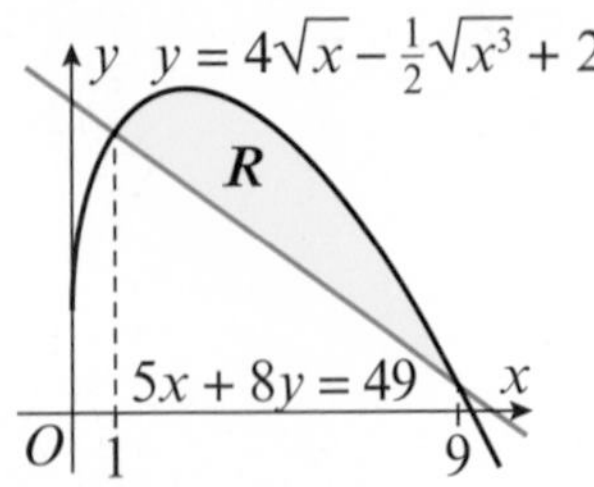

a Use integration to find the shaded region bounded by the curve, the lines $x = 1$, $x = 9$ and the x-axis.

b Find the shaded region R.

Hint

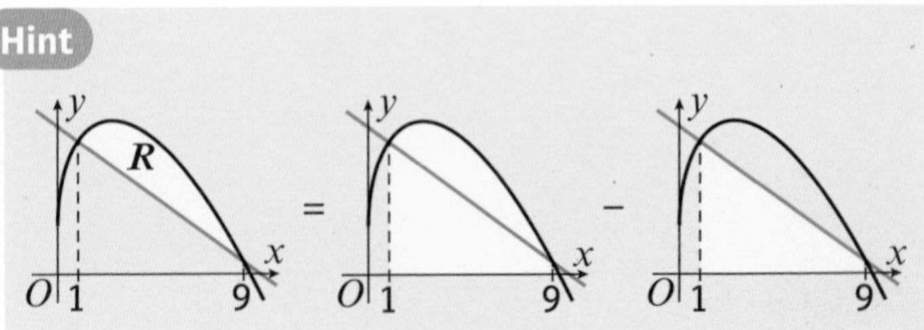

P **4** The shaded region R is bounded by the curve with equation $y = -\frac{1}{2}x^2 + \frac{9}{2}x - 4$ and the line with equation $x + 2y - 13 = 0$ as shown in the diagram. The line intersects the curve at the points A and B.

a Find the coordinates of the points A and B. **(3 marks)**

b Find the area of the shaded region R. **(5 marks)**

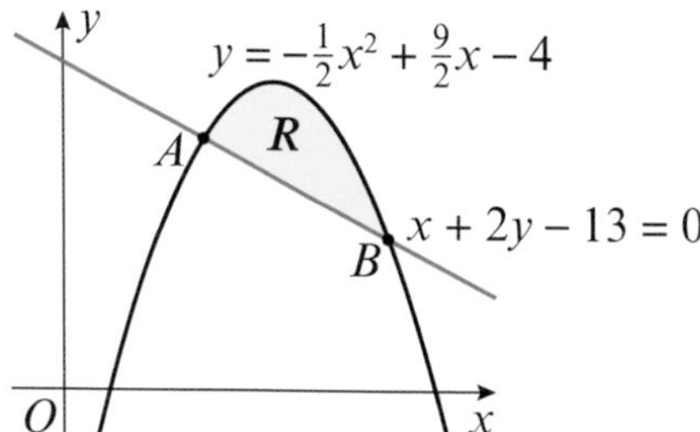

P **5** $h(x) = x^2(x - 4)^2$. The graph shows part of the curve with equation $y = h(x)$.

The straight line $y = 7x + 2$ intersects the curve at the points P and Q with x-coordinates 1 and 2 respectively. Find the area of the finite region bounded by the curve and the segment PG. **(7 marks)**

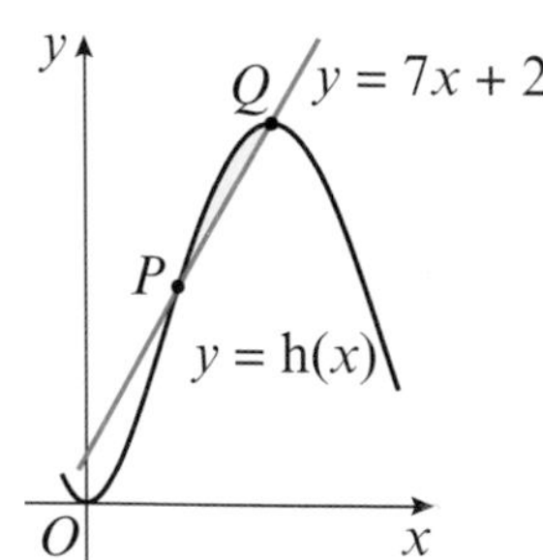

P **6** The line with equation $5x + 4y - 21 = 0$ crosses the curve with equation $y = \frac{4}{x^2}$ at the points A, B and C.

a Show that at points A, B and C, $5x^3 - 21x^2 + 16 = 0$. **(2 marks)**

Given that the x-coordinate of B is 1, find:

b the coordinates of the point C **(2 marks)**

c the area of the shaded region R. **(7 marks)**

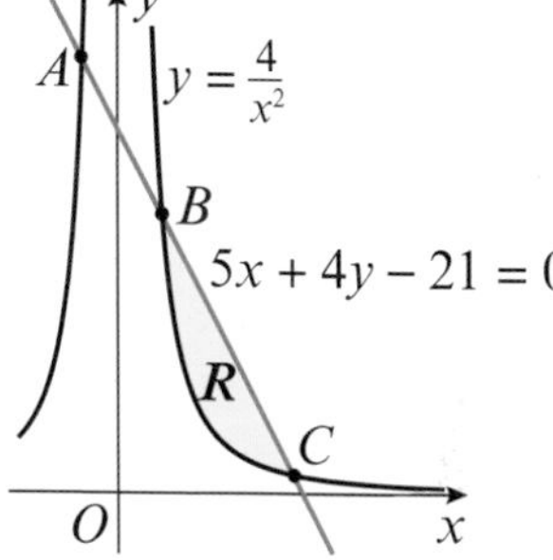

P **7** The line L with equation $3x + y + 6 = 0$ crosses the curve C with equation $3y = -x^3 - 2x^2 + 16x + 32$ at the points A, B and D.

a Show that at points A, B and C, $x^3 + 2x^2 - 25x - 50 = 0$. **(2 marks)**

Given that B has coordinates $(-2, 0)$, find:

b the coordinates of the point D **(2 marks)**

c the area of the shaded region R. **(7 marks)**

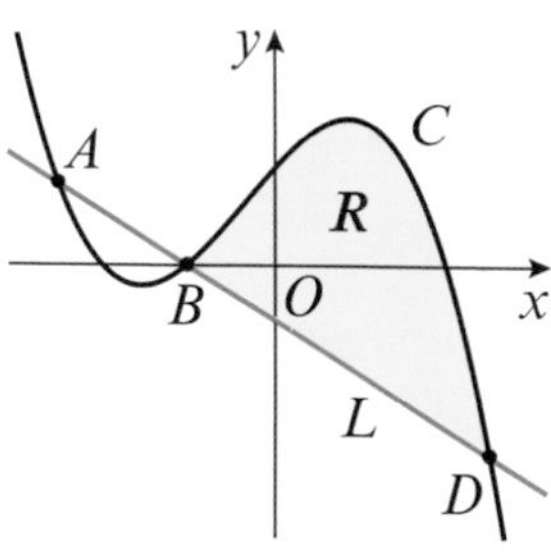

Problem solving Set A

Bronze

$$f'(x) = \frac{-5x^2 + 7x^{\frac{2}{3}}}{\sqrt{x}}$$

a Write $f'(x)$ in the form $ax^m + bx^n$, where a, b, m and n are rational numbers to be found. **(2 marks)**

b Given that $f(1) = -10$, find an expression for $f(x)$. **(4 marks)**

Silver

$f'(x) = \left(1 - 2x^{\frac{1}{3}}\right)^3$. Given that $f(8) = 24$, find $f(1)$. **(6 marks)**

Gold

$f'(x) = 3x^2 + px + \frac{8}{x^2}$. Given that $f(-4) = -85$ and $f(2) = 11$, find $f(x)$. **(6 marks)**

Problem solving Set B

Bronze

The diagram shows part of the curve with equation $y = \frac{1}{2}x^2 + 1$ and the line with equation $y = 9$. The line cuts the curve at the points A and B.

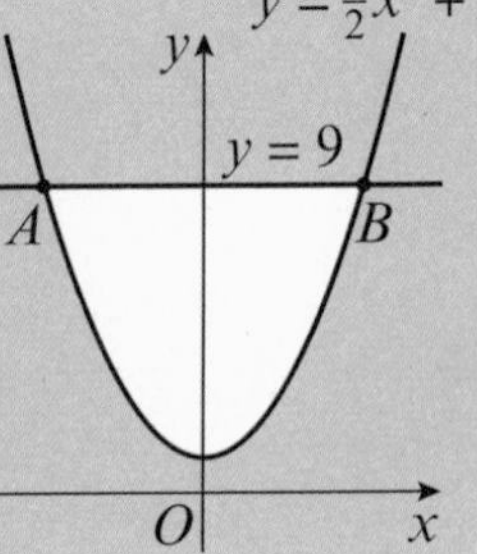

a Find the coordinates of the points A and B. **(2 marks)**

b Find the area of the finite region bounded by line AB and the curve. **(6 marks)**

Silver

The diagram shows part of the curve with equation $y = -x^2 - x + 12$ and the line with equation $x + y = 8$. The line cuts the curve at the points A and B.

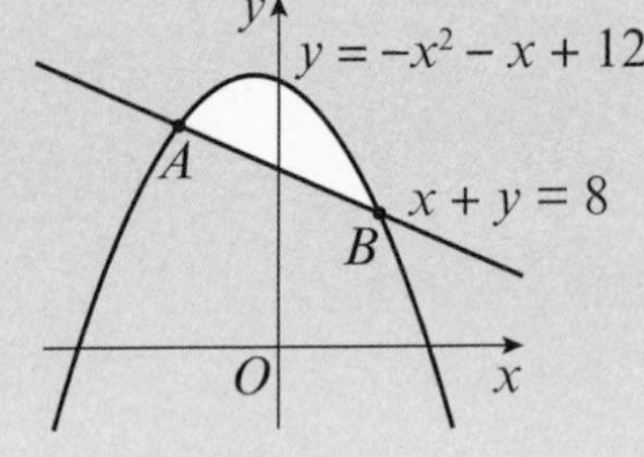

Find the area of the finite region bounded by line AB and the curve. **(9 marks)**

Gold

The diagram shows part of the curve with equation $y = -x^3 - 2x^2 + 4x + 8$ and the line with equation $y = 3x + 6$. The line cuts the curve at the points A, B and C.

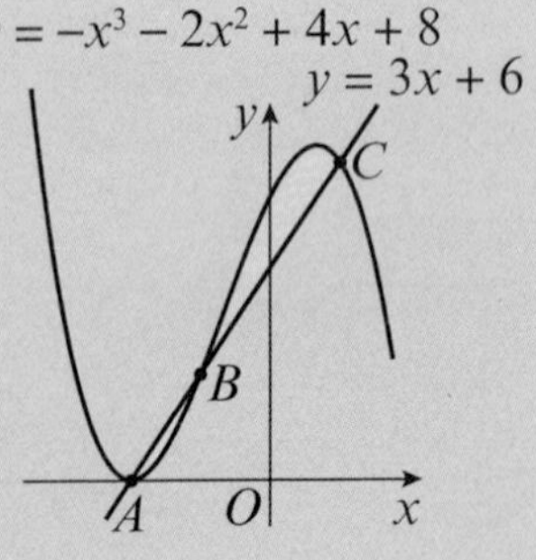

a Show that at points A, B and C, $x^3 + 2x^2 - x - 2 = 0$. **(2 marks)**

b Given that A has coordinates $(-2, 0)$, find the total area of the region bounded by the curve and the line segment BC. **(9 marks)**

Now try this → Exam question bank Q6, Q8, Q26, Q44, Q53, Q79

14.1 Exponential functions

1 Write each of the following as exact fractions:

a $\left(\frac{2}{3}\right)^3$ b 4^{-2}

c $\left(\frac{1}{5}\right)^{-2}$ d $\left(\frac{4}{3}\right)^{-3}$

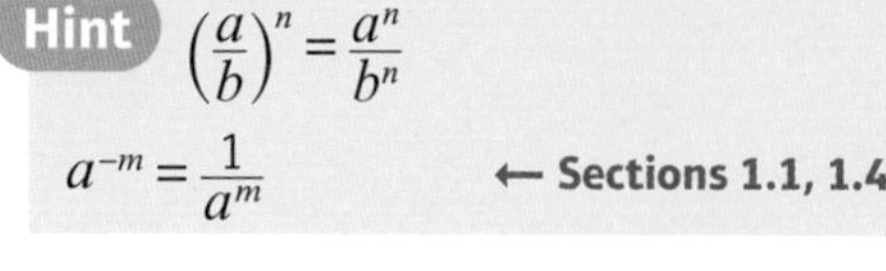

Hint $\left(\frac{a}{b}\right)^n = \frac{a^n}{b^n}$

$a^{-m} = \frac{1}{a^m}$

← Sections 1.1, 1.4

2 a Copy and complete the table for $y = 1.2^x$, giving each answer to 2 decimal places.

x	−3	−2	−1	0	1	2	3
y							

b Use the table to draw an accurate graph of $y = 1.2^x$.

Hint Use your calculator to work out each y-value in the table. Then plot the points.

3 a Copy and complete the table for $y = \left(\frac{2}{3}\right)^x$, giving each answer as an exact fraction.

x	−3	−2	−1	0	1	2	3
y							

b Use the table to draw the graph of $y = \left(\frac{2}{3}\right)^x$.

c Use your graph to solve $\left(\frac{2}{3}\right)^x = 2$.

Hint Draw the line $y = 2$ onto your graph. Read off the x-coordinate of the point of intersection.

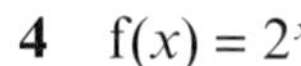

4 $f(x) = 2^x$

The diagram shows a sketch of the graph $y = f(x)$.

Use the graph to draw a sketch of the following, labelling the new coordinates of point A and any points of intersection with the coordinate axes:

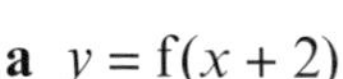

a $y = f(x + 2)$ b $y = f(x) - 2$

c $y = 2f(x)$ d $y = -f(x)$

y, $y = 2^x$, $A(0, 1)$, O, x

Hint If the graph has an asymptote other than one of the coordinate axes, you should draw and label it in your sketch.

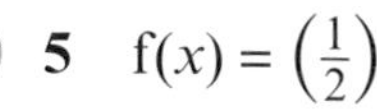

5 $f(x) = \left(\frac{1}{2}\right)^x$

a Sketch the graph of $y = f(x)$, labelling any points of intersection with the coordinate axes and giving the equations of any asymptotes. **(2 marks)**

b Use your answer to part **a** to sketch (labelling the equation of any asymptotes):

i $y = f(x + 3) - 4$

ii $y = -2f(x)$ **(4 marks)**

(E) **6** $g(x) = \left(\frac{3}{2}\right)^x$

On the same set of axes, sketch the following two graphs, labelling any points of intersection with the coordinate axes and giving the equation of any asymptotes.

a $y = g(x)$ and $y = g(-x)$ **(2 marks)**

b $y = g(x)$ and $y = g\left(\frac{1}{2}x\right)$ **(2 marks)**

(E/P) **7** The graph of $y = ka^x$ passes through the points (1, 2) and (–2, 128).

Find the values of the constants k and a. **(3 marks)**

14.2 $y = e^x$

1 On the same set of axes, sketch the graphs of $y = e^x$ and:

a $y = e^{-x}$ **b** $y = e^{x+5} - 6$

c $y = -3e^x - 5$ **d** $y = e^{4x}$

e $y = 6 - 5e^x$ **f** $y = e^{\frac{1}{2}x} - 2$

Hint You will need six different sketches. Each sketch should show $y = e^x$ and one other function.

2 Find f′(x) where f(x) equals:

a $e^{\frac{1}{2}x}$ **b** e^{-5x}

c $-4e^{6x}$ **d** $7e^{\frac{1}{3}x}$

Hint For all real values of x, and for any constant k, if $f(x) = e^{kx}$ then $f'(x) = ke^{kx}$

3 Find $\frac{dy}{dx}$ where y equals:

a $3e^{2x} + e^{4x}$ **b** $6e^{-2x} + \frac{5}{e^x}$

c $e^{2x}(e^{-x} - 4)$ **d** $\frac{1}{e^{3x}}(2 - e^{5x})$

Hint First expand the brackets or simplify the fraction.

For all real values of x, and for any constant k, if $y = e^{kx}$ then $\frac{dy}{dx} = ke^{kx}$

4 Write each of the following in the form $f(x) = Ae^{bx}$, where A and b are constants whose values are to be found.

Hint Use the laws of indices: $e^{m+n} = e^m \times e^n$

You can give exact values as powers of e.

a e^{5x+4} **b** e^{7x-1} **c** $e^{3-\frac{1}{2}x}$

(E) **5** Differentiate with respect to x:

a $8x^2 - \frac{1}{e^{6x}}$ **(2 marks)**

b $2e^{-x}(4e^{5x} - 3)$ **(3 marks)**

(E) **6** Given that $y = 5\sqrt{x} + \frac{10}{x^2} - 4e^{-2x}$, find $\frac{dy}{dx}$ **(3 marks)**

(E) **7** **a** Show that $e^{2(3x-1)}$ can be written in the form form Ae^{bx}, where A and b are constants whose values are to be found. **(2 marks)**

b Hence, or otherwise, sketch the graph of $y = e^{2(3x-1)}$. In your sketch, label any asymptotes and the coordinates of any intercepts of the y-axis. **(3 marks)**

(E/P) **8** Given $f(x) = e^{4x}$, find the exact equation of the tangent at the point where $x = \frac{1}{2}$ on the curve with equation $y = f(x)$. **(4 marks)**

14.3 Exponential modelling

1 $P = 20\,000\,e^{0.01t}$

a Find $\frac{dP}{dt}$

b Find the value of $\frac{dP}{dt}$ when $t = 10$.

Hint First differentiate and then substitute $t = 10$ into your answer.

2 $V = 32\,000e^{-0.08t}$

a Sketch the graph of V against t.

b Find $\frac{dV}{dt}$

c Find the value of $\frac{dV}{dt}$ when $t = 20$.

d Given that V is the value of a car and t is the time since purchase, interpret the significance of the sign of your answer to part **c**.

Hint A positive slope means that the function is increasing.

A negative slope means that the function is decreasing.

3 The number of deer, D, in a population after t years is modelled by the formula $D = 1300e^{\frac{1}{8}t}$

a Use this model to estimate the number of deer after:

i 0 years **ii** 2 years **iii** 30 months

b Use your answer to part **a i** to interpret the meaning of the constant 1300 in the model.

c Find $\frac{dD}{dt}$

d Show that after 8 years, the deer population is growing by approximately 440 deer per year.

e By substituting $t = 80$ into $\frac{dD}{dt}$, explain why this model is not valid for large values of t.

4 The value, V, of a car, in £s, is given by $V = 29\,000e^{-0.12t}$, where t is the time in years since the car was purchased.

a Find the value of the car after 6 years. **(1 mark)**

b Find the value of $\frac{dV}{dt}$ when:

i $t = 2$ **ii** $t = 6$ **(3 marks)**

c Use you answer to parts **a** and **b** to describe how the value of the car decreases over the first 6 years of ownership. **(2 marks)**

5 In 1980, the population, P, of a town was 18 000. The population in subsequent years can be modelled $P = P_0e^{0.02t}$, where t is the time in years since 1980.

a State the value of P_0. **(1 mark)**

b Find the rate, $\frac{dP}{dt}$, in people per year to the nearest person, at which the population P is increasing in 2005. **(3 marks)**

c Explain why this model is not valid for large values of t. **(1 mark)**

(E/P) **6** A child takes a dose of 250 mg of paracetamol. The level of paracetamol, P, present in the body at time t hours after it is taken can be given by the formula $P = 250e^{-0.15t}$.

a Use the model to estimate the level present after 6 hours. **(1 mark)**

b Show that $\frac{dP}{dt} = kP$, where k is a constant to be found. **(2 marks)**

c Interpret the sign of k. **(1 mark)**

(E/P) **7** The value, V, in £s, of a house t years after it reached a low value due to a property crash, can be modelled by the equation $V = 150\,000e^{0.06t}$.

a State the value of the house at time $t = 0$. **(1 mark)**

b Use the model to estimate the value of the house after 7 years. **(1 mark)**

c Find $\frac{dV}{dt}$ **(2 marks)**

A new property crash is likely to happen when the value of the house is increasing by £15 000 per year.

d Show that between $t = 8$ and $t = 9$ the conditions are correct for a new crash. **(3 marks)**

14.4 Logarithms

1 Complete the table by writing the equivalent logarithm or power.

Hint $a^x = n$ is equivalent to $\log_a n = x$ $(a \neq 1)$

	Logarithm	Power
a		$10^3 = 1000$
b	$\log_2 8 = 3$	
c		$7^2 = 49$
d		$6^{-3} = \frac{1}{216}$
e	$\log_4 \frac{1}{256} = -4$	
f	$\log_9 9 = 1$	
g		$25^{\frac{1}{2}} = 5$
h	$\log_{27} 3 = \frac{1}{3}$	

2 Without using a calculator, find the value of:

Hint For example, if $\log_8 64 = x$ then $8^x = 64$.

a $\log_8 64$ **b** $\log_2 \frac{1}{16}$

c $\log_5 5$ **d** $\log_3 \sqrt[3]{3}$

e $\log_{10} 0.01$ **f** $\log_9 1$

3 Without using a calculator, find the value of x for which:

Hint For example, if $\log_3 x = 4$ then $3^4 = x$.

a $\log_3 x = 4$ **b** $\log_5(3x - 1) = 3$ **c** $\log_x 729 = 3$ **d** $\log_x(4x) = 2$

4 Evaluate each of these logarithms to four decimal places.

Hint Use your calculator.

a $\log_5 18$ **b** $\log_e 0.6$ **c** $2\log_8 9$

5 Find the exact value of x for which $\log_8 (2x - 1) = \frac{5}{3}$ **(3 marks)**

6 Find the exact value of x for which $\log_x 64 = \frac{3}{4}$ **(3 marks)**

7 Solve the equation $\log_{125} x = -\frac{4}{3}$ **(3 marks)**

P 8 Find the exact value of $\log_a(\sqrt[3]{a^2})$. **(3 marks)**

14.5 Laws of logarithms

1 Write as a single logarithm:

Hint $\log_a (x^k) = k \log_a x$
$\log_a x + \log_a y = \log_a xy$
$\log_a x - \log_a y = \log_a\left(\frac{x}{y}\right)$

a $4\log_7 2$ **b** $\log_3 16 - \log_3 2$
c $\log_5 6 + \log_5 7$ **d** $3\log_4 5 + 2\log_4 8$
e $3\log_7 4 - 4\log_7 3$ **f** $\log_6 3 + \log_6 4 - \log_6 5$

2 By writing as a single logarithm, evaluate the following without using a calculator:

a $\log_{10} 5 + \log_{10} 20$ **b** $4\log_8 2 + \log_8 4$ **c** $\log_2 48 - \log_2 6$
d $\log_5 4 - \log_5 20$ **e** $2\log_8 5 + 4\log_8 2 - 2\log_8 10$

3 Write in terms of $\log_a x$, $\log_a y$ and $\log_a z$:

Hint Use the power law, multiplication law and division law of logarithms.
Remember that $\log_a a = 1$.

a $\log_a a^3$ **b** $\log_a(x^2y^3)$
c $\log_a\left(\frac{x^3}{z^4}\right)$ **d** $\log_a\left(\frac{x\sqrt{y}}{z^2}\right)$
e $\log_a(\sqrt[3]{ax^2yz^4})$ **f** $\log_a\left(\frac{a^3x^4y}{\sqrt{z}}\right)$

4 Solve the following equations.

Hint First move the logarithms onto the same side of the equation and then simplify the equation using either the multiplication or division laws of logarithms.

a $\log_6 4 + \log_6 x = 2$ **b** $2\log_8 6 = \log_8 x - 1$
c $2\log_6 x = 2 + \log_6 4$ **d** $2\log_4 (x - 1) - \log_4 9 = 1$

5 Solve the equation $3\log_5 4 = 2 - 2\log_5 x$. **(3 marks)**

P 6 Solve the equation $\log_5 x + 2\log_5 x = 4$, giving your answer in the form $x = p(\sqrt[q]{r})$, where p, q and r are integers to be found. **(3 marks)**

7 Solve the equation $\log_3(2x + 5) - \log_3(4x - 1) = 2$. **(4 marks)**

E/P **8** **a** Given that $\log_2(x+5) - 2\log_2(x-1) = 1$, show that $2x^2 - 5x - 3 = 0$. **(4 marks)**

b Hence, or otherwise, solve $\log_2(x+5) - 2\log_2(x-1) = 1$. **(2 marks)**

14.6 Solving equations using logarithms

1 Solve, giving your answer to 3 significant figures:

a $6^x = 31$ **b** $3^x = 70$

c $5^{2x} = 640$

Hint Rewrite each exponential equation as a logarithmic equation and solve using your calculator.

2 Solve, giving your answer to 3 significant figures:

a $5^{x+3} = 0.3$ **b** $4^{2x-5} = 85$

c $7^{1-4x} = 11$

Hint You can use the $\boxed{\log_{\square}\square}$ button on your calculator to find logs to any base.

3 Solve, giving your answer to 3 significant figures:

a $5^{2x} - 11(5^x) + 30 = 0$

b $4^{2x} - 19(4^x) + 60 = 0$

c $5^{2x+1} - 19(5^x) + 18 = 0$

Hint Write $5^{2x} = (5^x)^2$. Use the substitution $u = 5^x$ to write a quadratic in terms of u. Then factorise and solve the resulting quadratic. Finally, use logarithms to find x for the given values of 5^x.

4 **a** Solve the quadratic equation $9u^2 - 3u - 20 = 0$.

b Substitute $u = 3^x$ to show that the equation $3^{2x+2} - 3(3^x) - 20 = 0$ can be written as $9u^2 - 3u - 20 = 0$.

c Hence solve the equation $3^{2x+2} - 3(3^x) - 20 = 0$, explaining clearly why there is only one real solution.

Hint Remember that you cannot evaluate the logarithm of a negative number.

5 Solve, giving your answer to 3 significant figures:

a $8^x = 5^{x+3}$ **b** $7^{3-x} = 3^{2x}$

c $6^{2x} = 4^{x+3}$

Hint Take the logarithm of both sides and then use the power law of logarithms to solve the resulting equation.

E **6** Solve each of the following equations, giving your answers to 3 significant figures where appropriate.

a $7^{2x-5} = 400$ **(2 marks)**

b $6^{2x} + 4(6^x) - 21 = 0$ **(3 marks)**

E **7** Solve each of the following equations, giving your answers to 3 significant figures where appropriate.

a $4^{x+2} = 5^{x-1}$ **(2 marks)**

b $\log_7(5 - x) = 0.8$ **(2 marks)**

E **8** **a** Sketch the graph of $y = -2^x + 7$, stating the coordinates of any point where the graph crosses the coordinate axes and the equations of any asymptotes. Write coordinates as a logarithm when necessary. **(3 marks)**

b Solve the equation $-2^x + 7 = -6$. **(2 marks)**

P **9** **a** Using the substitution $u = 3^x$, show that the equation $9^x - 3^{x+1} - 10 = 0$ can be written in the form $u^2 - 3u - 10 = 0$. **(2 marks)**

b Hence solve the equation $9^x - 3^{x+1} - 10 = 0$, giving your answer to 2 decimal places. **(3 marks)**

14.7 Working with natural logarithms

1 Find the values of x, giving your answers in the form $a + b \ln c$, where a, b and c are rational constants.

Hint Take the natural logarithm of both sides: $\ln(e^x) = x$.

a $e^x = 8$ **b** $e^{\frac{1}{2}x} = 6$ **c** $e^{x+5} = 81$

d $e^{5x-1} = 23$ **e** $4e^{1-x} = 29$ **f** $\frac{2}{3}e^{3x} = 7$

2 Solve these equations, giving your answers in exact form.

Hint Use the fact that $e^{\ln x} = x$.

a $\ln x = 11$ **b** $\ln 6x = 3$

c $\ln(3 - 5x) = -1$ **d** $\ln(5 - 2x) = \frac{2}{5}$

e $2\ln(3x - 5) = 7$ **f** $\ln(x^2 - 9x + 21) = 0$

3 Solve the equation $e^{2x} - 5e^x + 4 = 0$, giving your answer in the form $\ln k$, where k is an integer to be found.

Hint Use the substitution $u = e^x$ to form a quadratic in u.
Solve the quadratic in u and hence find the value of x.

4 Solve the equation $2e^{4x} - 11e^{2x} + 14 = 0$, giving your answers in the form $a \ln b$, where a and b are rational constants to be found.

Hint Use the substitution $u = e^{2x}$.

5 Find the exact solution to the equation $e^x - 12 = e^{\frac{x}{2}}$. **(4 marks)**

6 Find the exact solutions to the equation $2e^{4x} + 15e^{-4x} = 13$. **(5 marks)**

P **7** Solve $e^{1-6x} = 4(12^x)$, giving your answer in the form $\frac{a + \ln b}{c + \ln d}$ **(4 marks)**

8 The graph of $y = 3 - 2\ln(x + 2)$ is shown in the diagram.

Find the exact coordinates, leaving your answer in terms of e or a natural logarithm where necessary, of:

a A **(2 marks)**

b B **(3 marks)**

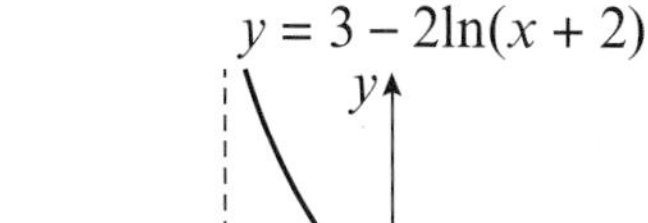

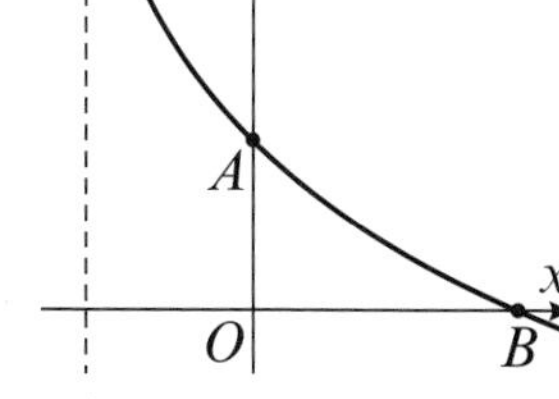

(E/P) **9** The temperature, T, in degrees Celsius, of bath water can be modelled by the equation $T = 31e^{-0.05x} + 7$, $x \geqslant 0$, where x is the time in minutes since the bath tub was full.

a Sketch a graph of $T = 31e^{-0.05x} + 7$, clearly labelling any points of intersection with the coordinate axes and the equation of any asymptotes. **(3 marks)**

b Interpret the meaning of the 7 in the equation. **(1 mark)**

Jack will add more hot water once the temperature reaches 25 °C.

c Given that Jack enters the bath at $x = 0$, find the time when Jack will first add more hot water. **(3 marks)**

14.8 Logarithms and non-linear data

1 Two variables V and x satisfy the formula $V = 5x^3$.

Hint Begin by taking logs of both sides of the equation.

a Show that $\log_{10} V = \log_{10} 5 + 3\log_{10} x$.

b Sketch the straight line graph of $\log_{10} V$ against $\log_{10} x$, labelling the intercept on the vertical axis.

c State the gradient of the line.

2 Two variables H and t satisfy the formula $H = 3\left(\frac{3}{2}\right)^t$

Hint Begin by taking logs of both sides of the equation.

a Show that $\log_{10} H = \log_{10} 3 + t\log_{10} \frac{3}{2}$

b Sketch the straight line graph of $\log_{10} H$ against t, labelling the intercept on the vertical axis.

c State the exact gradient of the line.

3 The data in the table follows a trend of the form $y = ab^x$, where a and b are constants.

x	0.5	2	5	7
y	847.5	1878.5	9229.1	26 672.0

a Show that $\log_{10} y = \log_{10} a + x\log_{10} b$.

b Copy and complete the table of values of x and $\log y$, giving your answers to 2 decimal places.

x	0.5	2	5	7
$\log y$				

c Plot the graph of $\log y$ against x and draw in a line of best fit.

d Use your graph to estimate the values of a and b to one decimal place.

4 The value, V, of a car can be modelled using an exponential function $V = ab^t$, where t is the time in years since the car was purchased and a and b are constants.

The car was purchased at the beginning of 2008.

The table shows value at the beginning of subsequent years.

Year	2008	2010	2012	2014	2016	2018
Value, £V	25 000	14 823	8788	5211	3089	1832

a Copy and complete the table below. Give your answers to 2 d.p.

Time in years since 2008, t	0	2	4	6	8	10
$\log_{10} V$						

b Show that $V = ab^t$ can be rearranged into the form $\log_{10} V = \log_{10} a + t\log_{10} b$.

c Plot the graph of $\log_{10} V$ against t using the values in the table.

d Draw a line of best fit and calculate the gradient.

e Use the graph to estimate the values of a and b to 3 significant figures.

P **5** The price of a stock, P, in pence, of a company can be modelled using an exponential function $P = ab^t$, where t is the time in days since the company was taken over and a and b are constants.

The takeover was complete on Friday 9/3/18 and the new stock began trading on Monday 12/3/18. The table shows the price of the stock over five days.

Date	12/3/18	13/3/18	14/3/18	15/3/18	16/3/18
Price, P pence	27.43	29.76	32.29	35.04	38.01

a Copy and complete the table below. Give your answers to 2 d.p. **(1 mark)**

Time in days since 12/3/18, t	0	1	2	3	4
$\log_{10} P$					

b Show that $P = ab^t$ can be rearranged into the form $\log_{10} P = \log_{10} a + t\log_{10} b$. **(2 marks)**

c Plot the graph of $\log P$ against t and draw a line of best fit. **(2 marks)**

d Use the graph to estimate the values of a and b to 4 significant figures. **(2 marks)**

E/P 6 The number of people, P, left in a concert venue t minutes after the band stops playing can be modelled by an equation of the form $P = ab^t$. The diagram shows the graph of $\log_4 P$ against t.

$\log_4 P$ 5.2 (20, 3.7) O t

a Write down the equation of the line shown on the graph. **(2 marks)**

b Find the number of people in the venue when $t = 0$, correct to 3 significant figures. **(1 mark)**

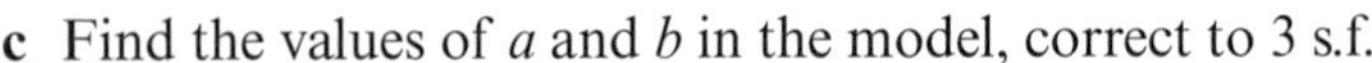

c Find the values of a and b in the model, correct to 3 s.f. **(2 marks)**

d Find the number of people half an hour after the band stopped playing. **(1 mark)**

E/P 7 A scientist recorded the number of bacteria, N, growing in a petri dish after t minutes. The diagram shows a graph of $\log_2 N$ against t.

$\log_2 N$ (10, 11.9) 9.3 O t

a Write down the equation of the line. **(2 marks)**

A scientist says the data can be modelled by an equation in the form $N = ab^t$

b Find the values of a and b to 1 decimal place. **(4 marks)**

c Interpret the meaning of the constant a in this model. **(1 mark)**

Problem solving Set A

Bronze

Solve the equation $6^{3x-2} = 23$, giving your answer to 2 decimal places. **(3 marks)**

Silver

a Using the substitution $u = 3^x$, show that the equation $9^x - 3^{x+2} + 20 = 0$ can be written in the form $u^2 - 9u + 20 = 0$. **(2 marks)**

b Hence solve the equation $9^x - 3^{x+2} + 20 = 0$, giving your answer in the form $\log_a b$. **(3 marks)**

Gold

Find the exact values of x for which $4^{2x} - 2^{2x+3} + 12 = 0$, leaving your answer in terms of logarithms where necessary. **(6 marks)**

Problem solving Set B

Bronze

The value, V, in £s, of a mobile phone can be modelled by the formula $V = 50 + 700\,e^{-0.4t}, t \geqslant 0$, where t is the time in years since the phone was purchased.

a Sketch the graph of V against t, labelling any asymptotes and any points of intersection with the coordinate axes. **(3 marks)**

b Calculate the value of the phone after 5 years. **(1 mark)**

c Find the time when the phone will be worth £100, giving your answer in the form $a \ln b$, where a and b are constants to be found. **(3 marks)**

Silver

The number of computers (in thousands), C, infected by an email virus can be modelled by the equation $C = e^{\frac{1}{4}t} - 1, 0 \leqslant t \leqslant 24$, where t is the time in hours since the email was sent.

a Calculate the number of computers, to the nearest whole number, infected after 11 hours. **(1 mark)**

b Find $\frac{dC}{dt}$ **(2 marks)**

c Find the time when the number of computers being infected is increasing at a rate of 12 000 per hour, leaving your answer in the form $a \ln b$, where a and b are constants to be found. **(3 marks)**

Gold

The temperature, T °C, of a bowl of soup is given by $T = ae^{kt} + 25, t \geqslant 0$, where t is the time in minutes since the soup was served.

a Explain briefly why k must be negative. **(1 mark)**

Given that initially the temperature of the soup is 70 °C and after 5 minutes the temperature is 55 °C:

b state the value of a **(1 mark)**

c find the value of k in the form $c \ln d$, where c and d are constants to be found **(3 marks)**

d find the time, to 2 decimal places, when the temperature of the soup is decreasing at a rate of 1.5 °C per minute. **(3 marks)**

Now try this → Exam question bank Q3, Q18, Q25, Q54, Q59, Q68

Exam question bank

This bank of exam-style questions have not been ordered by topic. Read each question carefully to work out which skills and techniques you will need to apply.

1 **a** Find the value of $27^{\frac{5}{3}}$ **(2 marks)**

b Simplify fully $\dfrac{(4x^{\frac{1}{2}})^3}{8x^2}$ **(3 marks)**

2 Simplify $\dfrac{4\sqrt{3}-2}{7-\sqrt{3}}$, giving your answer in the form $p\sqrt{3}-q$, where p and q are positive rational numbers to be found. **(4 marks)**

3 $f(x) = -2^x + 3$

Sketch the graph of $y = f(x)$, labelling the exact coordinates of any points of intersection with the coordinate axes and the equations of any asymptotes. **(4 marks)**

4 **a** Simplify $\sqrt{98}-\sqrt{50}$ giving your answer in the form $a\sqrt{2}$, where a is an integer to be found. **(2 marks)**

b Hence, or otherwise, simplify

$$\frac{16\sqrt{3}}{\sqrt{98}-\sqrt{50}}$$

giving your answer in the form $b\sqrt{c}$, where b and c are integers and $b \neq 1$. **(3 marks)**

5 Given that $\mathbf{a} = 6\mathbf{i} - 3\mathbf{j}$, find:

a the unit vector in the same direction of **a** **(3 marks)**

b the angle that **a** makes with the positive x-axis. **(2 marks)**

6 Given that $f(x) = 9\sqrt[3]{x^2} + \dfrac{4}{x^2} - 3x^3 + 2$, find $\int f(x)\,dx$. **(4 marks)**

7 Show that $\dfrac{2\sin^2 x - 1}{\cos^2 x} \equiv \tan^2 x - 1$. **(3 marks)**

8 Find $\int_{\frac{1}{2}}^{1} (4-3x)^3\,dx$, leaving your answer as a fraction in its simplest terms. **(5 marks)**

9 In the binomial expansion of $(1+3x)^n$, where $n > 0$, the coefficient of x^2 is 6 times the coefficient of x. Work out the value of n. **(4 marks)**

10 The diagram shows part of curve C with equation $y = x^2 - 8x + 18$ and part of the line L with equation $y = 2x + 5$.

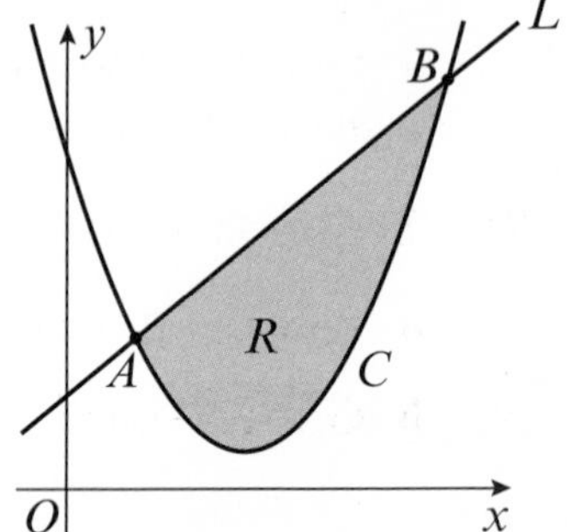

a Using an appropriate algebraic method, find the coordinates of A and B. **(4 marks)**

b Define the shaded region R. **(2 marks)**

11 $f(x) = 8x^2 - \frac{80}{3}x^{\frac{3}{2}} + 5x;\ x \geqslant 0$

Find the x-coordinates of the points on the curve with equation $y = f(x)$ where the gradient is equal to -4. **(5 marks)**

12 The diagram shows a sketch of part of the curve $y = f(x)$, $x \in \mathbb{R}$, where $f(x) = (2x - 3)^2(x + 2)$.

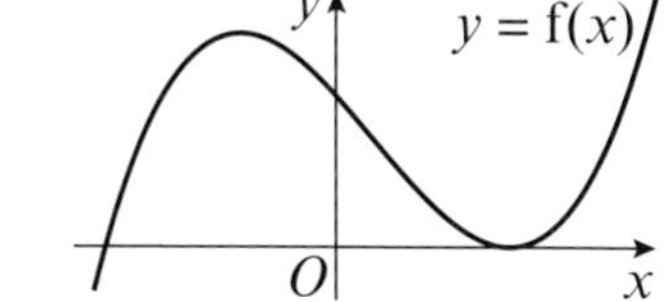

a i Given that the curve with equation $y = f(x) - k$, $x \in \mathbb{R}$, passes through the origin, find the value of the constant k.

ii Given that the curve with equation $y = f(x + c)$, $x \in \mathbb{R}$, has a local minimum point at the origin, find the value of the constant c. **(3 marks)**

b Show that $f'(x) = 12x^2 - 8x - 15$. **(3 marks)**

13 $f(x) = x^2 + (k - 4)x + \left(\frac{1}{2}k + 3\right)$

a Find the discriminant of $f(x)$ in terms of k. **(3 marks)**

b Given that the equation $f(x) = 0$ has two equal roots, find the possible values of k. **(2 marks)**

c Show that when $k = 8$, $f(x) > 0$ for all values of x. **(3 marks)**

14 The resultant of the forces $\mathbf{F}_1 = 6\mathbf{i} + 3p\mathbf{j}$ and $F_2 = 4p\mathbf{i} - 5\mathbf{j}$ acts parallel to $\mathbf{i}$. Find:

a the value of p **(2 marks)**

b the magnitude of the resultant force. **(2 marks)**

15 $f(x) = \frac{1}{3}\sqrt{x} + \frac{3}{\sqrt{x}};\ x > 0$

a Find $f'(x)$. **(2 marks)**

b Find the coordinates of the stationary point on the curve $y = f(x)$. **(2 marks)**

c By first finding $f''(x)$, determine whether the stationary point found in **b** is a local minimum or a local maximum. **(2 marks)**

16 At time $t = 0$, a ship, S, and a boat, B, sail from position O. The ship sails with velocity vector $(-\mathbf{i} + 3\mathbf{j})$ km h^{-1} and the boat sails with velocity vector $(6\mathbf{i} - 2\mathbf{j})$ km h^{-1}.

a Work out the position vector of the ship and the boat after 2 hours. **(2 marks)**

b Calculate the distance of the ship from the boat after 2 hours, leaving your answer as a simplified surd. **(3 marks)**

17 The graph of $y = \tan(x + k)$ passes through the point $(60°, 1)$.

a Find one positive and one negative possible value of k. **(3 marks)**

b Give the equations of two different asymptotes. **(3 marks)**

18 Find the exact value of x such that $\log_{16}(1 - 3x) = \frac{3}{4}$ **(3 marks)**

19 The point B lies on the line with equation $3x + 2y = 12$. Given that $|OB| = 10$, find the vector $\overrightarrow{OB}$ in the form $p\mathbf{i} + q\mathbf{j}$, where $p > 0$. **(4 marks)**

20 **a** Expand $\left(1 + \frac{2}{x}\right)^2$, simplifying each term. **(2 marks)**

b Use the binomial expansion to find, in ascending powers of x, the first four terms in the expansion of $\left(1 + \frac{2}{5}x\right)^6$, simplifying each term. **(4 marks)**

c Hence find the coefficient of x in the expansion of $\left(1 + \frac{2}{x}\right)^2\left(1 + \frac{2}{5}x\right)^6$. **(2 marks)**

21 The line l_1 has the equation $2y - 9 = 5x$. The point A with x-coordinate 3 lies on l_1. The line l_2 is perpendicular to l_1 and passes through the point A.

a Find an equation for l_2 in the form $ax + by + c = 0$. **(4 marks)**

The lines l_1 and l_2 cross the x-axis at the points B and C respectively.

b Calculate the area of the triangle ABC. **(4 marks)**

22 The diagram shows a sketch of part of the curve $y = \mathrm{g}(x)$ where $\mathrm{g}(x) = (x + 1)(2x - 3)^2$

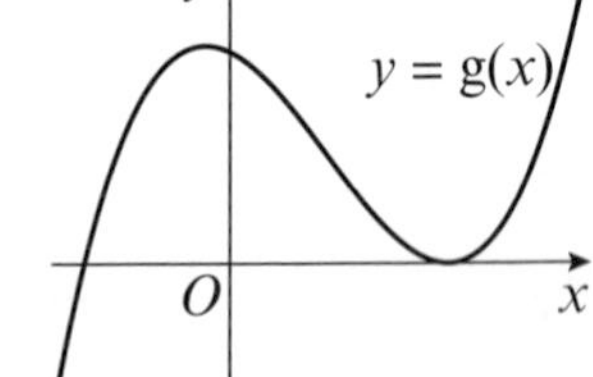

Deduce the values of x for which:

a $\mathrm{g}(x) \leqslant 0$ **(1 mark)**

b $\mathrm{g}(2x) = 0$ **(3 marks)**

23 The equation $kx^2 + 5kx + 5 = 0$, where k is a constant, has no real roots. Prove that $0 < k < \frac{4}{5}$ **(4 marks)**

24 **a** Factorise completely $x^3 + 12x^2 + 36x$ **(2 marks)**

b Sketch the curve with equation $y = x^3 + 12x^2 + 36x$, showing the coordinates of the points where the curve cuts or touches the x-axis. **(2 marks)**

25 Find the exact solutions to the equation $6e^{6x} - 38e^{3x} + 40 = 0$ **(5 marks)**

26 Given that $\int_9^k \frac{8}{\sqrt{x}}\,dx = 32$, find the value of k. **(4 marks)**

27 $\mathrm{f}(x) = ax^3 + bx^2 + cx$

The curve with equation $y = \mathrm{f}(x)$ has a gradient of 7 when $x = -1$ and a gradient of 5 when $x = 2$. Additionally, $\mathrm{f}''(1) = 2$. Find the values of the constants a, b and c. **(6 marks)**

28 Two circles, C_1 and C_2, have equations $x^2 + (y + 2)^2 = 16$ and $x^2 + (y + 5)^2 = 16$ respectively.

a For each of these circles, state the radius and the coordinates of the centre. **(3 marks)**

b Sketch the circles C_1 and C_2 on the same diagram. **(2 marks)**

c Find the exact distance between the points of intersection of C_1 and C_2. **(3 marks)**

29 **a** Find the first 4 terms, in ascending powers of p, of the binomial expansion of $(q + p)^{100}$ **(3 marks)**

Let p represent the probability that any one lightbulb in a batch of 100 is faulty and let q represent the probability it is not faulty. Given that $p = 0.02$,

b use your answer from part **a** to work out the probability that no more than 3 lightbulbs in a batch of 100 are faulty. Give your answer correct to 5 decimal places. **(5 marks)**

30 **a** Prove that when p and q are positive real numbers with $p > q$,

$\sqrt{p^2 - q^2} > p - q$ **(3 marks)**

b Prove by counter-example that this is not true for all values of p and q. **(1 mark)**

31 A parallelogram has sides of lengths 9 cm and 7.5 cm.
The size of the larger interior angles is θ.
Given that the area of the parallelogram is 60 cm^2, find the exact value of $\tan\theta$. **(4 marks)**

32 Find the set of values of x for which:

a $3x - 7 > 3 - 7x$ **(2 marks)**

b $x^2 - 4x \leqslant 21$ **(4 marks)**

c **both** $3x - 7 > 3 - 7x$ **and** $x^2 - 4x \leqslant 21$. **(1 mark)**

33 The diagram shows a sketch of the curve with equation $y = \mathrm{g}(x)$.

The curve has a single turning point, a local minimum, at the point $M(5, -1.5)$.
The curve crosses the x-axis at two points, $P(3, 0)$ and $Q(8, 0)$.
The curve crosses the y-axis at a single point $R(0, 9)$.

a State the coordinates of the turning point on the curve with equation $y = 3\mathrm{g}(x)$. **(1 mark)**

b State the largest root of the equation $\mathrm{g}(x - 2) = 0$. **(1 mark)**

c State the range of values of x for which $\mathrm{g}'(x) \geqslant 0$. **(1 mark)**

d Given that the equation $\mathrm{g}(x) + k = 0$, where k is a constant, has at least one real root, state the range of possible values of k. **(1 mark)**

34 **a** Use a counter-example to show that the following statement is false.
'$5^x \geqslant 2^x$ for all values of x' **(2 marks)**

b Alison chooses two consecutive odd numbers. She squares them, adds the results, then halves the total. Prove that the result is always an odd number.
For example, $\frac{1}{2}(9^2 + 11^2) = 101$ **(3 marks)**

35 Solve the simultaneous equations

$x + y = 2$

$3y^2 - x^2 = 12$ **(6 marks)**

36 Solve the following equation algebraically, showing each step of your working:
$(4^{x-1})^2 - 10(4^{x-1}) + 16 = 0$ **(5 marks)**

37 In the triangle ABC, $AB = 14$ cm, $BC = 6$ cm and $\angle ABC = \theta$. The area of the triangle is 24 cm^2.

a Find the two possible values of θ correct to 1 decimal place. **(3 marks)**

b Given that AC is the longest side of the triangle, find the length of AC. **(2 marks)**

38 The expression $16 - 6x - x^2$ can be written in the form $q - (x + p)^2$, where p and q are integers.

a Find the value of p and the value of q. **(3 marks)**

b Calculate the discriminant of $16 - 6x - x^2$. **(1 mark)**

c Sketch the curve with equation $y = 16 - 6x - x^2$, showing clearly the coordinates of any points where the curve crosses the coordinate axes. **(3 marks)**

39 **a** On the same set of axes, sketch the graphs of $y = \cos x$ and $y = \cos 3x$ for $0 \leqslant x \leqslant 360°$. **(3 marks)**

b Describe the transformation from the graph of $y = \cos x$ to $y = \cos 3x$. **(2 marks)**

c Given that $\cos 3x = \cos 60°$, use your graph to find all the solutions to the equation $\cos 3x = \cos 60°$ for $0 \leqslant x \leqslant 360°$. **(4 marks)**

40 **a** Find the first 3 terms, in ascending powers of x, of the binomial expansion of $\left(2 - \frac{x}{15}\right)^{10}$

Give each term in its simplest form. **(4 marks)**

$f(x) = (p + qx)\left(2 - \frac{x}{15}\right)^{10}$, where p and q are constants. Given that the first two terms, in ascending powers of x, in the series expansion of $f(x)$ are 512 and $-\frac{256x}{3}$,

b find the value of p **(2 marks)**

c find the value of q. **(2 marks)**

41 $f(x) = 6x^3 + 5x^2 - 3x - 2$

a Factorise $f(x)$ completely. **(6 marks)**

b Show that $\frac{6x^3 + 5x^2 - 3x - 2}{2x^3 + 3x^2 + x}$ can be written in the form $A + \frac{B}{x}$, where A and B are integers to be found. **(3 marks)**

42 The circle C has centre $(2, -1)$ and passes through the point $P(8, 5)$.

a Find an equation for C. **(4 marks)**

The circle crosses the positive y-axis at the point Q.

b Find the area of the triangle OPQ, where O is the origin, giving your answer in its simplest surd form. **(4 marks)**

43 $f(x) = x^2 - 6x + 16, x \in \mathbb{R}$.

a Show that $f(x)$ can be written in the form $(x - p)^2 + q$, where p and q are constants to be found. **(2 marks)**

b Hence sketch the graph of $y = f(x)$ showing the coordinates of the turning point. **(3 marks)**

c Given that the equation $f(x) = k$, where k is a constant, has two distinct real solutions, find the range of possible values of k. **(3 marks)**

44 The diagram shows a sketch of part of the curve C with equation $y = -x^2 + 2x + 4$.

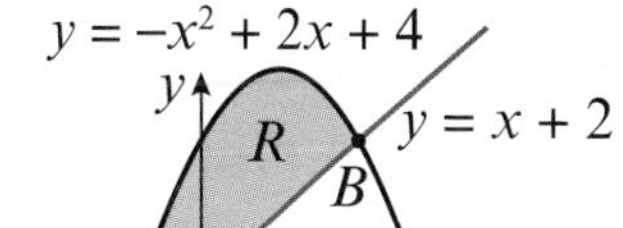

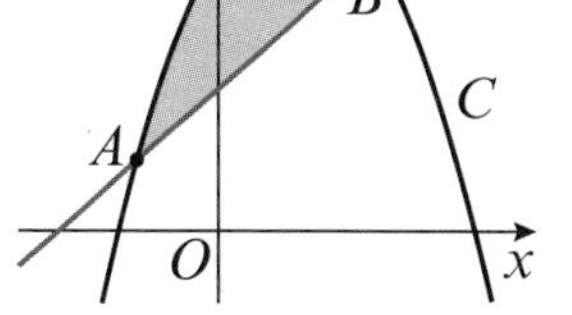

The straight line $y = x + 2$ crosses the curve at the points A and B.

a Find the coordinates of points A and B. **(2 marks)**

b The the area of the finite region R. **(7 marks)**

45 The diagram shows triangle ABC.

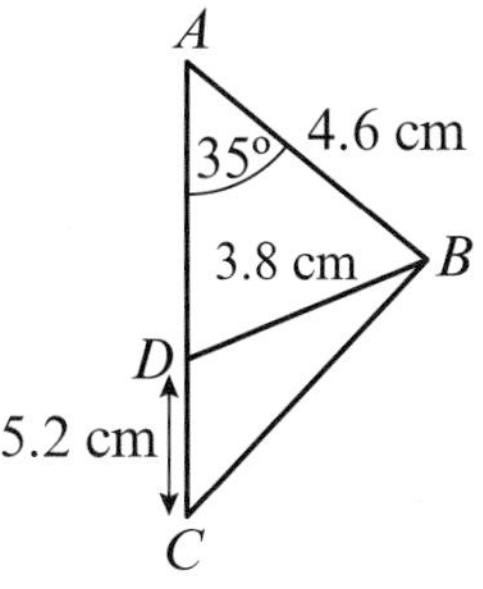

Calculate the area of triangle ABC. **(6 marks)**

46 The path of a toy rocket launched into the air can be modelled by the equation $y = x - 0.05x^2$, where x is the horizontal distance the rocket travels (in metres) and y is the vertical height (in metres) measured from where the rocket was fired.

a Sketch the graph of $y = x - 0.05x^2$ to show the path of the rocket. **(2 marks)**

b State the horizontal distance travelled by the rocket when it lands on the ground. **(3 marks)**

In reality the rocket achieves a maximum height of 4.5 m.

c Evaluate the model in light of this information. **(2 marks)**

47 Solve $6\tan^2 x = 11\tan x + 10$ for $0 \leqslant x < 360°$, giving your answers to 1 decimal place. You must show each step of your working. **(6 marks)**

48 A curve has equation $y = x^3 + 3x^2 - 9x + 6$.

a Find $\frac{dy}{dx}$ **(2 marks)**

b Hence find the range of values of x for which y is increasing. Write your answer using set notation. **(4 marks)**

49 **a** Factorise completely $36x - 25x^3$. **(3 marks)**

b Sketch the curve C with equation $y = 36x - 25x^3$. Show on your sketch the coordinates at which the curve meets the x-axis. **(3 marks)**

The points A and B lie on C and have x-coordinates of -1 and 1 respectively.

c Show that the length of AB is $2\sqrt{k}$ where k is a constant to be found. **(4 marks)**

50 $f(x) = 4 - 2\sqrt{x} + \frac{1}{2}x^2$

a Find the coordinates of the stationary point on the curve with equation $y = f(x)$. **(3 marks)**

b Find the equation of the normal to the curve at $P(4, 8)$. **(5 marks)**

The normal to the curve at P intersects the x-axis at A.

c Find the area of the triangle OAP. **(3 marks)**

51 **a** Fully expand $(1 + x)^5$ **(3 marks)**

b Use your answer to part **a** to show that:

i $(1 + \sqrt{5})^5 = 176 + 80\sqrt{5}$

ii $\log_2(1 + \sqrt{5})^5 = p + \log_2(11 + 5\sqrt{5})$ where p is an integer. **(6 marks)**

52 $f(x) = x^2 - 6x + 13, x \in \mathbb{R}$.

a Express $f(x)$ in the form $(x - a)^2 + b$ where a and b are constants. **(2 marks)**

The curve C with equation $y = f(x)$ crosses the y-axis at the point P and has a minimum point at the point Q.

b Sketch the graph of C, showing the coordinates of points P and Q. **(3 marks)**

c Find the length of the line segment PQ, writing your answer as a simplified surd. **(3 marks)**

53 The diagram shows part of the curve with equation

$y = \sqrt{x^3} + 2x - 8\sqrt{x}$

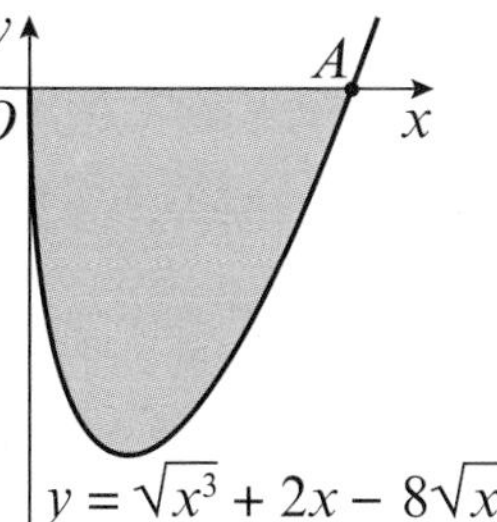

The curve intersects the x-axis at the origin and at the point A.
Find:

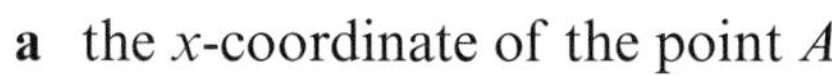

a the x-coordinate of the point A **(3 marks)**

b the area of the shaded region. **(6 marks)**

54 **a** Given that $\log_3(7x - 18) - 2\log_3(x - 2) = 1$, show that $3x^2 - 19x + 30 = 0$. **(4 marks)**

b Hence, or otherwise, solve $\log_3(7x - 18) - 2\log_3(x - 2) = 1$. **(2 marks)**

55 The straight line with equation $y = 2x - 4$ does not cross or touch the curve with equation $y = x^2 + px - 2p$, where p is a constant.

a Show that $p^2 + 4p - 12 < 0$. **(3 marks)**

b Hence find the set of possible values of p. **(3 marks)**

56 In the triangle DEF, $\overrightarrow{DE} = 3\mathbf{i} + 7\mathbf{j}$ and $\overrightarrow{DF} = 8\mathbf{i} - \mathbf{j}$.

a Prove that triangle DEF is scalene. **(4 marks)**

b Find the size of $\angle DFE$, correct to 1 decimal place. **(4 marks)**

57 $g(x) = 4x^3 - 16x^2 - 35x + 147$

a Using the factor theorem, explain why $g(x)$ is divisible by $(x + 3)$. **(2 marks)**

b Hence show that $g(x)$ can be written in the form
$g(x) = (x + 3)(ax + b)^2$ where a and b are integers to be found. **(4 marks)**

The diagram shows a sketch of part of the curve with equation $y = g(x)$.

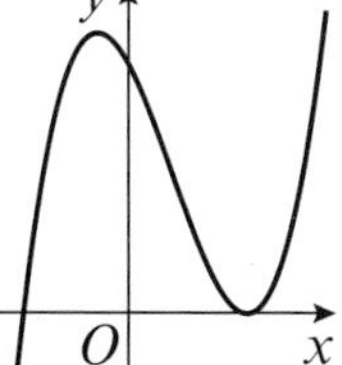

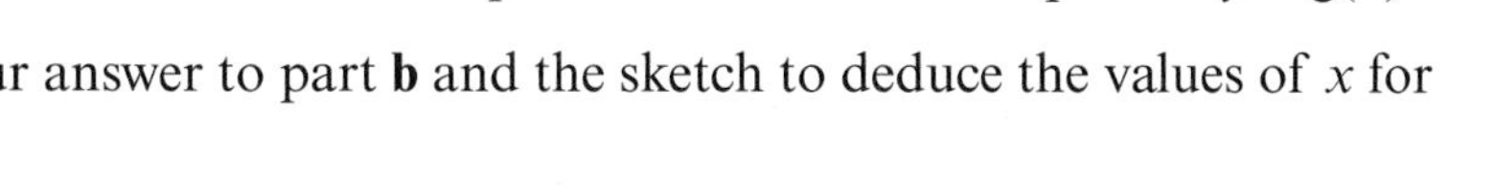

c Use your answer to part **b** and the sketch to deduce the values of x for which:

i $g(x) \leqslant 0$ **ii** $g(2x) \leqslant 0$ **(3 marks)**

58 The circle C has equation $x^2 + y^2 - 8x + 12y + 16 = 0$.

a Find:

i the coordinates of the centre of C **ii** the radius of C **(3 marks)**

The line with equation $y = kx$, where k is a constant, does not intersect C.

b Find the range of possible values of k. **(6 marks)**

59 The amount of money, A, in a bank account t years after it is deposited can be given by the equation $A = Pe^{\frac{rt}{100}}$, where P is the amount deposited and $r\%$ is the interest rate.

Given that £700 was deposited 4 years ago and that the interest rate on the account is 3.5%, find:

a the current value of the account **(2 marks)**

b $\frac{dA}{dt}$ **(2 marks)**

c the time taken for the money to double. **(3 marks)**

60 **a** Show that the equation $\sin^2 x = 8\cos^2 x - 6\cos x$ can be written in the form $(3\cos x - 1)^2 = 2$ **(3 marks)**

b Hence solve $\sin^2 x = 8\cos^2 x - 6\cos x$ for $0 \leqslant x < 360°$, giving your answers to 2 decimal places. **(5 marks)**

61 **a** Find the binomial expansion of $(1 + 3x)^4$. **(4 marks)**

b Hence show that $(1 + 3x)^4 + (1 - 3x)^4 = 2 + 108x^2 + 162x^4$ **(3 marks)**

c Use your answer to part **b** to show that the curve with equation $f(x) = (1 + 3x)^4 + (1 - 3x)^4$ has only one stationary point and find the coordinates. **(4 marks)**

62 The line with equation $mx - y - 3 = 0$ touches the circle with equation $x^2 - 4x + y^2 - 6y = 7$. Find the two possible values of m. **(7 marks)**

63 The diagram shows triangle ABC, with $AB = 6$ cm, $BC = (2x - 5)$ cm and $AC = (x + 2)$ cm.

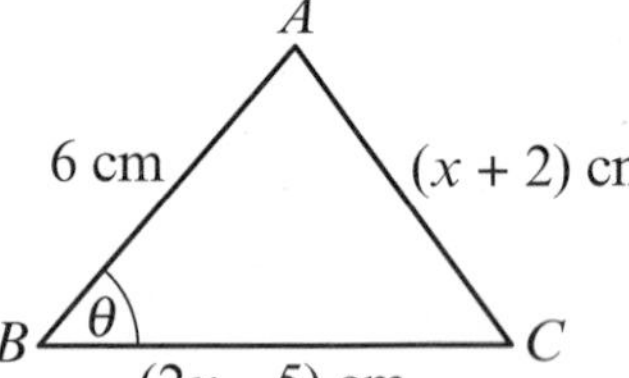

Given that $\angle ABC = \theta$ where $\cos\theta = \frac{3}{4}$

a show that $x^2 - 14x + 34 = 0$ **(3 marks)**

b By completing the square or otherwise, find the exact values of the two possible values of x. **(3 marks)**

64 **a** Prove that, if $1 + x^2 < (1 + x)^2$, then $x > 0$. **(3 marks)**

b Show, by means of a counter-example, that the inequality $1 + x^2 < (1 + x)^2$ is not true for all values of x. **(2 marks)**

65 The line $y = k - x$ is a tangent to the curve $y = x^2 + 5x - 2$. Find the value of k. **(3 marks)**

66 Solve, for $0 \leqslant x \leqslant 360°$:

a $\sin(x + 15°) = \frac{\sqrt{3}}{2}$ **(4 marks)**

b $\cos 2x = -\frac{1}{\sqrt{2}}$ giving your answers to 1 decimal place. **(6 marks)**

67 A company makes a particular type of portable DVD player.
The annual profit made by the company is modelled by the equation
$P = 6.25(30x - x^2) - 1256.25$, where P is the profit measured in thousands of pounds and x is the selling price of the DVD player, in pounds.

a i Write the model in the form $P = c - a(x - b)^2$, where a, b and c are constants to be found.

ii Hence, or otherwise, find the values of x which give $P = 0$.
Give your answer to 3 significant figures.

iii Sketch the graph of P against x, clearly labelling the intersections with the x-axis and the maximum point on the graph. **(7 marks)**

b Using the model, explain why £25 is not a sensible selling price for the DVD player. **(2 marks)**

The company wishes to maximise its annual profit.

State, according to the model:

c i the maximum possible annual profit

ii the selling price of the disc player that maximises the annual profit. **(2 marks)**

68 The number of reported cases of a disease in a country are shown in the table.

Year	Cases, C
1990	5128
1995	3943
2000	3031
2005	2331
2010	1792
2015	1378

This data can be modelled using an exponential function of the form $C = ab^t$, where t is the time in years since 1990 and a and b are constants.

a Show that $C = ab^t$ can be written as $\log_{10}C = \log_{10}a + t\log_{10}b$. **(2 marks)**

b Plot a graph of $\log_{10}C$ against t and draw in a line of best fit. **(3 marks)**

c Use your graph to estimate the values of a and b to 2 significant figures. **(3 marks)**

69 $f(x) = (x^2 + k)(2x + 5) + 7$ where k is a constant.

a Write down the remainder when $f(x)$ is divided by $(2x + 5)$. **(1 mark)**

Given that $(x - 1)$ is a factor of $f(x)$,

b prove that $k = -2$. **(3 marks)**

c Find all the solutions to $f(x) = 0$. **(5 marks)**

70 A circle C_1 has equation $x^2 + y^2 - 10x + 6y - 11 = 0$.

a Find:

i the coordinates of the centre of C_1 **ii** the radius of C_1 **(3 marks)**

The circle C_1 cuts the x-axis at the points P and Q.

b Find an equation of the circle C_2 with diameter PQ. **(6 marks)**

71 **a** Show that $x^2 - 8x + 17 > 0$ for all real values of x. **(3 marks)**

b "If I add 2 to a number and square the sum, the result is greater than the square of the original number."
State, giving a reason, whether the above statement is always true, sometimes true or never true. **(2 marks)**

72 The graph shows part of the curve C with equation $y = -x^3 + 2x^2 + 15x$.

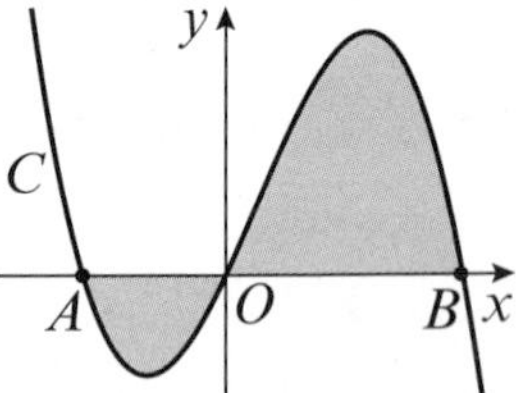

The curve C crosses the x-axis at the origin O and at points A and B.

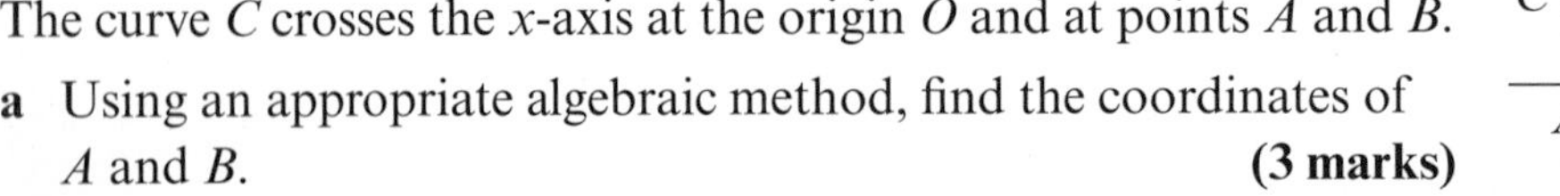

a Using an appropriate algebraic method, find the coordinates of A and B. **(3 marks)**

b The finite region shown shaded is bounded by the curve C and the x-axis. Use calculus to find the total area of the shaded region. **(8 marks)**

73 The diagram shows three points P, Q and R in the same horizontal plane.

R is 200 m from P on a bearing of 068°. R is 95 m from Q on a bearing of 310°.
Calculate:

a the distance PQ **(5 marks)**

b the bearing of Q from P. **(5 marks)**

74 A horse riding facility wishes to create a riding space using 300 m of fencing. The riding space must comprise a rectangular area and two semi-circles, as shown in the diagram.

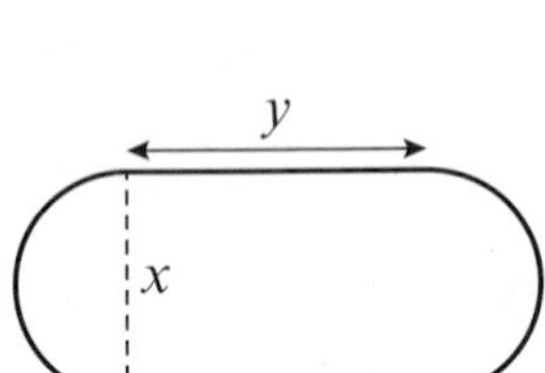

a Show that the area of the riding space can be expressed by the formula $A = 150x - \frac{\pi}{4}x^2$ **(4 marks)**

b Given that x can vary, find the exact maximum area of the riding space. **(4 marks)**

75 $f(x) = 2x^3 + kx^2 - 24x + 45$ where k is a constant.

Given that $(x - 5)$ is a factor of $f(x)$,

a show that $k = -7$ **(2 marks)**

b factorise $f(x)$ completely. **(4 marks)**

Given that $g(y) = 2(5^{3y}) - 7(5^{2y}) - 24(5^y) + 45$,

c find the values of y that satisfy $g(y) = 0$, giving your answers to 2 decimal places where appropriate. **(3 marks)**

76 Solve $\cos(3x - 20°) = -0.6$ for $0 \leqslant x < 180°$, giving your answers to 1 decimal place. You should show each step in your working. **(7 marks)**

77 The diagram shows the line segment joining the points $P(0, 4)$ and $Q(6, 2)$.

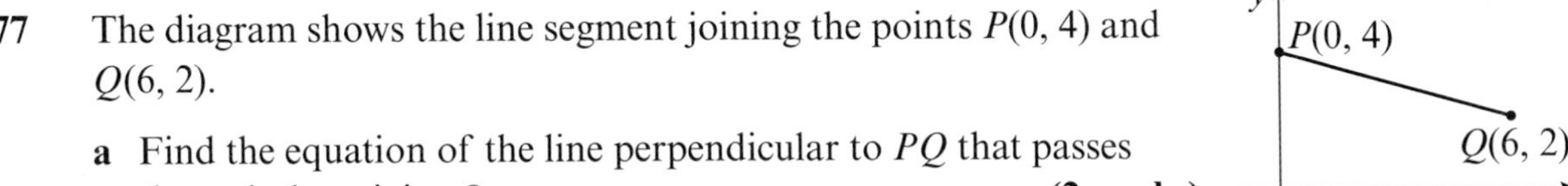

a Find the equation of the line perpendicular to PQ that passes through the origin, O. **(2 marks)**

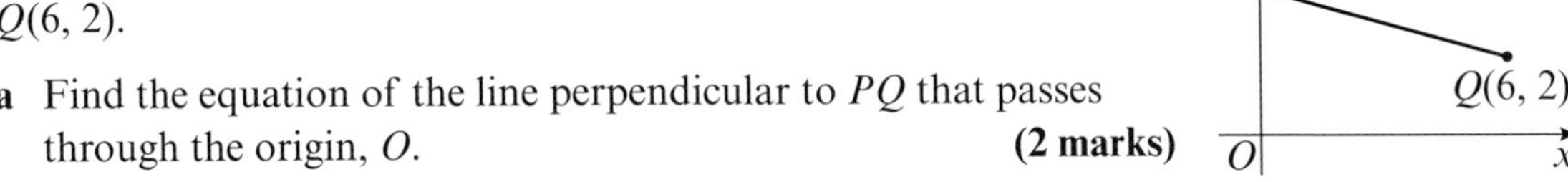

b Find the coordinates of the point where this perpendicular meets PQ. **(4 marks)**

c Show that the perpendicular distance of PQ from the origin is $\frac{6\sqrt{10}}{5}$ **(2 marks)**

d Find the length of PQ giving your answer in the form $a\sqrt{10}$. **(2 marks)**

e Find the area of triangle OPQ. **(2 marks)**

78 **a** On the same set of axes sketch the graphs of $y = \cos(x - 60°)$ and $y = -\cos x$, in the interval $-90° \leqslant x \leqslant 360°$. **(5 marks)**

b State the coordinates of the points of intersection of the graph of $y = \cos(x - 60°)$ with the x- and y-axes for $-90° \leqslant x \leqslant 360°$. **(3 marks)**

c Use your graph to find the solutions to the equation $\cos(x - 60°) + \cos x = 0$ in the interval $-90° \leqslant x \leqslant 360°$. **(3 marks)**

79 $f(x) = 4x^{\frac{3}{2}} - \frac{1}{2}x^3$. The diagram shows part of a sketch of the curve with equation $y = f(x)$.

The shaded region R is bounded by the curve and the x-axis.

a Find the coordinates of point A in the form $(\sqrt[3]{p}, q)$ where p and q are integers to be found. **(3 marks)**

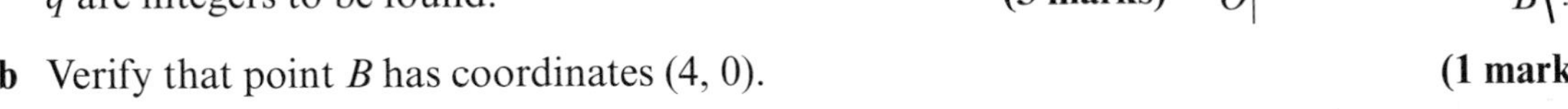

b Verify that point B has coordinates $(4, 0)$. **(1 mark)**

c Find the exact area of the region R. **(6 marks)**

80 $OABC$ is a parallelogram. $\overrightarrow{OA} = \mathbf{a}$ and $\overrightarrow{OC} = \mathbf{b}$. The point Z divides OB in the ratio 3 : 2.

a Express the vector $\overrightarrow{CZ}$ in terms of $\mathbf{a}$ and $\mathbf{b}$. **(2 marks)**

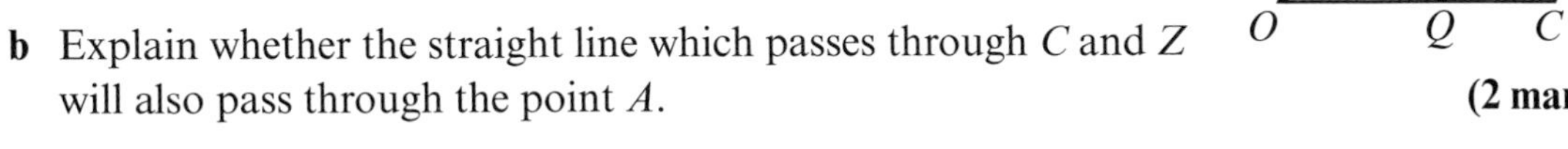

b Explain whether the straight line which passes through C and Z will also pass through the point A. **(2 marks)**

Given that OA, QP and CB are parallel,

c show that $AP : PB = 3 : 2$ and $PZ : ZQ = 2 : 3$. **(3 marks)**

81 The line l_1 with equation $y = 2x + 5$ intersects the line l_2 with equation $4x + 3y - 35 = 0$ at the point P.

a Find the coordinates of P. **(5 marks)**

The lines l_1 and l_2 cross the line $y = 1$ at the points Q and R respectively.

b Find the area of triangle PQR. **(4 marks)**

The distance of the point $A(10, k)$ from P is less than 10.

c Find the range of possible values for k. **(4 marks)**

82 The circle C has centre $P(5, 3)$ and passes through the point $Q(13, 9)$.

a Find an equation for C. **(4 marks)**

The line l_1 is the tangent to C at Q.

b Find an equation for l_1 in the form $ax + by + c = 0$, where a, b and c are integers. **(4 marks)**

The line l_2 is parallel to l_1 and passes through the midpoint of PQ.
Given that l_2 intersects C at A and B,

c find the length of AB, giving your answer in its simplest surd form. **(3 marks)**

83 $f(x) = x^3 - 2x^2 + x + 18$

a Show that $(x + 2)$ is a factor of $f(x)$. **(2 marks)**

b Hence show that -2 is the only real root of $f(x)$. **(4 marks)**

c Prove that $f(x) \geqslant 18$ for all $x \geqslant 0$. **(3 marks)**

84 $f(x) = 6x^3 + 11x^2 - 3x - 2$

a Given that $(x + 2)$ is a factor of $f(x)$, factorise $f(x)$ completely. **(6 marks)**

b Given that $g(y) = 6\sin^3 y + 11\sin^2 y - 3\sin y - 2$, find the values of y that satisfy $g(y) = 0$ for $0 < \theta \leqslant 360°$, giving your answers to 1 decimal place where appropriate. **(3 marks)**

85 **a** Solve the equation $5\sin(2\theta + 60°) = 4$ for $0 \leqslant \theta < 360°$, giving your answers to 1 decimal place. You must show each step of your working. **(6 marks)**

b Solve the equation $8\tan x - 3\cos x = 0$ for $360° \leqslant x < 540°$, giving your answers to 2 decimal places. **(5 marks)**

86 The line l_1 passes through the points $A(4, -1)$ and $B(-2, 7)$.

a Find an equation of l_1 giving your answer in the form $ax + by + c = 0$. **(4 marks)**

b Find an equation of the line l_2 which passes through B and is perpendicular to l_1. **(3 marks)**

The point C has coordinates $(k, 1)$ and angle ABC is a right angle.

c Find the value of k. **(2 marks)**

d Find the equation of the circle that passes through the points A, B and C. **(5 marks)**

87 The points P and Q lie on a circle with centre N as shown in the diagram.

The point P has coordinates $(2, -3)$ and the midpoint M of PQ has coordinates $(5, 1)$.

The line l passes through the points M and N.

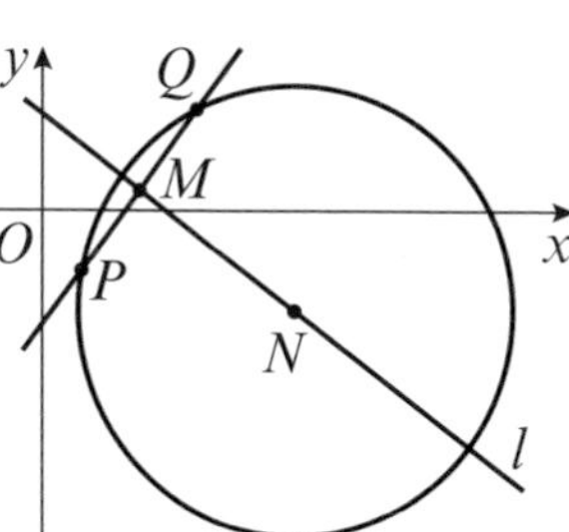

a Find an equation for l. **(4 marks)**

Given that the x-coordinate of N is 13,

b use your answer to part **a** to show that the y-coordinate of N is -5 **(1 mark)**

c find an equation for the circle. **(4 marks)**

Given that QR is a diameter of the circle,

d find the area of the triangle PQR, giving your answer in its simplest surd form. **(4 marks)**

88 **a** On separate sets of axes, sketch the graphs of:

i $y = -5x + c$, where c is a positive constant **ii** $y = \frac{1}{x} + 6$

On each sketch, show the coordinates of any point at which the graph crosses the y-axis and the equation of any horizontal asymptote. **(4 marks)**

Given that $y = -5x + c$ meets the curve $y = \frac{1}{x} + 6$ at two distinct points,

b show that $(6 - c)^2 > 20$ **(3 marks)**

c hence find the range of possible values for c. **(4 marks)**

89 The points P, Q and R have coordinates $(6, 2)$, $(k, 8)$ and $(9, 3)$ respectively. Given that the distance between points P and Q is twice the distance between points P and R,

a calculate the possible values of k. **(7 marks)**

Given that the line passing through P and Q has equation $3x - y - 16 = 0$,

b find the coordinates of the midpoint of PQ. **(4 marks)**

90 A sheet of cardboard has dimensions 60 cm by 36 cm.

A square of side length x cm is removed from each of the four corners as shown in the diagram. The sheet of cardboard is then folded to make a box.

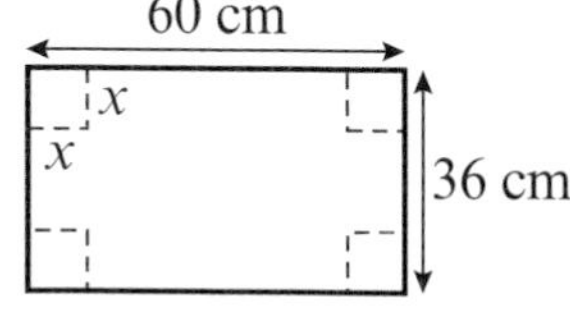

a Show that the volume of the box is given by the equation $V = 4x^3 - 192x^2 + 2160x$ **(4 marks)**

b Find the value of x for which the volume is a maximum. Leave your answer in the form $a + b\sqrt{19}$, where a and b are constants to be found. **(4 marks)**

c Find the maximum volume of the box, giving your answer to one decimal place. **(2 marks)**

91 A running track can be modelled by two straight lines, which are opposite sides of a rectangle, joined by two semicircular arcs each of radius r metres, as shown in the diagram.

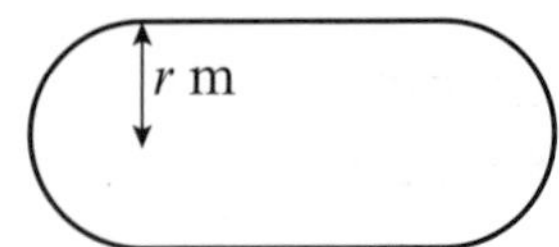

The distance around the entire track is 500 m, and the track itself can be assumed to be a thin line.

a The area inside the track is $A\,\text{m}^2$. Show that $A = 500r - \pi r^2$. **(5 marks)**

b Hence find, in terms of π, the maximum value of A and the value of r at which this maximum occurs.
(You do not have to justify that the value is a maximum.) **(6 marks)**

92 The points $A(8, 10)$ and $B(16, -2)$ lie on the circle with centre $C(3, k)$. Find:

a the value of k **(5 marks)**

b an equation for the circle **(3 marks)**

c the area of triangle ABC. **(5 marks)**

93 The diagram shows a trapezium $PQRS$.

The coordinates of the vertices are $P(-10, -6)$, $Q(15, 9)$, $R(10, 16)$ and $S(-5, 7)$.

a Verify that the lines PQ and SR are parallel. **(3 marks)**

b Prove that the trapezium is not isosceles. **(3 marks)**

c The diagonals of the trapezium meet at M. Find the coordinates of M. **(4 marks)**

d Show that neither of the diagonals of the trapezium bisects the other diagonal. **(3 marks)**

Answers

CHAPTER 1

1.1 Index laws

1 a b^9 b a^2 c $15x^9$ d $2x$
2 a $6x^5 - 7x^3$ b $-15x^2 + 40x^3$
c $3x^2 - 31x + 14$
3 a $x^6 - x^2 - x^4$ b $3x^2 + 2x^6$ c $3x + 6 - 8x^7$
4 a $21x^5y^5$ b $5x^9y$
5 a $9r^2s^7$ b $16a^{11}b^4$
6 a 5 b 4
7 $p = 4, q = 2, r = 4$
8 a $5x$ b $4x$ c 1

1.2 Expanding brackets

1 a $x^2 + 7x + 12$ b $x^2 - 49$ c $x^2 - 9x + 18$
2 a $x^2 + 2xy + y^2$ b $x^2 - 8x + 16$
c $4x^2 - 20xy + 25y^2$
3 a $20x^2 - 23xy - 21y^2$ b $x^3 - 9x^2y - x + 9y$
c $x^2 + 2x + 2y - 2xy - 3y^2$
4 a $10x^3 - 18x^2 - 26x^2y + 12xy^2 + 36xy$
b $27x^3 - 54x^2y + 36xy^2 - 8y^3$
c $8x^3 + 16x^2 - 22x + 6$
5 $a^3 + b^3 + 3a^2b + 3ab^2$
6 $a = 3, b = 4$
7 a $(4x^2 + 4x + 1)$ cm^2 b $(8x^3 + 12x^2 + 6x + 1)$ cm^3
8 a $(4x^2 + 8x + 2)$ m^2 b $x = 3$
9 $p = 30, q = 1, r = 11, s = 2$

1.3 Factorising

1 a $6(x + 3)$ b $9x(3x - 1)$ c $x^2(x - 4)$
2 a $7ab^2(1 + 3a)$ b $8b(a - 8b^2)$
c $5b^2(ab^2 + 4a^3c + 3b^3c^2)$
3 a $(x + 3)(x + 2)$ b $(x - 5)(x + 2)$
c $(x + 7)(x - 7)$ d $(y + 4x)(y - 4x)$
4 a $3x^2(x^2 + 2x - 6)$ b $4(x + 3)(x + 4)$
c $(2x + 3)(x - 1)$
5 $x(1 - 5x)(1 + 5x)$
6 $4x(3 + 2x)(3 - 2x)$
7 $4x(2x + 1)(x + 2)$
8 $\frac{2x(x + 2)}{(x + 3)}$

1.4 Negative and fractional indices

1 a x^{-5} b x^{-1} c $\frac{1}{2}x^{-13}$
2 a $10x^{\frac{11}{12}}$ b $3x^{\frac{11}{6}}$ c $4x^2$ d $14x^{0.25}$
3 a $x^{\frac{7}{10}}$ b $x^{-\frac{5}{4}}$ c $x^{\frac{25}{6}}$
4 a 64 b $-\frac{1}{64}$ c $\frac{9}{25}$
5 a $\frac{1024}{3125}x^{\frac{15}{2}}$ b $x^{-5} + 8x^{-7}$ c $5x^{-5} - 3x^{-2}$
6 a 512 b $64x^5$
7 a $\frac{1}{5}$ b $\frac{6561}{256}x^{-1}$
8 a 27 b $3x^{-\frac{1}{2}}$
9 a $\frac{1}{3}x$ b $18x^{-2}$
10 5^{4x-10}

1.5 Surds

1 a $3\sqrt{2}$ b $3\sqrt{7}$ c $5\sqrt{10}$
2 a $\sqrt{3}$ b $\sqrt{2}$ c $\sqrt{6}$
3 a $10\sqrt{2}$ b $6\sqrt{3}$ c $\sqrt{2}$
4 a $9 - \sqrt{3}$ b $-1 + \sqrt{2}$ c $24 - 4\sqrt{7} - 6\sqrt{3} + \sqrt{21}$
5 a $5\sqrt{5}$ b $4\sqrt{7}$
6 $\sqrt{15} + \sqrt{3}$
7 a $-9 + 3\sqrt{7}$ b $37 + 12\sqrt{7}$
8 a 99 b $11 - 5\sqrt{3}$
9 $a = \frac{11}{2}$

1.6 Rationalising denominators

1 a $\frac{\sqrt{3}}{3}$ b $7\sqrt{5}$ c $\sqrt{3}$
2 a $\frac{\sqrt{3}}{3}$ b $\frac{\sqrt{6}}{6}$ c $\frac{\sqrt{15}}{5}$
3 a $\sqrt{2} - 1$ b $\frac{1 + \sqrt{7}}{3}$ c $3\sqrt{6} + 3\sqrt{5}$
4 a $5 + 2\sqrt{6}$ b $\frac{5 - 2\sqrt{7}}{5}$ c $51 - 36\sqrt{2}$
5 $4 + 2\sqrt{3}$
6 a $7 + 3\sqrt{5}$ b $4 - \sqrt{5}$
7 $p = \frac{9}{2}, q = 2$
8 $A = 49, B = 4, C = 28$
9 $\frac{2 + \sqrt{14}}{2}$

Problem solving: Set A

B a $2\sqrt{3}$ b $4\sqrt{6}$
S a $16 + 12\sqrt{5}$ b $4 + 3\sqrt{5}$
G $\frac{13}{11}\sqrt{3} - \frac{1}{11}$

Problem solving: Set B

B a 2^{2x+4} b $x = \frac{1}{2}$
S a $y^2 - 6y + 8 = 0$ b $x = \frac{3}{2}$ or $x = 2$
G $x = \frac{3}{2}$ or $x = \frac{5}{2}$

CHAPTER 2

2.1 Solving quadratic equations

1 a $x = -5$ or $x = -2$ b $x = 8$ or $x = -3$
c $x = 0$ or $x = -6$
2 a $x = 0$ or $x = 3$ b $x = 3.5$ or $x = -3.5$
c $x = 2.5$ d $x = 2$ or $x = 0.5$
3 a $x = 4$ b $x = 3.5$ or $x = -3$
c $x = -3$ or $x = -4$
4 a $x = -4 \pm \sqrt{10}$ b $x = -2 \pm \sqrt{5}$ c $x = \frac{6 \pm \sqrt{6}}{2}$

5 **a** $x = 1$ or $x = 9$ **b** $x = \pm\frac{5}{4}$
c $x = \frac{5}{3}$ or $x = -5$ **d** $x = 3.53$ or $x = 0.472$
e $x = 0$ or $x = 5$ **f** $x = -0.177$ or $x = -2.82$

6 **a** $(2x + 3)(3x - 5) = 20 \Rightarrow 6x^2 - x - 15 = 20$
$\Rightarrow 6x^2 - x - 35 = 0$
b Length = 8 m, width = 2.5 m

7 $-2 \pm \sqrt{3}$

8 $x = 0.485$ and $x = -16.5$

9 **a** 3.12 seconds **b** 1 and 2 seconds

2.2 Completing the square

1 **a** $(x + 1)^2 - 1$ **b** $(x - 4)^2 - 16$ **c** $(x - 6)^2 - 36$

2 **a** $(x - 4)^2 - 4$ **b** $\left(x - \frac{1}{2}\right)^2 - \frac{49}{4}$ **c** $\left(x - \frac{3}{2}\right)^2 - \frac{25}{4}$

3 **a** $3(x - 2)^2 + 5$ **b** $5(x - 1)^2 + 7$ **c** $3\left(x - \frac{7}{6}\right)^2 - \frac{25}{12}$

4 **a** $x = 1 \pm \sqrt{11}$ **b** $x = \frac{-6 \pm \sqrt{6}}{5}$ **c** $x = \frac{7 \pm \sqrt{73}}{6}$

5 **a** $(x - 3.5)^2 - 14.25$ **b** -14.25

6 **a** $-3\left(x + \frac{1}{3}\right)^2 + \frac{13}{3}$ **b** $x = \frac{-1 \pm \sqrt{13}}{3}$

7 **a** $7 - (x - 3)^2$ **b** $x = 3 \pm \sqrt{7}$

2.3 Functions

1 **a** 0 **b** 5 **c** 15.75 **d** 5

2 $x = 2$ and $x = -2$

3 **a** -2 and -7 **b** 9 and -9
c $\sqrt[3]{-3}$ and $\sqrt[3]{-4}$ **d** 0, -3 and 7

4 **a** $(x + 3)^2 - 7$ **b** $-3 \pm \sqrt{7}$ **c** -7, when $x = -3$

5 **a** $(x + 3)^2 + 4$
b Smallest value of $(x + 3)^2$ is 0 so minimum value is 4.

6 **a** 1 and -1 **b** $\sqrt[5]{\frac{1}{3}}$ and $\sqrt[5]{\frac{1}{2}}$ **c** 1 and 4

7 **a** $(2^x - 2)(2^x - 4)$ **b** 1 and 2

2.4 Quadratic graphs

1 **a**

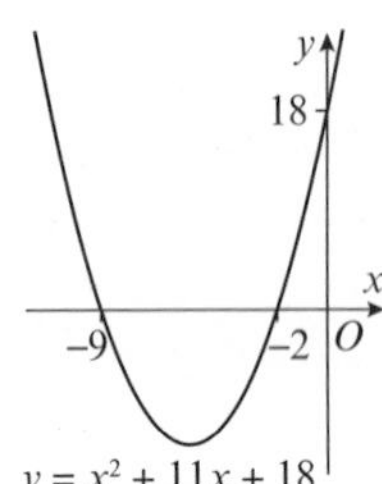

b

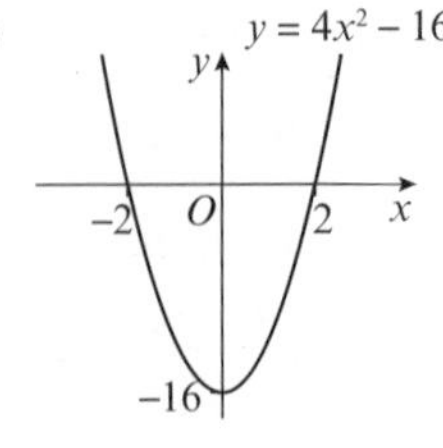

c

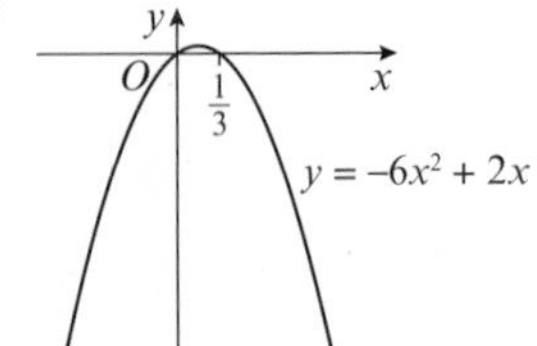

2 **a** $(1, 9)$ **b** $\left(-\frac{1}{2}, -\frac{25}{4}\right)$ **c** $\left(-\frac{13}{2}, \frac{1}{4}\right)$

3 **a**

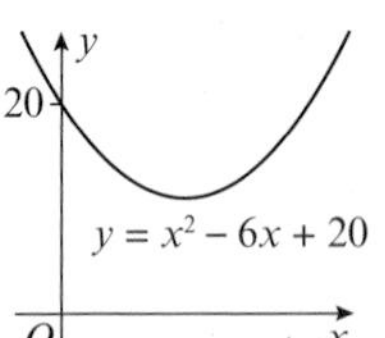

Turning point (3, 11), line of symmetry $x = 3$

b

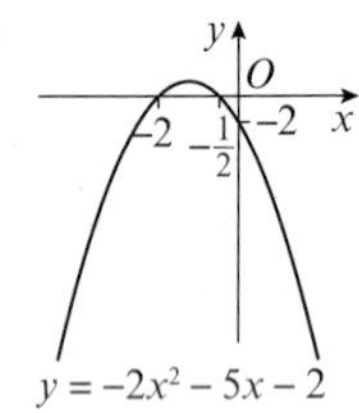

Turning point $\left(-\frac{5}{4}, \frac{9}{8}\right)$, line of symmetry $x = -\frac{5}{4}$

c

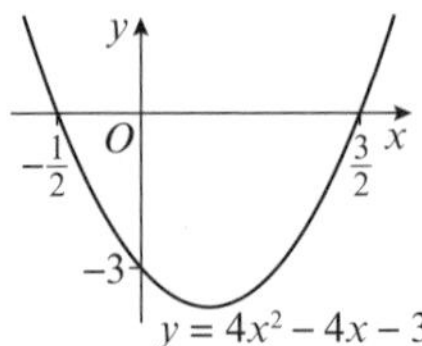

Turning point $\left(\frac{1}{2}, -4\right)$, line of symmetry $x = \frac{1}{2}$

4 **a**

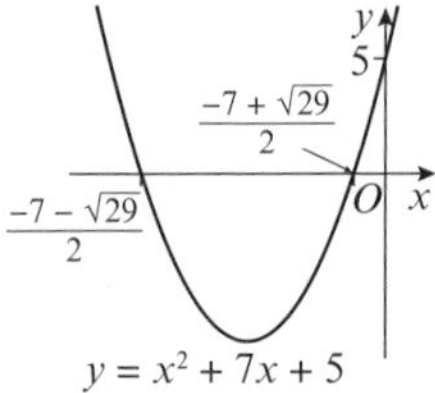

Turning point $\left(-\frac{7}{2}, -\frac{29}{4}\right)$, line of symmetry $x = -\frac{7}{2}$

b

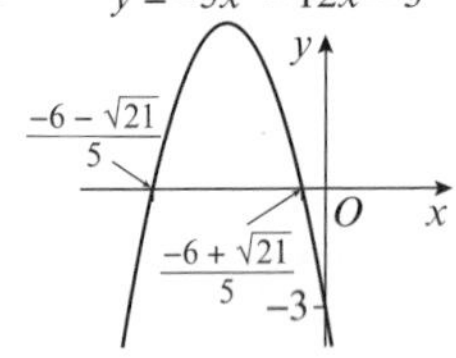

Turning point $\left(-\frac{6}{5}, \frac{21}{5}\right)$, line of symmetry $x = -\frac{6}{5}$

c

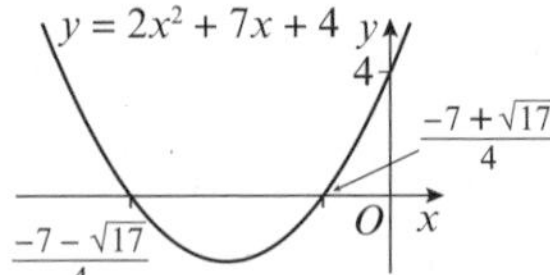

Turning point $\left(-\frac{7}{4}, -\frac{17}{8}\right)$, line of symmetry $x = -\frac{7}{4}$

5 **a** $p = 4$, $q = 9$
b

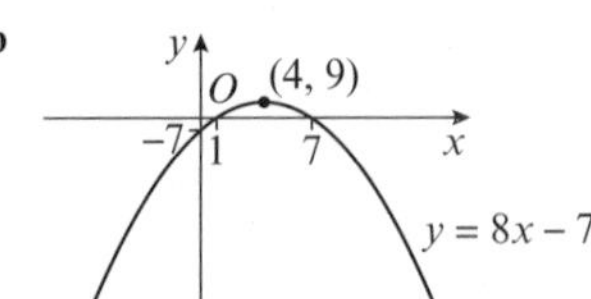

6 **a** $(x + 3)^2 - 5$
b $f(x) = x^2 + 6x + 4$

y
P (0, 4)
O x
Q (−3, −5)

c The smallest y-value of the curve in part **b** is -5, so $f(x) = -6$ cannot have real solutions.

7 **a** $x = 10$ or $x = -1$

b

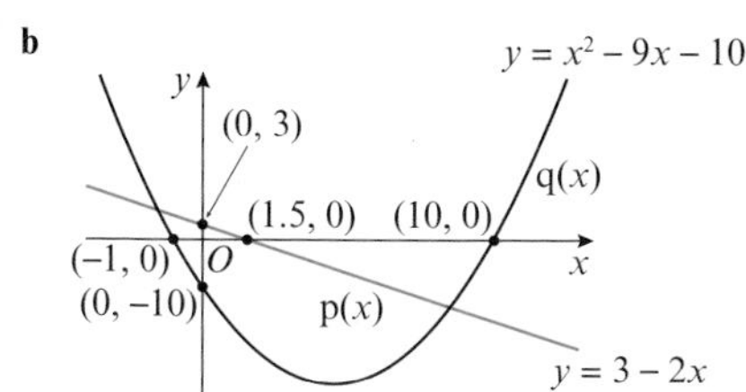

8 $a = 5, b = -20, c = 15$

2.5 The discriminant

1 **a** $b^2 - 4ac > 0$ **b** $b^2 - 4ac = 0$ **c** $b^2 - 4ac < 0$

2 **a** **i** 25 **ii** 2 roots
b **i** 0 **ii** 1 root
c **i** –72 **ii** no roots

3 **a** 4, ii **b** 0, iii **c** –12, i

4 **a** $k^2 - 100$ **b** $k = \pm 10$

5 **a** $k^2 + 2k + 9$ **b** $k^2 + 2k + 9 = (k + 1)^2 + 8$
c $(k + 1)^2 + 8 \geqslant 8 > 0$

6 $k > -\frac{9}{20}$

7 $p > \frac{49}{8}$

2.6 Modelling with quadratics

1 **a** 25 m **b** $-0.04(x - 12.5)^2 + 6.25$
c 6.25 m, 12.5 m

2 **a** $P = 73.5x - 5.25x^2 - 107.5 \Rightarrow P = -5.25(x^2 - 14x) - 107.5$
$\Rightarrow P = 149.75 - 5.25\,(x - 7)^2$
b **i** £149 750 **ii** £7

3 **a** **i** (1.25, –3) **ii** $y = 1.92(x - 1.25)^2 - 3$
b The canal is only 2.5 m wide so the boat is too wide

4 **a** 0 workers produce 0 cars
b 420 **c** 500 cars, 2500 workers

5 **a** The ball was thrown from a height of 120 m.
b $\frac{4085}{32} - 4.9\left(t - \frac{5}{4}\right)^2$
c **i** 6.35 seconds (3 s.f.)
ii 128 metres (3 s.f.), 1.25 seconds

6 **a** Negative profit can be interpreted as a loss, so the model may still be valid.
b $P = 160 - 6.25(x - 10)^2$ **c** £7.47
d **i** £160 000 **ii** £10

7 2000, £24

Problem solving: Set A

B **a** $1 - (x - 1)^2$ **b** 2 m

S **a** The stone was thrown from a height of 114 m
b **i** $10.4x - 5.2x^2 = -5.2(x^2 - 2x) \Rightarrow h(x) = 114 - 5.2(x^2 - 2x)$
ii $h(x) = 119.2 - 5.2(x - 1)^2$
c **i** 5.79 m (3 s.f.) **ii** 119.2 m, 1 m

G **a** The stone was thrown from a height of 125 m
b 134 m (3 s.f.) **c** 125 m (3 s.f.)

Problem solving: Set B

B **a** $4^2 - 8k^2 = 0 \Rightarrow 16 - 8k^2 = 0$ **b** $\pm\sqrt{2}$

S 24

G 6 and 2

CHAPTER 3

3.1 Linear simultaneous equations

1 **a** $x = 2, y = -3$ **b** $x = 1, y = 2$
c $x = -1, y = -2$

2 **a** $x = 1, y = 1$ **b** $x = 2.25, y = 6.5$
c $x = 0.5, y = 1$

3 **a** $x = 3, y = 5$ **b** $x = 3, y = -5$
c $x = 5, y = -2$

4 $x = 0.5, y = 1.5$

5 $x = \frac{3}{7}, y = \frac{44}{7}$

6 **a** $(6ky + 9x = 12) - (6ky - 6x = 27)$ gives $15x = -15$, so $x = -1$
b $k = 0.5$

7 Nisha. You can rearrange each equation to get the other, so any pair of values that satisfies one equation will satisfy both.

3.2 Quadratic simultaneous equations

1 **a** $x = -1, y = -64$ or $x = 16, y = 4$
b $x = 1, y = 3$ or $x = 3, y = 1$
c $x = 2, y = 3$ or $x = -\frac{2}{3}, y = \frac{1}{3}$

2 **a** $x = 1, y = -4$ or $x = 4, y = 2$
b $x = -4, y = -6$ or $x = 3, y = 1$
c $x = 1, y = 2$ or $x = 2, y = -1$

3 **a** $x = 1.63, y = -2.12$ or $x = 7.37, y = 15.12$
b $x = -1.00, y = 2.00$ or $x = 1.75, y = -0.75$
c $x = 0.59, y = -4.66$ or $x = 4.51, y = 11.02$

4 $x = 2, y = 2$ or $x = 8.67, y = -4.67$ (3 s.f.)

5 $x = -0.631, y = 0.261$ or $x = -0.227, y = -0.547$ (3 s.f.)

6 $\frac{-2 \pm 3\sqrt{2}}{4}$

7 **a** $y = 8 + 2x \Rightarrow x^2 + 2k(8 + 2x) + 4k = 0$
$\Rightarrow x^2 + 4kx + 20k = 0$
b $k = 5$ **c** $x = -10, y = -12$

3.3 Simultaneous equations on graphs

1 **a**

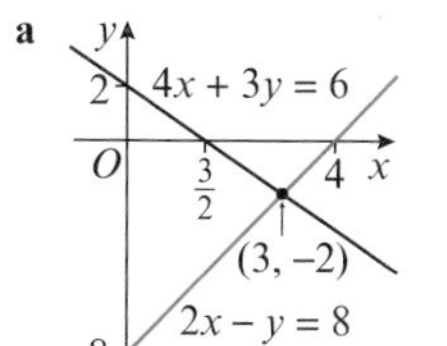

b

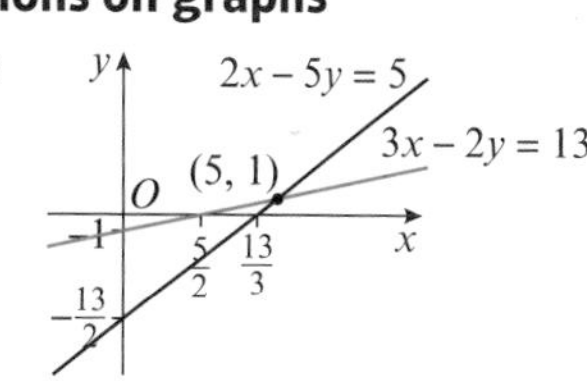

2 **a**

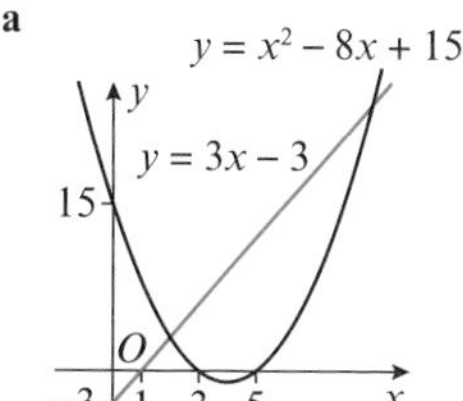

b (2, 3) and (9, 24)

c $x = 2 : 2^2 - 8 \times 2 + 15 = 3 \times 2 - 3 = 3$
$x = 9 : 9^2 - 8 \times 9 + 15 = 3 \times 9 - 3 = 24$

3 **a** $y = 3x^2 + 8x - 3$, $y = 7x - 2$; x-intercepts -3, $\frac{1}{3}$; y-intercepts $\frac{2}{7}$, -2, -3; O

b $(-0.77, -7.37)$ and $(0.43, 1.04)$

4 $A(1, 13)$ and $B(4, 16)$

5 $A(-2, 5)$ and $B(2, -3)$

6 **a** $y = 1 - \frac{1}{2}x$, $y = x^2 - 4x - 10$; 1, 2, -10, $2 - \sqrt{14}$, $2 + \sqrt{14}$; O

b $(-2, 2)$ and $(5.5, -1.75)$

7 $x = 5.18$ (3 s.f.)

8 **a** $\left(\frac{4}{9}, \frac{34}{9}\right)$

b $y = f(x)$, $y = g(x)$; 6, 4, $\frac{6}{5}$, 8; O

3.4 Linear inequalities

1 **a** $x \geqslant 8$ **b** $x > -11$ **c** $x \geqslant \frac{3}{10}$

2 **a** $2 \leqslant x \leqslant 4$ **b** $x < 27$ **c** $x \geqslant \frac{1}{3}$

3 **a** $\left\{x : -3 < x < \frac{2}{3}\right\}$ **b** $\{x : 5 \leqslant x \leqslant 12\}$

c $\left\{x : x < \frac{1}{2}\right\} \cup \{x : x > 38\}$

4 **a** $x > 3$ **b** $x < 4$

5 **a** $\{x : x > -1\}$ **b** $\left\{x : x \leqslant \frac{4}{5}\right\}$

6 **a** $\{x : x \geqslant -1\}$ **b** $\{x : x < 3\}$

7 **a** $2(x + 30) + 2x < 400$ **b** $x < 85$ m

8 **a** $\{x : -1 \leqslant x < 2\}$

b $\{x : x < 2.25\} \cup \{x : x > 101\}$

3.5 Quadratic inequalities

1 **a** $-5 < x < 5$ **b** $-\frac{3}{2} < x < 1$

c $x < -1, x > \frac{1}{3}$ **d** $x \leqslant 0, x \geqslant \frac{5}{2}$

2 **a** $\{x : 2 < x < 7\}$ **b** $\{x : -4 < x < -1\}$

c $\{x : -5 < x < -3\} \cup \{x : -2 < x < 1\}$

3 **a** $x < -\frac{1}{6}, x > \frac{1}{6}$ **b** $0 < x < \frac{5}{3}$

c $x \leqslant -\frac{1}{4}, x > 0$

4 $x^2 - 6x + 11 = (x - 3)^2 + 2 \Rightarrow x^2 - 6x + 11 \geqslant 2 > 0$

5 **a** $x > 3$ **b** $-4 \leqslant x \leqslant 10$

c $3 < x \leqslant 10$

6 **a** $(k - 2)^2 - 4 \times 1 \times (4 - 2k) > 0$, therefore $k^2 + 4k - 12 > 0$

b $k < -6, k > 2$

7 **a** $22x + 4 > 103 \Rightarrow x > 4.5$

b $21x^2 + 9x < 702 \Rightarrow -6 < x < \frac{39}{7}$

c $4.5 < x < \frac{39}{7}$

8 $0 < k < \frac{8}{9}$

9 $\frac{10 - 2\sqrt{7}}{9} < p < \frac{10 + 2\sqrt{7}}{9}$

10 $x < 1$ or $x > 2.5$

3.6 Inequalities on graphs

1 **a** **i**

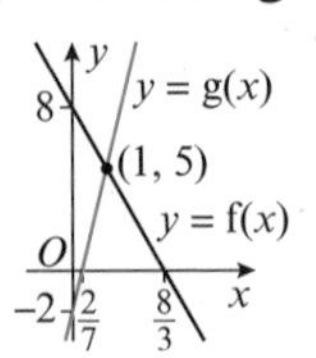

ii $x > 1$

b **i**

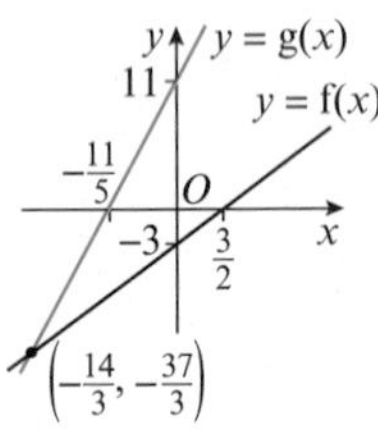

ii $x > -\frac{14}{3}$

c **i**

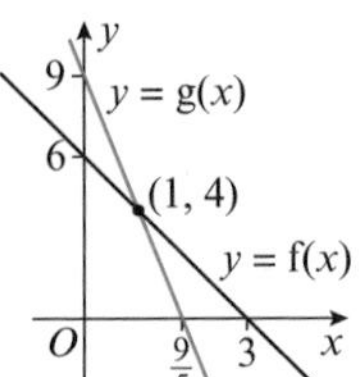

ii $x < 1$

2 **a** **i**

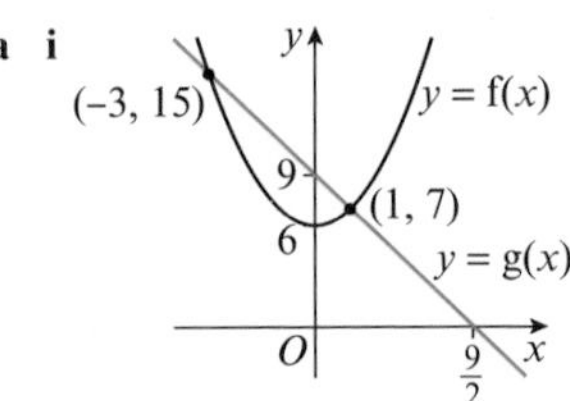

ii $-3 \leqslant x \leqslant 1$

b **i**

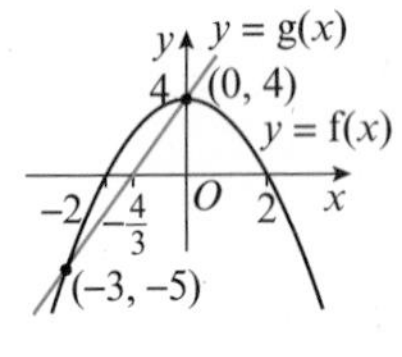

ii $x \leqslant -3$ and $x \geqslant 0$

c **i**

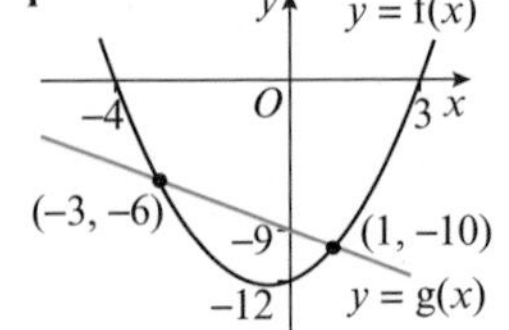

ii $-3 \leqslant x \leqslant 1$

3 **a** **i**

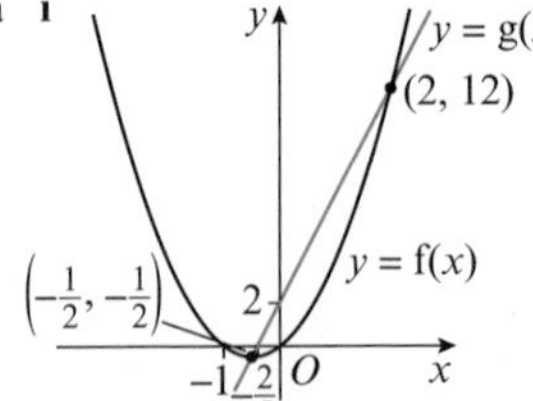

ii $\{x : x < -0.5\} \cup \{x : x > 2\}$

b **i**

ii $\{x : 0 < x < 2\}$

c **i** $y = f(x)$, $y = g(x)$; $(-4, 6)$, $(0, 2)$; 2, -2, -1, 2; O

ii $\{x : x < -4\} \cup \{x : x > 0\}$

4 **a**

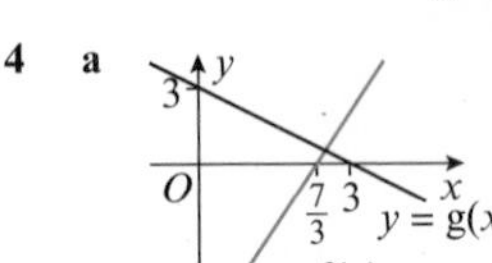

b $\left(\frac{5}{2}, \frac{1}{2}\right)$ **c** $x > \frac{5}{2}$

5 **a** $y = f(x)$, $y = g(x)$

b (1, −5) and (5, 3)

c $1 < x < 5$

6 **a** $x = -2$ and 5

b $y = q(x)$, $y = p(x)$

c (1, −12) and $\left(\frac{11}{2}, \frac{15}{4}\right)$ **d** $\left\{x : x < 1\right\} \cup \left\{x : x > \frac{11}{2}\right\}$

3.7 Regions

1 **a**

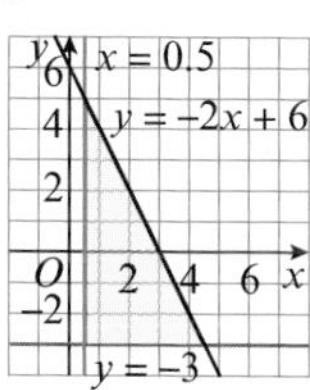

b (0.5, 5), (4.5, −3) and (0.5, −3)

c 16

2 **a**

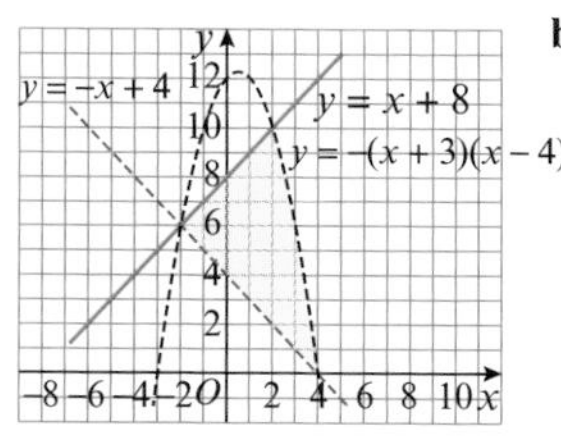

b (4, 0), (−2, 6), (2, 10)

c They all lie on one of the two lines with strict inequalities.

3 **a** (0, 2), (−1, 0), (−1.5, 0.5) and (−2, 0)

b $y \geqslant x^2 + 3x + 2$, $y \geqslant 2x^2 + 4x + 2$ and $y \geqslant x + 2$

c (−1.5, 0.5) and (0, 2)

4 **a**

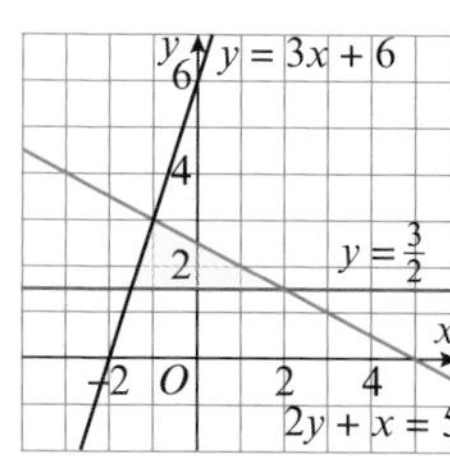

b $\frac{21}{8}$

5 **a**

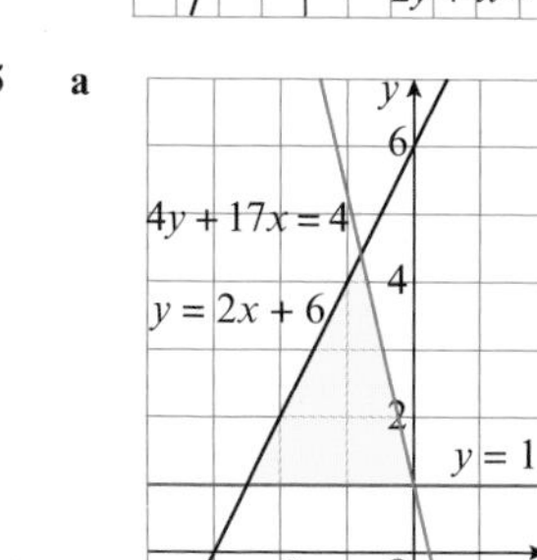

b $\frac{17}{4}$

6 **a, c**

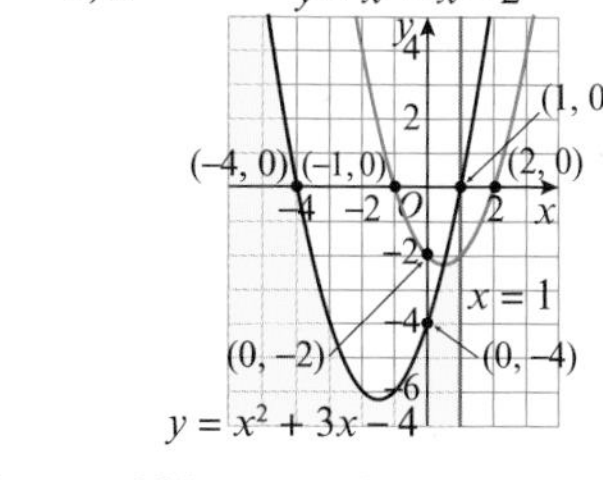

b (0.5, −2.25), (1, 0) and (1, −2)

7 **a** 6.25 **b** $y \geqslant 2\sqrt{x} - 3$, $y \leqslant 10 - 1.28x$, $x \geqslant 2.25$

Problem solving: Set A

B **a** $x^2 - (2x + 3) + 2k = 0 \Rightarrow x^2 - 2x + (2k - 3) = 0$

b $k = 2$ **c** $x = 1, y = 5$

S **a** $x^2 - 4k\left(4x - \frac{1}{4}\right) + 3k = 0 \Rightarrow x^2 - 16kx + 4k = 0$

b $x = \frac{1}{2}, y = \frac{7}{4}$

G $x = \frac{1}{2}, y = 2$

Problem solving: Set B

B **a** $x = -2$ and $x = -3$ **c** (−4, 2) and (0, 6)

b

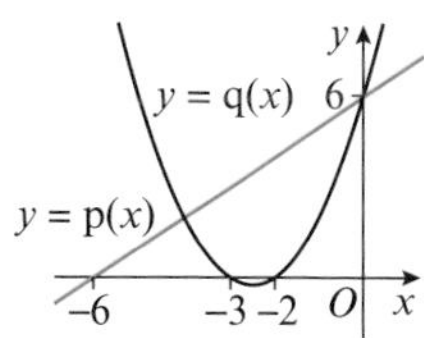

d $x < -4$ or $x > 0$

S **a, d**

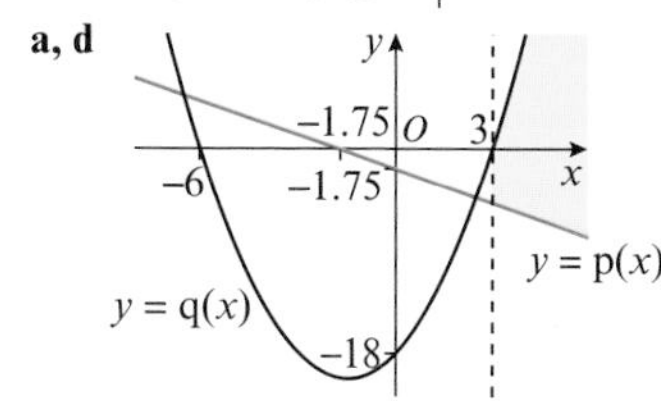

b (−6.5, 4.75) and (2.5, −4.25) **c** $-6.5 < x < 2.5$

G **a, c**

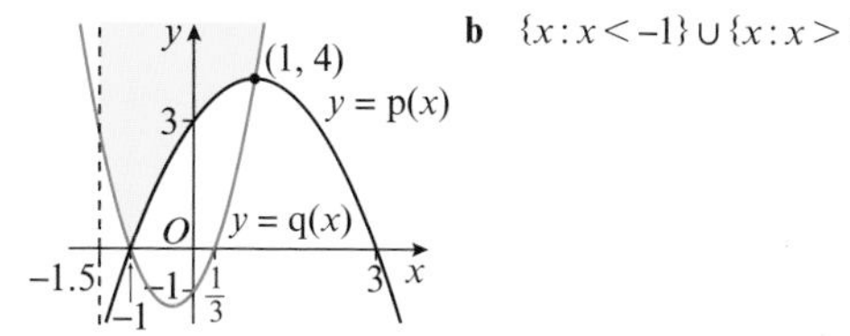

b $\{x : x < -1\} \cup \{x : x > 1\}$

CHAPTER 4

4.1 Cubic graphs

1 **a** $y = (x + 1)(x - 2)(x + 3)$

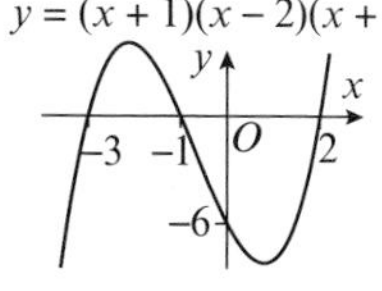

b $y = x(2x - 1)(2x + 1)$

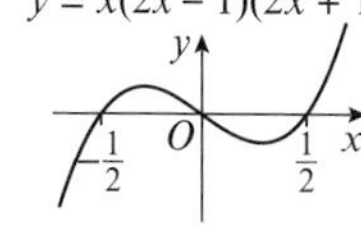

c $y = (x - 2)(x + 4)^2$

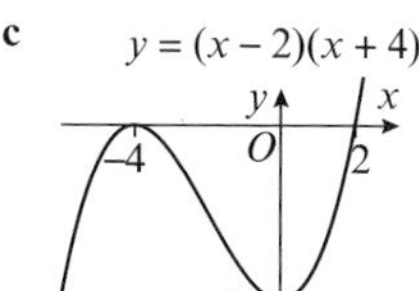

2 **a** $y = x(x-1)(x+3)$

b $y = x(1-x)(x+3)$

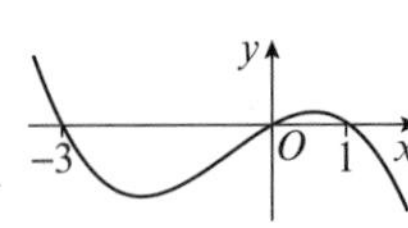

c $y = 3x^2(2x-1)$

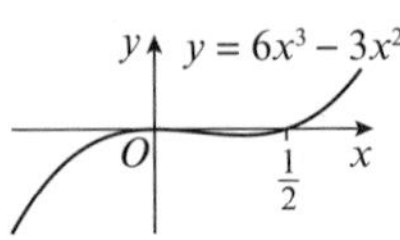

3 **a**

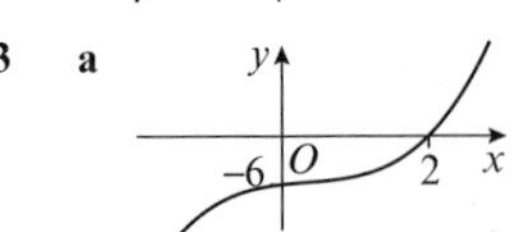

b

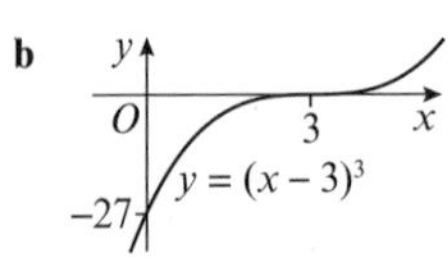

c

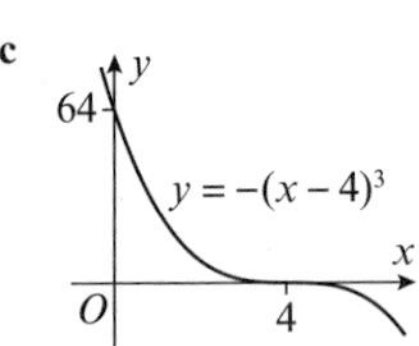

4 **a** $x(x+4)^2$

b

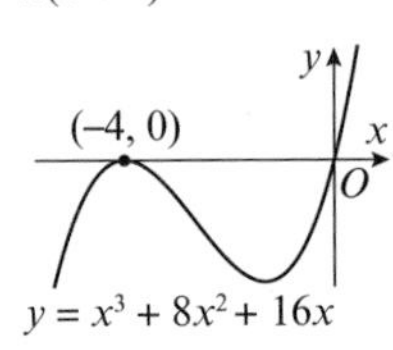

5 **a** $x(5-2x)(5+2x)$

b

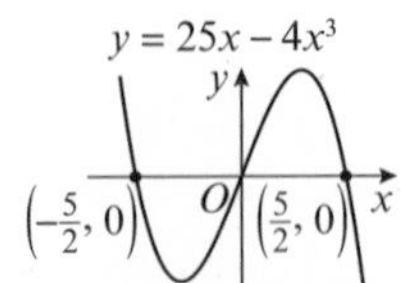

6 **a** Assumed that the coefficient of x^3 is 1 when it isn't.

b $y = \frac{1}{8}x^3 - \frac{3}{4}x^2 - \frac{9}{2}x + 27$

7 **a** $a = 6, b = 9, c = 4$ **b** $(0, -4)$

8 $a = 3, b = 18, c = 9, d = 30$

4.2 Quartic graphs

1 **a** $y = (x+1)(x+3)(x-1)(x-3)$

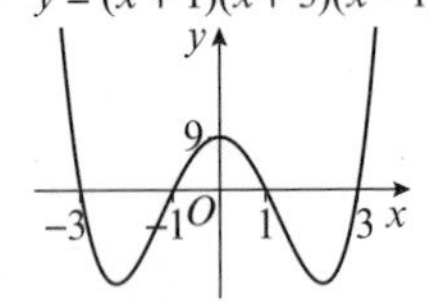

b

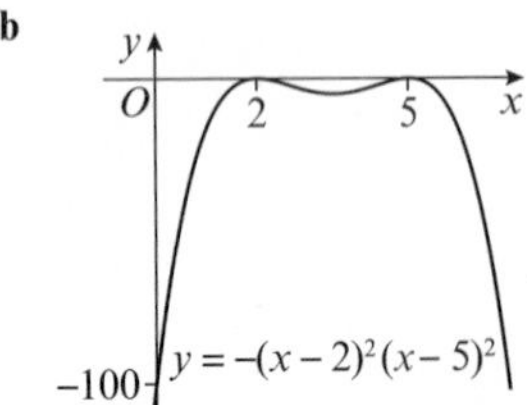

c $y = (2x+1)^3(x-3)$

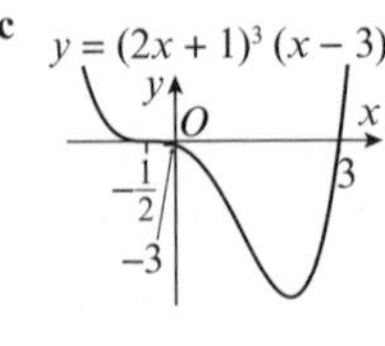

2 **a**

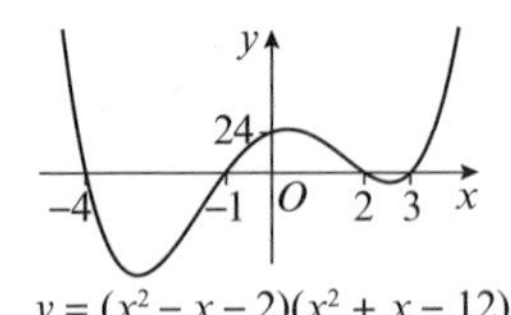

$y = (x^2 - x - 2)(x^2 + x - 12)$

b **c**

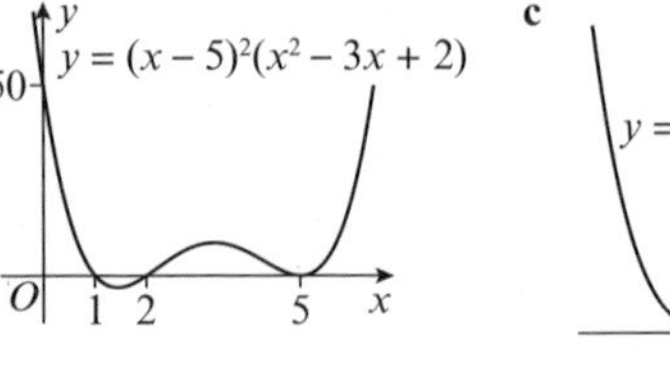

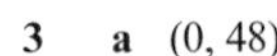

3 **a** $(0, 48)$ **b** $c = 16, d = 4, e = 48$

4

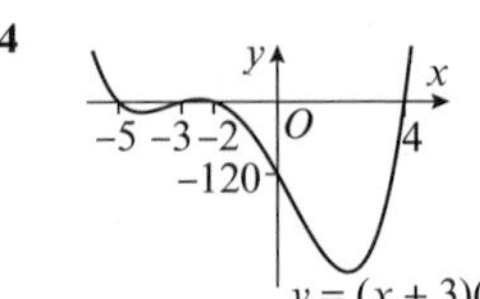

5

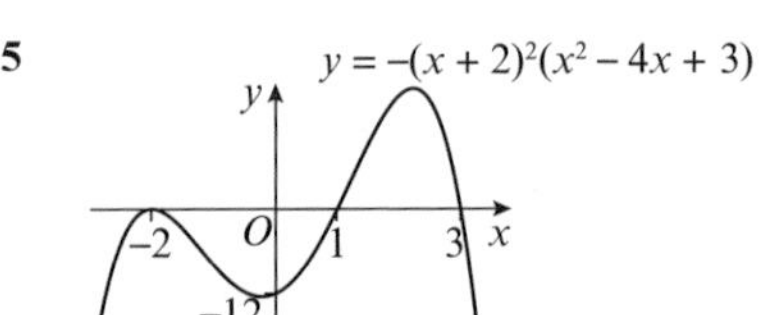

6 **a** $(0, -12)$ **b** $b = 2, c = 7, d = 8, e = 12$

7 **a** $(0, 36)$ **b** $b = 2, c = 11, d = 12, e = 36$

4.3 Reciprocal graphs

1

$y = \frac{2}{x}$ $y = \frac{6}{x}$

2

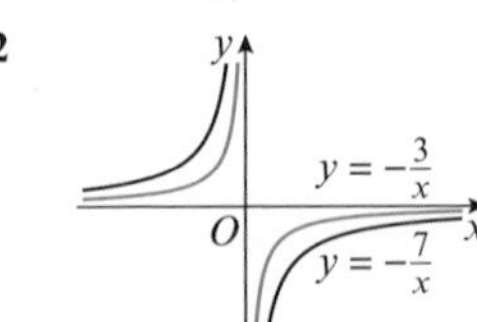

3

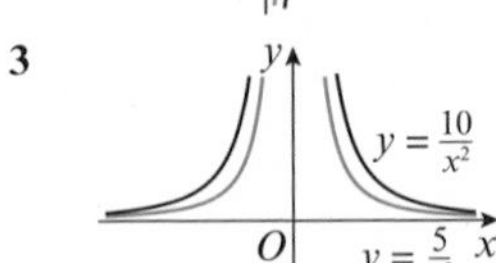

4

$y = -\frac{1}{x^2}$ $y = -\frac{9}{x^2}$

5 **a** 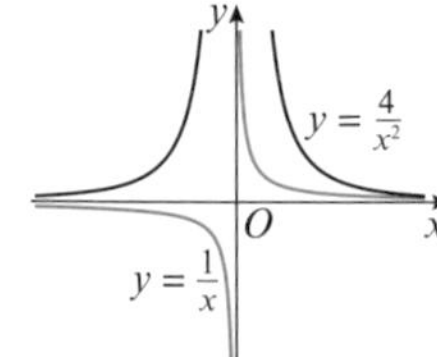

b Both graphs have asymptotes $y = 0$ and $x = 0$.

6 **a** 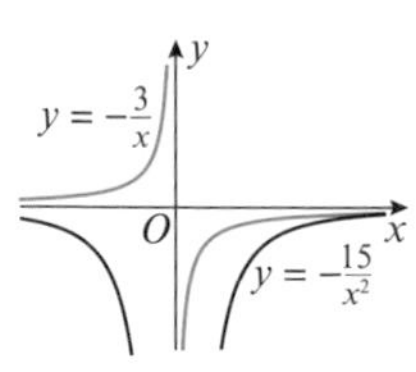

b Both graphs have asymptotes $y = 0$ and $x = 0$.

4.4 Points of intersection

1 **a** **i** 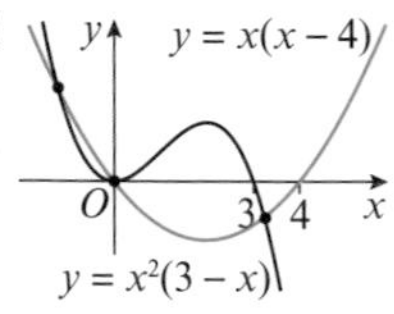

ii $(0, 0)$, $(1 + \sqrt{5}, 2 - 2\sqrt{5})$ and $(1 - \sqrt{5}, 2 + 2\sqrt{5})$

b **i** $y = x^2 + x - 2$ 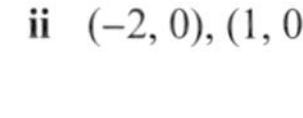

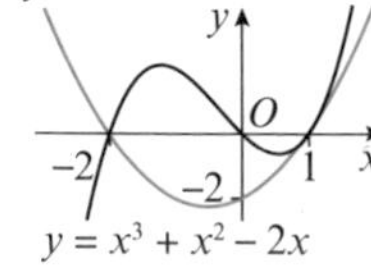

ii $(-2, 0)$, $(1, 0)$

2 **a** 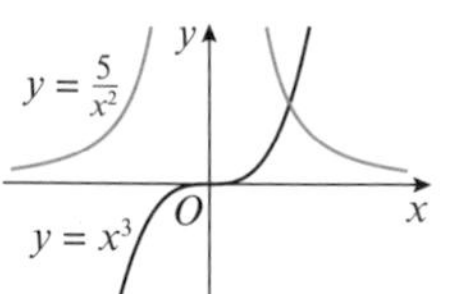

b $x^3 = \frac{5}{x^2} \Rightarrow x^5 - 5 = 0$

c 1

3 **a** 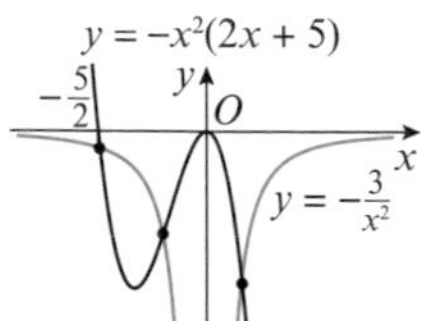

b $-\frac{3}{x^2} = -x^2(2x + 5) \Rightarrow x^4(2x + 5) - 3 = 0$

c 3

4 **a** **i** 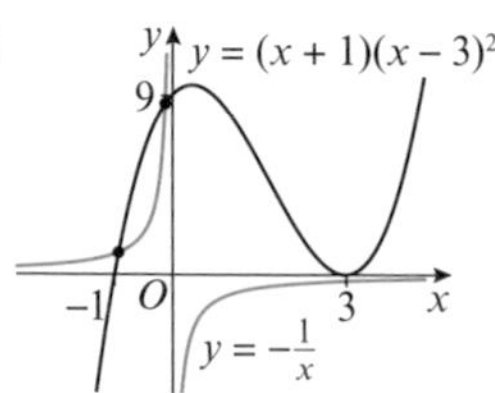

ii 2 intersections $\Rightarrow$ 2 real solutions

b **i** 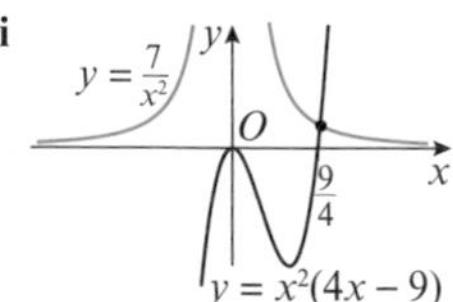

ii 1 intersection $\Rightarrow$ 1 real solution

5 $(-1, -7)$, $(1, -3)$

6 **a** 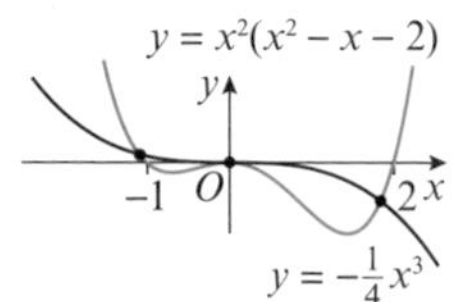

b 3

c $-4x^2(x^2 - x - 2) = x^3 \Rightarrow x^3 + 4x^2(x^2 - x - 2) = 0$
$\Rightarrow 4x^4 - 3x^3 - 8x^2 = 0 \Rightarrow x^2(4x^2 - 3x - 8) = 0$

d $x = 0$, $x = \frac{3 \pm \sqrt{137}}{8}$

7 **a** **i** $(-0.5, 0)$ **ii** $x = 0$, $y = 2$

b 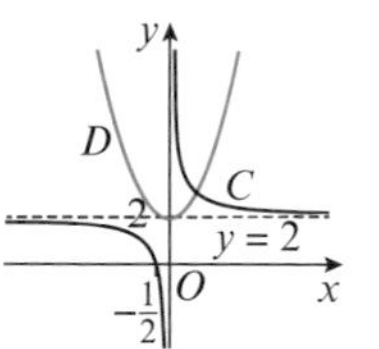

c $(1, 3)$

8 **a** 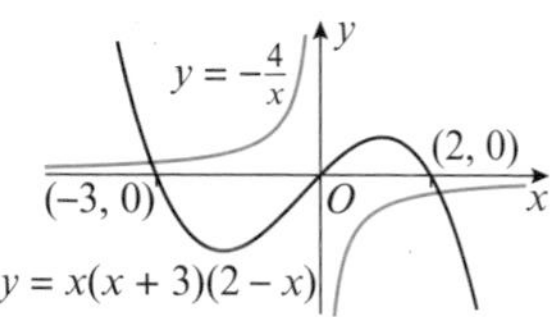

b 2 intersections $\Rightarrow$ 2 solutions

9 **a** 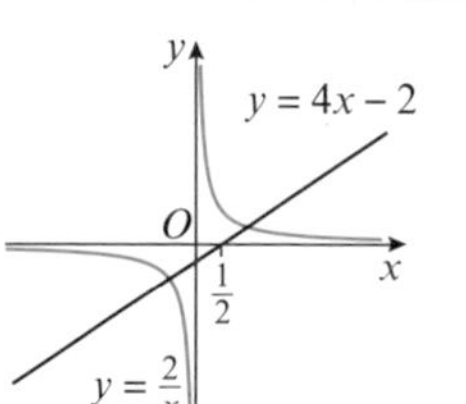

b $(-0.5, -4)$ and $(1, 2)$

10 $(-5, 60)$ and $(5, -40)$

11 **a** $a > -6.25$ **b** $(2.15, 9.29)$ and $(8{,}85, 22.7)$

4.5 Translating graphs

1 **a** **i** 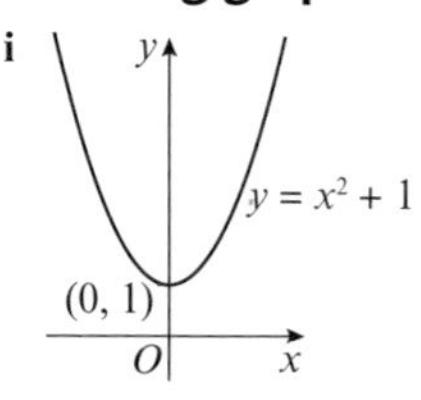

ii 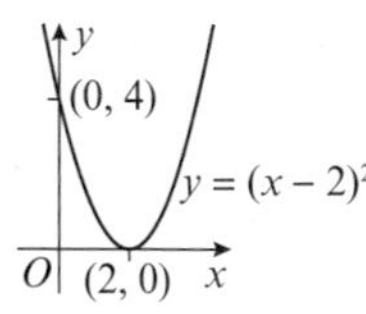

iii $y = (x + 3)^2$ 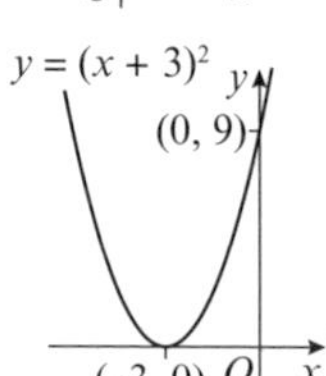

iv 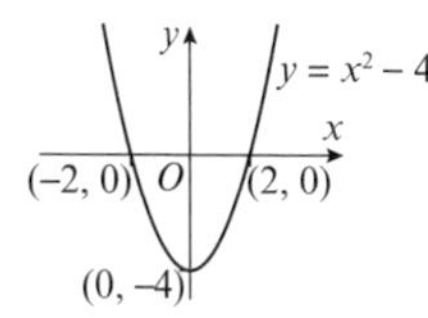

b **i** 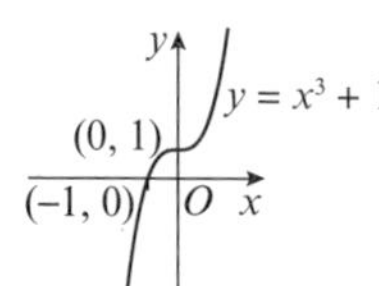

ii 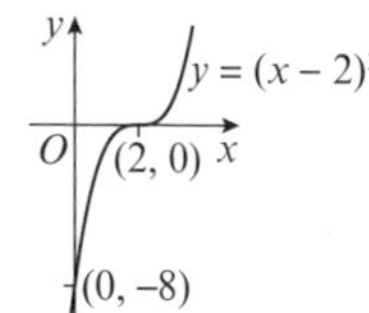

iii (0, 27) $y = (x + 3)^3$ (−3, 0) O x y

iv y x O $(\sqrt[3]{4}, 0)$ (0, −4) $y = x^3 - 4$

c i y $y = \frac{1}{x} + 1$ $y = 1$ (−1, 0) O x $x = 0$

ii $y = \frac{1}{x-2}$ $y = 0$ O $\left(0, -\frac{1}{2}\right)$ $x = 2$ x

iii $x = -3$ y $\left(0, \frac{1}{3}\right)$ $y = 0$ O x $y = \frac{1}{x+3}$

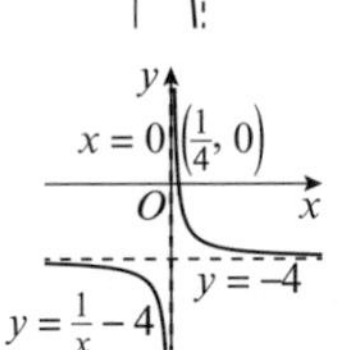

iv y $x = 0$ $\left(\frac{1}{4}, 0\right)$ O x $y = -4$ $y = \frac{1}{x} - 4$

2 a

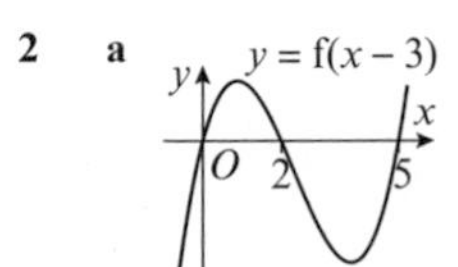

b

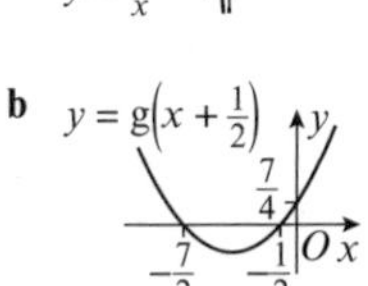

c

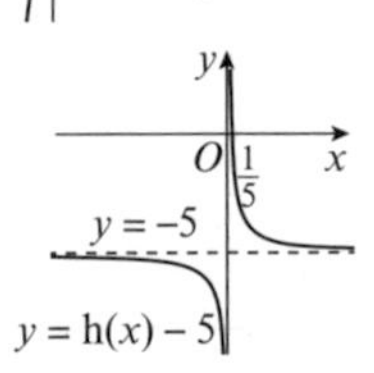

3 a

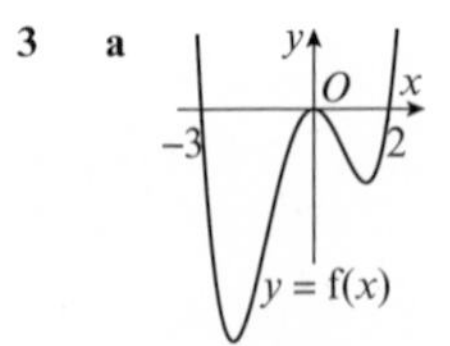

b

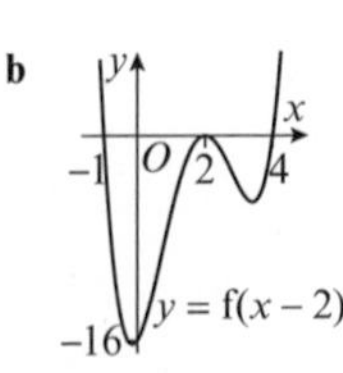

4 a

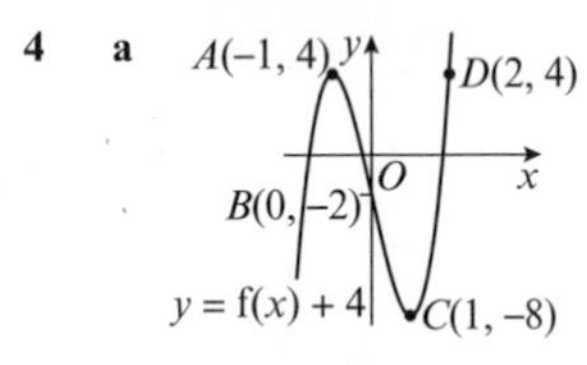

b

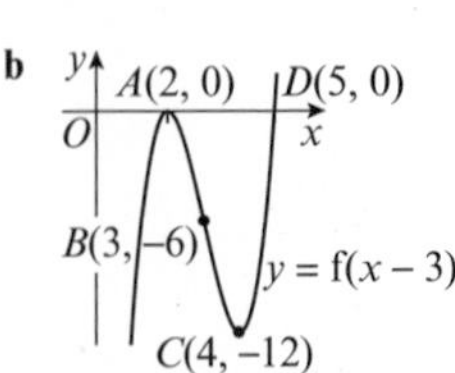

5 a $y = (x + 4)^2(x - 2)$

b

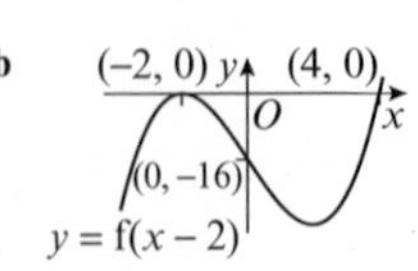

6 a (3, 0) **b** 7

7 2 and −3

8 $-7 < a < -3$

4.6 Stretching graphs

1 a i

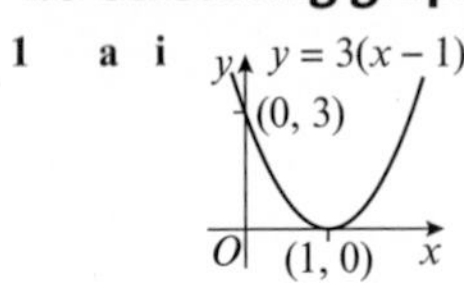

ii $y = \left(\frac{1}{2}x - 1\right)^2$ y (0, 1) O (2, 0) x

iii y (1, 0) O x (0, −1) $y = -(x - 1)^2$

iv

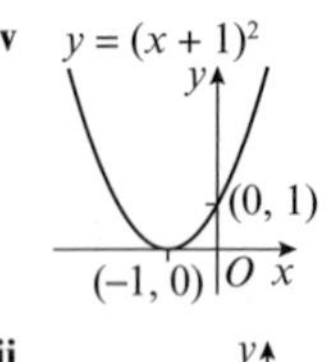

b i $y = 3x^3 + 3$ y (0, 3) (−1, 0) O x

ii y (0, 1) (−2, 0) O x $y = \left(\frac{1}{2}x\right)^3 + 1$

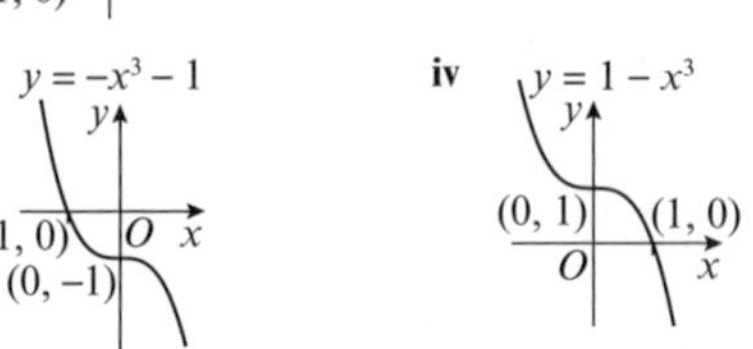

c i

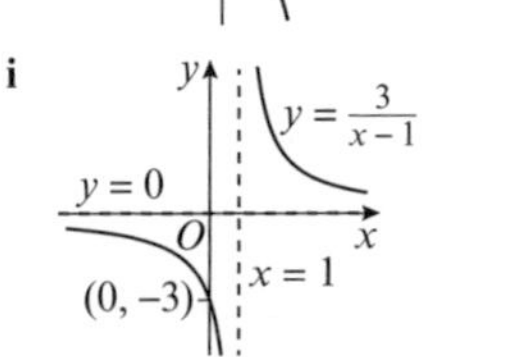

ii

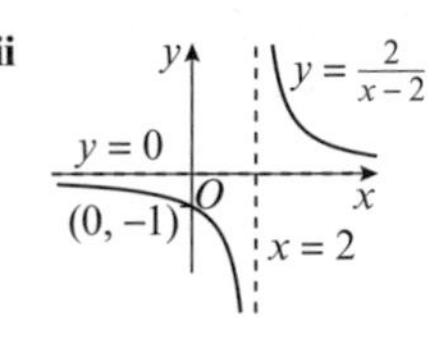

iii

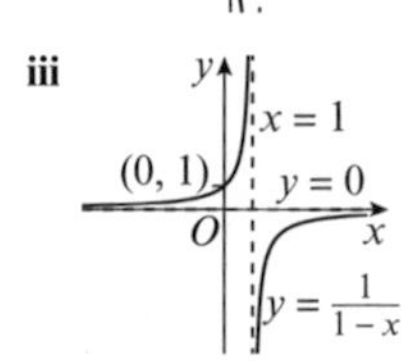

iv

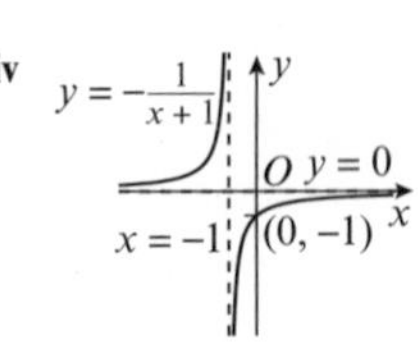

2 a

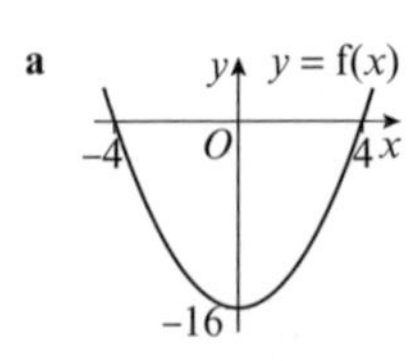
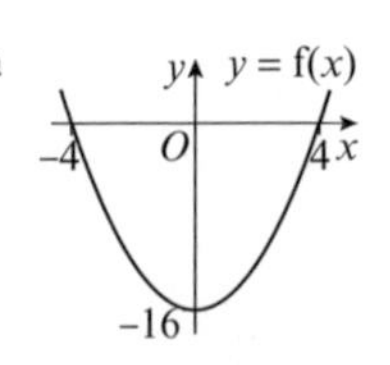

b i $y = \frac{1}{4}x^2 - 16$

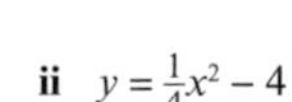

ii $y = \frac{1}{4}x^2 - 4$

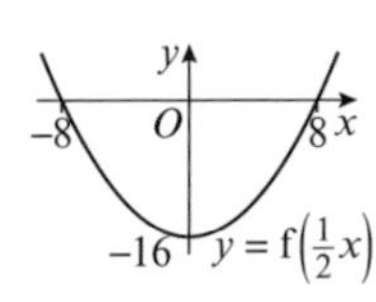

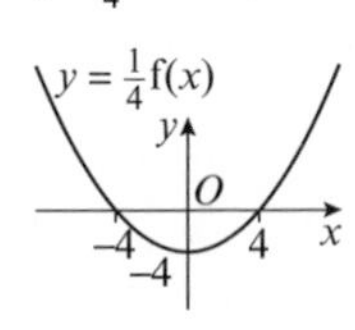

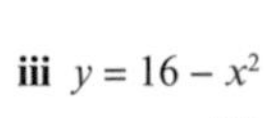

iii $y = 16 - x^2$

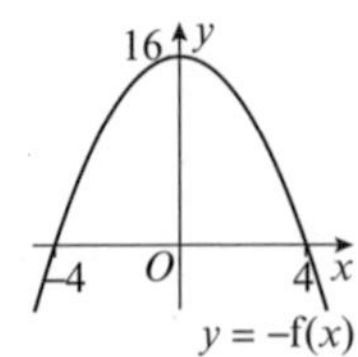

c i (0, −16) **ii** (0, −4) **iii** (0, 16)

3 a, b

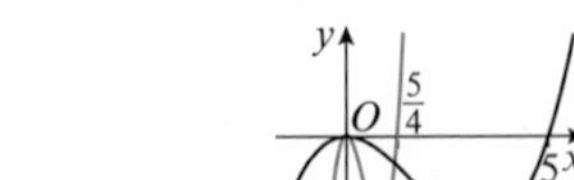
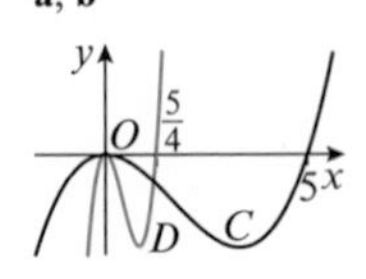

c Stretch by scale factor $\frac{1}{4}$ in the horizontal direction

4 $B(-2, 4)$ y $C(0, 2)$ $y = \mathrm{f}\left(\frac{1}{2}x\right)$ $A(-4, 0)$ O $D(2, 0)$ x

5 a 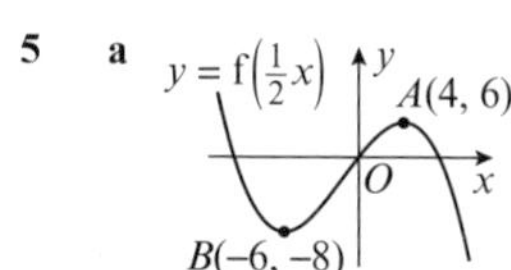

b 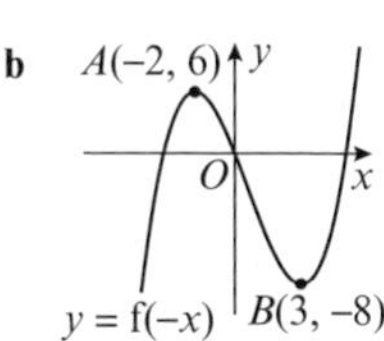

6 a 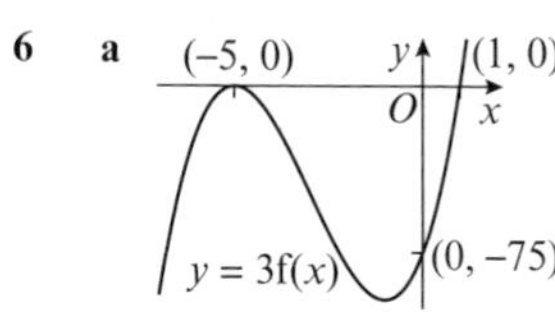

b $y = 3(x + 5)^2(x - 1)$

7 a $a = 8$ b $b = \frac{1}{2}$

4.7 Transforming functions

1 a 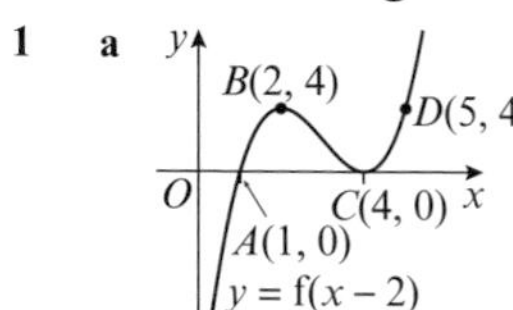

b 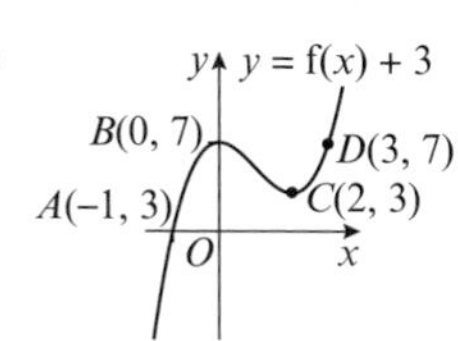

c 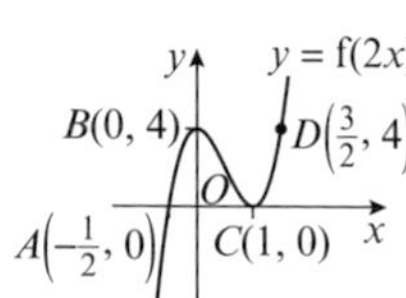

2 a b c 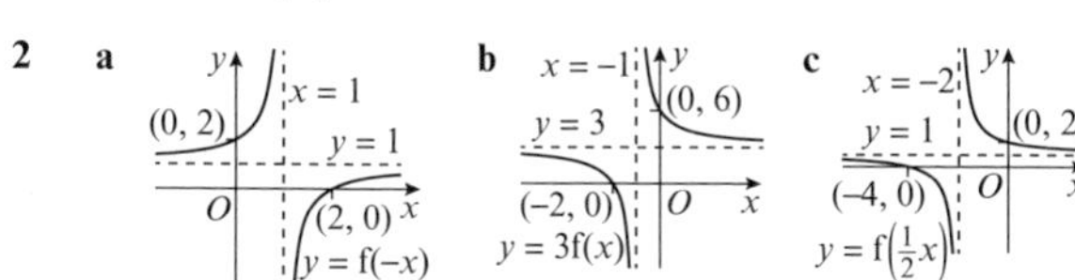

3 a $A(-1, 0)$, $B(0, -4.5)$, $C(1, -8)$, $D(3, 0)$

b $A(-1, 0)$, $B(0, -36)$, $C(1, -64)$, $D(3, 0)$

c $A(-1, -1)$, $B(0, -10)$, $C(1, -17)$, $D(3, -1)$

4 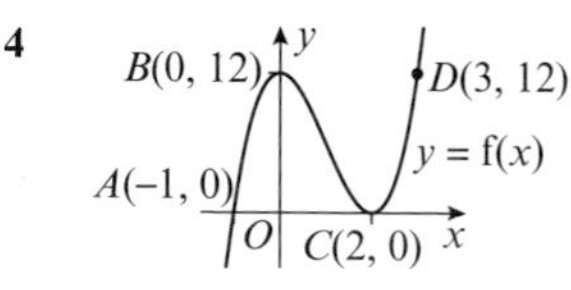

5 a 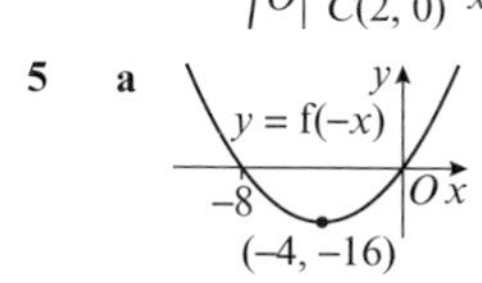

b 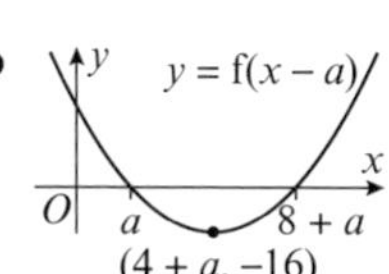

6 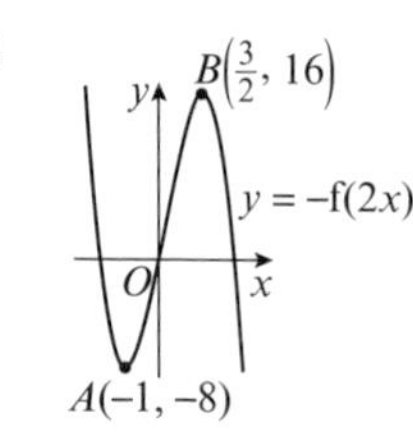

7 a $(3, 0)$

b i y 80 $y = \mathrm{f}(x - 2)$ $(4, 16)$ x O 2 5

ii 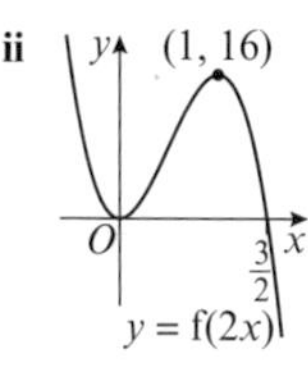

c -11

8 $\{-6\} \cup \{a : a \geqslant -2\}$

Problem solving: Set A

B a $b = -2$, $c = -5$, $d = 6$ b $(0, 6)$

S $b = -\frac{13}{2}$, $c = \frac{13}{2}$, $d = 5$

G a $m = -8$, $n = 20$, $p = -16$

b $a = -2$; $y = -2x^4 + 16x^3 - 40x^2 + 32x$

Problem solving: Set B

B a 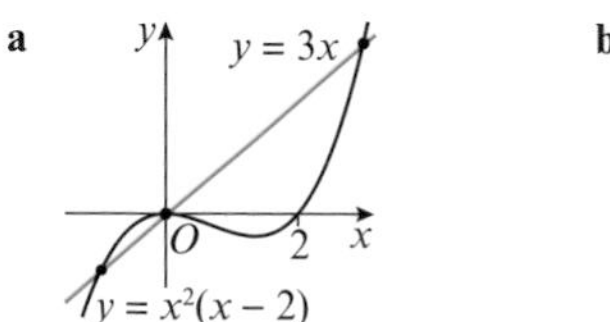

b 3

c $x^2(x - 2) = 3x \Rightarrow x^3 - 2x^2 - 3x = 0 \Rightarrow x(x + 1)(x - 3) = 0$

d $(-1, -3)$, $(0, 0)$, $(3, 9)$

S a 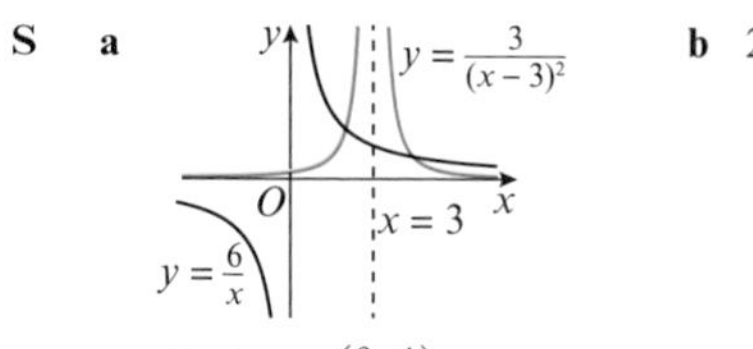

b 2

c $(2, 3)$ and $\left(\frac{9}{2}, \frac{4}{3}\right)$

G a $(4, 1)$ b $y = 5 - x$ y $y = \frac{4}{(x-2)^2}$ $(0, 5)$ $(1, 4)$ $(0, 1)$ $(4, 1)$ O $(5, 0)$ x $x = 2$

CHAPTER 5

5.1 $y = mx + c$

1 a $\frac{3}{4}$ b $-\frac{3}{2}$ c $\frac{2}{5}$

2 a $y = \frac{1}{3}x + 7$ b $x - 3y + 21 = 0$

3 $-\frac{5}{3}$

4 a $\frac{4}{5}$ b $\left(0, \frac{12}{5}\right)$ c $(-3, 0)$

5 2

6 $\left(\frac{25}{19}, \frac{52}{19}\right)$

7 a 5 b $\frac{5}{3}$

8 a $2x - 3y - 12 = 0$ b 3 c -17

9 a -10 b -10 c $x + 2y + 20 = 0$

5.2 Equations of straight lines

1 a $y = -2x + 3$ b $4y = x + 26$
c $8y = -x - 19$ d $5y = -3x - 12$

2 a $5y = -6x + 9$ b $6y = -5x + 13$
c $15y = -12x + 1$

3 $\left(-\frac{3}{5}, 0\right)$

4 $\left(0, \frac{7}{3}\right)$

5 $y = \frac{4}{5}x - \frac{42}{5}$

6 $x + 3y - 3 = 0$

7 $(12, -1)$

8 a $x - 4y - 21 = 0$ b $(1, -5)$

9 a $2x + 3y - 5 = 0$ b 1 c -2

5.3 Parallel and perpendicular lines

1 a Parallel b Not parallel
c Parallel

2 a Not perpendicular b Not perpendicular
c Perpendicular

3 $3y = 2x + 21$

4 $y + 5x = 0$

5 $4x + 3y - 31 = 0$

6 a $2 \times (-4) = 1 - 3 \times 3 = -8$ b $2x - 3y - 18 = 0$

7 a $\frac{2}{5}$ b $5x + 2y - 19 = 0$

8 a 13 b $\frac{2}{5}$
c $5x + 2y - 11 = 0$ d $\left(\frac{11}{5}, 0\right)$

9 a $y = \frac{1}{2}x + 4$ b $(8, 8)$

5.4 Length and area

1 a 10 b 25 c $4\sqrt{17}$

2 Congruent. $AB = BC = \sqrt{41}$

3 10.5

4 5 and -9

5 a $y = 2 - 4x$ b $5\sqrt{17}$

6 a $y = \frac{1}{2}x + 4$ b $x = -2, y = \frac{1}{2} \times (-2) + 4 = 3$
c $AB = 3\sqrt{5}$
d $(p - 4)^2 + \left(\frac{1}{2}p - 2\right)^2 = 36$
$\Rightarrow p^2 - 8p + 16 + \frac{1}{4}p^2 - 2p + 4 = 36$
$\Rightarrow 5p^2 - 40p - 64 = 0$

7 a $x + 3y + 3 = 0$ b $5\sqrt{10}$

8 a $3x + 4y - 7 = 0$ b $\frac{49}{24}$

9 a $y = 3 - 0.5x$ b $(4, 1)$
c $\sqrt{(4 - 0)^2 + (1 + 7)^2} = 4\sqrt{5}$ d 60

5.5 Modelling with straight lines

1 a i $k = \frac{25}{2}$ ii cost per concert ticket
iii $C = \frac{25}{2}n$

b i $k = \frac{5}{3}$ ii m s^{-1} iii $d = \frac{5}{3}t$ or $3d = 5t$

c i $k = 45$ ii miles per gallon
iii $m = 45g$

2 a linear b 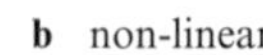 non-linear

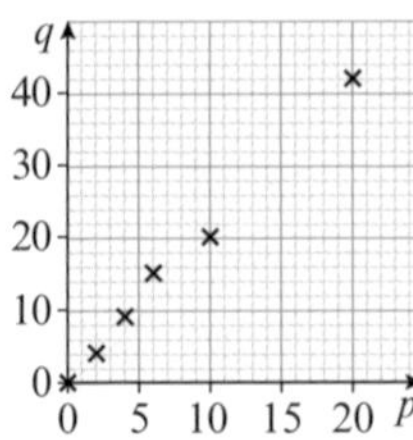

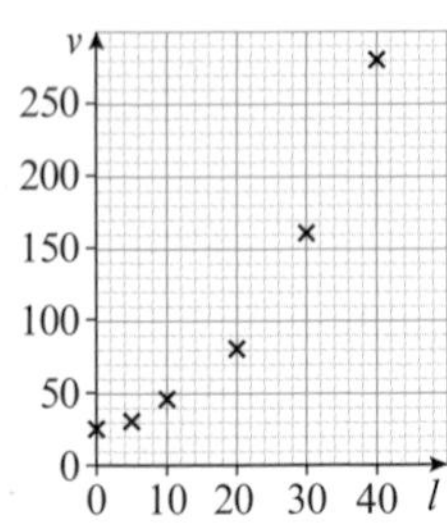

c  linear

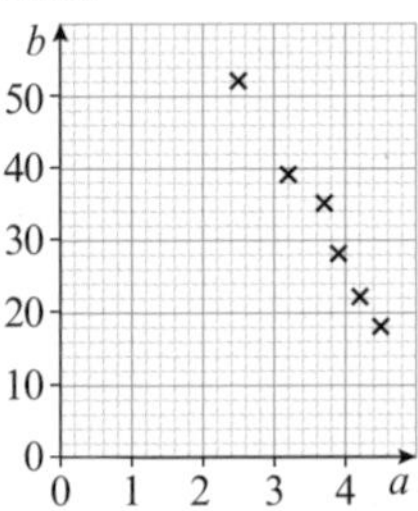

3 a 1.25 b $D = 1.25P$
c The number of US dollars per British pound.

4 a

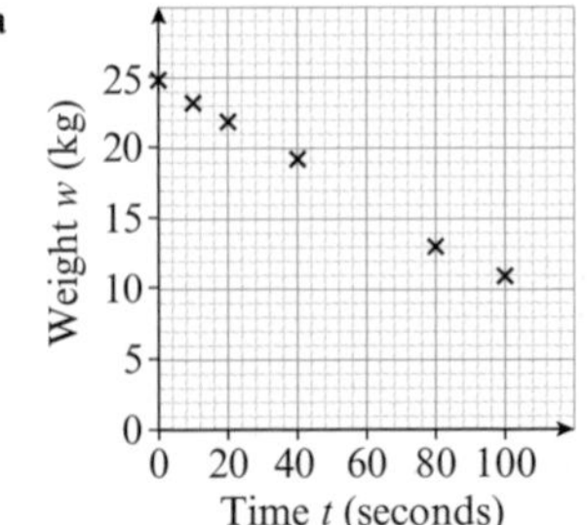

A linear model is appropriate.

b $w = -0.14t + 24.8$

Answers may vary slightly depending on which points of the graph were used.

c $a = -0.14$ is the number of kg the weight goes down by per second. $b = 24.8$ is the initial weight of the bag of sand

d 177 seconds (3 s.f.)

Answers may vary slightly depending on which points of the graph were used.

e The estimate assumes that rate of flow of sand is constant. As the bag empties, the reduced weight of sand in the bag may cause the sand to flow at a slower rate.

5 a $v = 8 + 3t$

b $a = 8\ \text{m s}^{-1}$ is the velocity when $t = 0$; $b = 3$ is the rate of change of velocity (i.e. the acceleration).

c $53\,\text{m}\,\text{s}^{-1}$

d The model suggests the speed will continue to increase indefinitely at a constant rate, which is not possible.

6 a

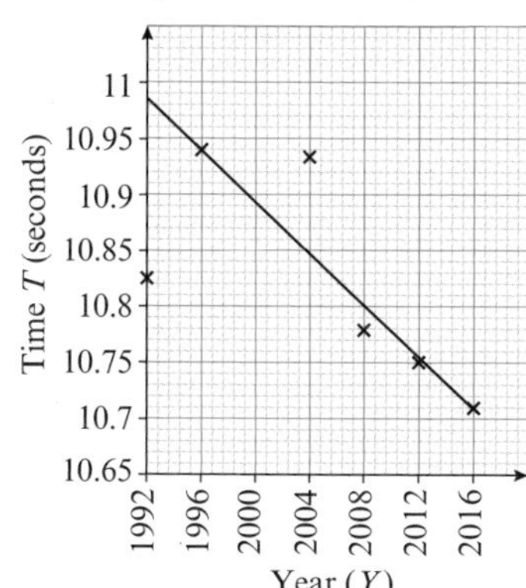

b $T = 33.894 - 0.0115Y$ c 10.66 seconds

d E.g. the model is not valid as it would eventually predict times $\leqslant 0$ to complete the race.

7 a $C = 3.6m + 2.4$

b a represents a cost of £3.60 per mile and b represents the £2.40 standing charge.

c £45.60

8 a $V = 5475 - 1095t$

b The rate of change of the value of the motorbike, i.e. a decrease of £1095 per year.

c £2737.50

d The model suggests that the value will be negative after some years ($t > 5$), which is not realistic.

Problem solving: Set A

B a 3 b $-\frac{1}{2}$ c $x + 2y - 15 = 0$

d (15, 0) e 45

S a (0, −1) b 15

G 45

Problem solving: Set B

B a $(10 - 7)^2 + (k - 1)^2 = 16 \Rightarrow 9 + k^2 - 2k + 1 = 16$
$\Rightarrow k^2 - 2k - 6 = 0$

b $k = 1 + \sqrt{7}$

S 11 and −5

G $(2 + \sqrt{14}, 2 + \sqrt{14})$

CHAPTER 6

6.1 Midpoints and perpendicular bisectors

1 a (−1, 1) b (−2, 2) c $\left(\frac{3}{5}, -\frac{1}{4}\right)$

d $\left(\frac{\sqrt{3}}{2}, -2\sqrt{5}\right)$ e $(2 + 2\sqrt{2}, 2 - 2\sqrt{2})$

2 a $x - 3y - 5 = 0$ b $x - 2y + 12 = 0$

c $4x + 3y + 26 = 0$

3 a (1, 6) b $2\sqrt{5}$

4 $a = 9, b = -1$

5 a $5y = 16 - 3x$ b $6y = x$ c $\left(\frac{96}{23}, \frac{16}{23}\right)$

6 $x + 3y - 7 = 0$

7 a (1, 2) b $4\sqrt{2}$

c $PRQS$ is a square with side length $\sqrt{(4\sqrt{2})^2 + (4\sqrt{2})^2}$

8 a 4 b 8

c $2x - 11y + 78 = 0$

6.2 Equation of a circle

1 a $(x + 5)^2 + (y - 3)^2 = 64$ b $(x - 7)^2 + (y + 8)^2 = 100$

c $(x + 1)^2 + (y + 4)^2 = 18$

2 a (−2, 9), radius 12 b (5, −2), radius $4\sqrt{2}$

c (−6, 0), radius $3\sqrt{5}$

3 Centre (−4, 6), radius $\sqrt{67}$

4 $(x - 3)^2 + (y - 14)^2 = 52$

5 10 and −4

6 a (3, −5) b $5\sqrt{2}$

7 a $(x - 1)^2 + (y - 6)^2 = 20$ b $(0, 6 + \sqrt{19}), (0, 6 - \sqrt{19})$

8 a $p = -9, q = -2$ b $\sqrt{74}$

c $x^2 + y^2 + 4x - 6y - 61 = 0$

9 a $(0, -2), k = 21$ b $5\sqrt{2}$

6.3 Intersections of straight lines and circles

1 a (8, 0) and (−4, 0)

b $(0, 4 + 4\sqrt{3})$ and $(0, 4 - 4\sqrt{3})$

2 (0, 7) and (6, −5)

3 a Quadratic reduces to $x^2 - 8x + 16 = 0$
$b^2 - 4ac = 64 - 64 = 0$, so meet at one point.

b (4, 7)

4 $b^2 - 4ac = -184 < 0$, so line does not meet circle.

5 a Circle with centre (0, 0) and radius 10

b (0, 10) and (6, −8)

6 a Substitute $y = x + k$ into the equation for the circle.

b 5 and −11

c The line is a tangent to the circle.

7 a (−2, 11) and (9, 10) b $y = 11x - 28$

c Centre is (3, 5). When $x = 3$, $y = 11 \times 3 - 28 = 5$ so perpendicular bisector passes through centre of circle.

8 $-\frac{40}{9} < k < 0$

9 a 25 b 10 and −2

6.4 Use tangent and chord properties

1 a Substituting $x = -2$ in the equation gives $y = 10$.

b $5x - 12y + 130 = 0$

2 a $(x - 8)^2 + (y - 2)^2 = 80$ b $x + 2y + 8 = 0$

3 a $y = 9x + 31$

b Centre is (−3, 4). When $x = -3$, $y = 9 \times (-3) + 31 = 4$, so perpendicular bisector passes through the centre of circle.

4 a $p = -4$; $k = 1$ or 5

b $3x + y + 9 = 0$
Centre is (−4, 3); $3(-4) + 3 + 9 = 0$

5 a $p = -6, q = -2$ b $\sqrt{13}$

c $x^2 + y^2 - 2x + 8y + 4 = 0$ d $2x - 5y - 51 = 0$

6 a Substitute $x = 8$ and $y = 4$ into equation for circle $(x + 5)^2 + (y - 9)^2 = 194$

b 51.6

7 a $y = 3x - 3$ b 6

c $(x - 3)^2 + (y - 6)^2 = 20$

d $(3 + \sqrt{2}, 6 + 3\sqrt{2})$ and $(3 - \sqrt{2}, 6 - 3\sqrt{2})$

8 a $(x - 4)^2 + (y + 1)^2 = 52$ b $y = 19 - 5x$

c 36.1

9 a $3x - y - 26 = 0$ b $x - 2y - 7 = 0$ c $(9, 1)$

6.5 Circles and triangles

1 a AB: $y = 2x + 3$, BC: $y = x - 1$, AC: $2y = 3x + 2$

b Solve $y = 2x + 3$ and $y = x - 1$ simultaneously to get $x = -4$ and $y = -5$.

These also satisfy $2y = 3x + 2$, so $(-4, -5)$ lies on all three perpendicular bisectors.

c $(x + 4)^2 + (y + 5)^2 = 130$

2 a $PR^2 = 392$, $RQ^2 = 72$, $PQ^2 = 464$, $PQ^2 = PR^2 + RQ^2$, so PQ is the hypotenuse of a right-angled triangle and the diameter of the circle.

b $(x + 2)^2 + (y - 5)^2 = 116$

3 a $AC^2 = 136$, $AB^2 = 68$, $BC^2 = 68$, $AB = AC$ and $AC^2 = AB^2 + BC^2$ so ABC is an isosceles right-angled triangle.

b $(x - 3)^2 + (y + 2)^2 = 34$

4 a $(6, 4)$ b $(x - 6)^2 + (y - 4)^2 = 100$

5 a $(5, -3)$ b 50

6 a Centre $(2, -4)$, radius $\sqrt{53}$

b E.g. $PQ^2 = (9 + 5)^2 + (-6 + 2)^2 = 212$; diameter $= 2 \times \sqrt{53}$, So diameter$^2 = 4 \times 53 = 212$

c $(0, 3)$

7 a $(-2, 3)$

b Equation can be written as $(x + 2)^2 + (y - 3)^2 = 36$
So radius is $\sqrt{36} = 6$

c 10.9 (3 s.f.)

8 a -15 b $\left(-\frac{5}{2}, -6\right)$

9 a 12 b $(x - 4)^2 + (y - 5)^2 = 65$

Problem solving: Set A

B a $x^2 + (y - 5)^2 = 100$ b $(-6, 13)$ and $\left(\frac{14}{5}, -\frac{23}{5}\right)$

S Equation of circle: $(x - 5)^2 + (y - 1)^2 = 25$

$$y = \frac{44 - 3x}{4} \Rightarrow x^2 - 10x + 25 + \left(\frac{44 - 3x - 4}{4}\right)^2 = 25$$

$$\Rightarrow x^2 - 10x + \left(\frac{1600 - 240x + 9x^2}{16}\right) = 0$$

$$\Rightarrow 16x^2 - 160x + 1600 - 240x + 9x^2 = 0$$

$$\Rightarrow 25x^2 - 400x + 1600 = 0$$

$$\Rightarrow x^2 - 16x + 64 = (x - 8)^2 = 0$$

Line meets circle at one point and is a tangent to the circle.

G $k > 19$ or $k < -63$

Problem solving: Set B

B a $(2, -3)$ b 5

c $A(-2, -6)$ and $B(6, -6)$ d $(2, -6)$

e $\angle APM = \sin^{-1}\left(\frac{4}{5}\right)$ f $AQ = 6.7$

S a $(4, -5)$ b 12.5

c $A(-8, -8.5)$ and $B(16, -8.5)$

d $\angle APB = 2 \times \sin^{-1}\left(\frac{12}{12.5}\right)$ e $AQ = 42.9$

G a $A(-4.5, -6)$ and $B(10.5, -6)$ b 15.9

CHAPTER 7

7.1 Algebraic fractions

1 a $3x^3 + 4x^2 - 2x$ b $4 - 8x + 3x^2$

c $-4x^4 + 2x^3 - x^2$

2 a $x + 3$ b $x - 1$ c $2x - 1$

3 a $\frac{2x + 1}{x + 3}$ b $\frac{3x - 2}{2x + 3}$ c $\frac{3x - 2}{x - 4}$

4 $\frac{(3x - 5)(5x + 3)}{x(3x - 5)} = \frac{5x + 3}{x} = 5 + \frac{3}{x}$

5 $k = -15$

6 $k = 2$

7 $a = 2, b = 3, c = 2, p = 2, q = -5$

7.2 Dividing polynomials

1 a $(x - 1)(2x^2 - 3x + 5)$ b $(x + 2)(3x^2 + 2x - 1)$

c $(x - 3)(2x^2 + 7x + 4)$ d $(x + 4)(4x^2 - 3x + 1)$

2 a $3x^3 + 2x^2 - 5x - 3$ b $2x^3 - 5x^2 + 3x - 2$

c $3x^3 + x^2 - 5x + 1$ d $2x^3 - 3x^2 - 7x - 3$

3 a $2x^2 + 4x - 4$ b $3x^2 - 2x + 6$

c $4x^2 + 8x + 5$ d $2x^2 - 3x + 15$

4 a -44 b 25

c -80

5 3

6 a -210 b $(x - 2)(2x + 3)(3x - 5)$

7 a $k = 15$ b $(x + 3)(2x^2 - 3x + 5)$

c One solution $x = -3$ when $x + 3 = 0$.
For $2x^2 - 3x + 5 = 0$, $b^2 - 4ac = -31 < 0$, so no other real solutions.

8 a 43

b $(x - 2)(3x^2 + 16x + 24) + 43$

9 a $(x - 1)(10x^2 - 19x - 15)$ b $(x - 1)(2x - 5)(5x + 3)$

c $x = 1, x = \frac{5}{2}, x = -\frac{3}{5}$

7.3 The factor theorem

1 a $f(-1) = 0$ b $f(-2) = 0$

c $f(3) = 0$ d $f(4) = 0$

2 a $f(2) = 0$, $(x - 2)(x + 3)(2x - 1)$

b $f(-3) = 0$, $(x + 3)(x + 5)(2x + 1)$

c $f(1) = 0$, $(x - 1)(3x - 2)(2x + 3)$

d $f(-4) = 0$, $(x + 4)(5x - 3)(3x + 2)$

3 a $(x - 1)(x - 3)(x + 6)$ b $(x + 2)(x + 5)(2x - 1)$

c $(x + 3)(x - 4)(3x + 5)$

4 a i $(x - 2)(x + 1)(2x - 9)$ b i $(x - 5)(x + 4)(2x - 1)$

ii

ii

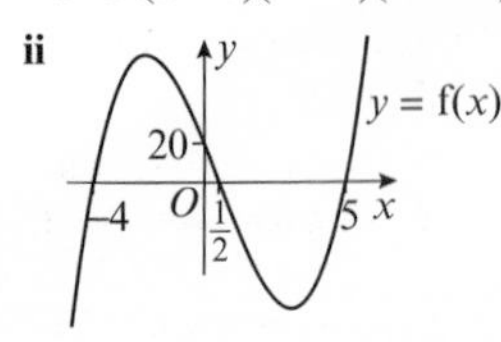

c i $(x+3)(2x+7)(3x-1)$

ii

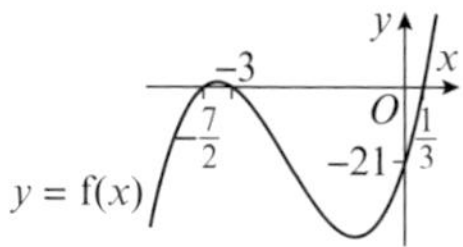

5 $x = 3, x = -\frac{3}{2}, x = \frac{4}{3}$

6 a $f(-4) = 0$ b $(x+4)(3x-2)^2$

7 $(x-1)(x-2)(2x+3)$

8 a $g(-1) = -1 + 2 + 19 + k = 0 \Rightarrow k = -20$

b $(x+1)(x-4)(x+5)$

c

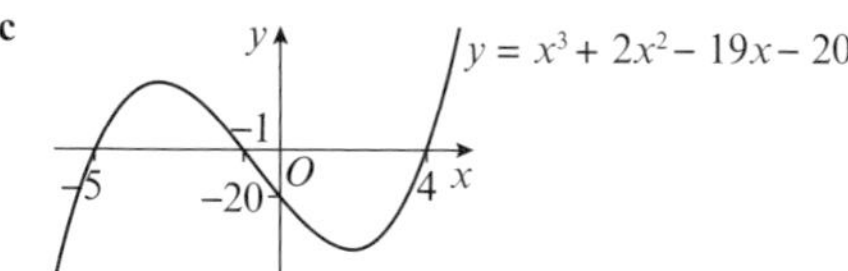

9 a $p(-3) = 0$ b $(x+3)(5x-2)^2$

c $x = -3$ and $x = \frac{2}{5}$

7.4 Mathematical proof

1 $\text{LHS} = (x-4)(2x^2+15x+18) = 2x^3+7x^2-42x-72 = \text{RHS}$

2 Using Pythagoras' theorem:
$PQ^2 = 16^2 + 12^2 = 400$, $QR^2 = 18^2 + 6^2 = 360$
$PR^2 = 2^2 + 6^2 = 40$
$QR^2 + PR^2 = 360 + 40 = 400 = PQ^2$
As $PQ^2 = QR^2 + PR^2$ using Pythagoras, triangle PQR is right-angled.

3 At point of intersection, $kx = 4x^2 - 5x + 4$, so
$4x^2 - x(5+k) + 4 = 0$
Only one real root, so $b^2 - 4ac = 0$ gives $k^2 + 10k - 39 = 0$
$\Rightarrow (k+13)(k-3) = 0 \Rightarrow k = 3$ and $k = -13$, so the lines with equations $y = 3x$ and $y = -13x$ are tangents to the curve.

4 If A, B and C lie on the same straight line, the gradients of the line segments AB, BC and AC are equal.

Gradient of $AB = \frac{1-3}{2+2} = -\frac{2}{4} = -\frac{1}{2}$

Gradient of $BC = \frac{-5-1}{14-2} = -\frac{6}{12} = -\frac{1}{2}$

As gradient of AB = gradient of BC and B is a common point, then A, B and C are collinear.

5 Completing the square,
$x^2 - 6x + 10 = (x-3)^2 - 9 + 10 = (x-3)^2 + 1$
As $(x-3)^2 \geqslant 0$ for all values of x, then $(x-3)^2 + 1 \geqslant 1$ and $x^2 - 6x + 10 > 0$ for all values of x.

6 $\text{LHS} = x^2 + 2xy + y^2 - (x^2 - 2xy + y^2) = 4xy = \text{RHS}$

7 If first odd number is $2n+1$, next consecutive odd number is $2n+3$.
$(2n+3)^2 - (2n+1)^2 = 4n^2 + 12n + 9 - (4n^2 + 4n + 1)$
$= 8n + 8 = 8(n+1)$
So the difference between the squares of two consecutive odd numbers is a multiple of 8.

8 Let odd number be $2n+1$.
$(2n+1)^3 - (2n+1)^2 = 8n^3 + 12n^2 + 6n + 1 - (4n^2 + 4n + 1)$
$= 8n^3 + 8n^2 + 2n$
As $8n^3 + 8n^2 + 2n = 2(4n^3 + 4n^2 + n)$, the difference between the cube and the square of an odd number is even.

9 No real roots means $b^2 - 4ac < 0$.
Solving $(k-3)^2 - 4 \times 1 \times (3-2k) = 0$ gives
$k^2 + 2k - 3 = 0 \Rightarrow (k+3)(k-1) = 0$
Critical values are -3 and 1, so for no real roots $-3 < k < 1$.

7.5 Methods of proof

1 a $n = 5$: $5^2 + 5 + 17 = 47$; $n = 6$: $6^2 + 6 + 17 = 59$
$n = 7$: $7^2 + 7 + 17 = 73$; $n = 8$: $8^2 + 8 + 17 = 89$
$n = 9$: $9^2 + 9 + 17 = 107$; $n = 10$: $10^2 + 10 + 17 = 127$
Proof by exhaustion, so for $5 \leqslant n \leqslant 10$, $f(n)$ is prime.

b When $n = 17$: $17^2 + 17 + 17 = 17(17 + 1 + 1)$, divisible by 17 so not prime. Counter-example, so $f(n)$ is not prime for all positive integers, n.

2 2 and 3 are both prime numbers. $3 - 2 = 1$ is not even, so the difference of two prime numbers is not always even.

3 a $(1+2x)^2 - 4x^2 - 1 = 4x > 0$ for all positive values of x.
So $(1+2x)^2 - 4x^2 - 1 > 0 \Rightarrow (1+2x)^2 > 1 + 4x^2$

b If $x = -1$, $(1+2x)^2 = 1$ and $1 + 4x^2 = 5$, counter-example, so $(1+2x)^2 > 1 + 4x^2$ is not true for all values of x.

4 a Cancelling a factor of x is not valid when $x = 0$.

b $x = 0$

5 When $x = -2$, $(-2+2)^2 = 0$ and $(-2)^2 = 4$, so $(x+2)^2 > x^2$ is not true when $x = -2$.

6 $n = 1$: $9 > 4$; $n = 2$: $25 > 8$; $n = 3$: $49 > 16$; $n = 4$: $81 > 32$
$n = 5$: $121 > 64$; $n = 6$: $169 > 128$

7 Sometimes true. For example, when $x = 4$, $4^2 = 16$ which is greater than 4, so true. But when $x = \frac{1}{2}$, $\left(\frac{1}{2}\right)^2 = \frac{1}{4}$, which is not greater than $\frac{1}{2}$, so false.

8 $(x+5)^2 - 4x - 9 = x^2 + 6x + 16 = (x+3)^2 + 7 \geqslant 0$ for all real values of x, proved by completing the square.
So $(x+5)^2 - 4x - 9 \geqslant 0 \Rightarrow (x+5)^2 \geqslant 4x + 9$

9 Proof by exhaustion:
$n = 2$: $2^2 + 2 = 6$ not divisible by 4
$n = 3$: $3^2 + 2 = 11$ not divisible by 4
$n = 4$: $4^2 + 2 = 18$ not divisible by 4
$n = 5$: $5^2 + 2 = 27$ not divisible by 4
$n = 6$: $6^2 + 2 = 38$ not divisible by 4
$n = 7$: $7^2 + 2 = 51$ not divisible by 4

Problem solving: Set A

B a -24 b $(x-4)(x^2 - 5x - 18) - 24$

c $(x+2)(x-3)(x-8)$

d Proof by exhaustion:
$n = 4$: $p(4) = 4^3 - 9 \times 4^2 + 2 \times 4 + 48 = -24 < 0$
$n = 5$: $p(5) = 5^3 - 9 \times 5^2 + 2 \times 5 + 48 = -42 < 0$
$n = 6$: $p(6) = 6^3 - 9 \times 6^2 + 2 \times 6 + 48 = -48 < 0$
$n = 7$: $p(7) = 7^3 - 9 \times 7^2 + 2 \times 7 + 48 = -36 < 0$
So for $3 < n < 8$, $x^3 - 9x^2 + 2x + 48 < 0$

S a $(2x+1)(x^2 - 8x + 20)$

b One root when $2x + 1 = 0$, i.e. $x = -\frac{1}{2}$
Using discriminant for $x^2 - 8x + 20$,
$b^2 - 4ac = 64 - 4 \times 1 \times 20 = -16 < 0$ indicates no real roots so $p(x) = 0$ has exactly one real root.

c $2x^3 - 15x^2 + 32x + 20 = (2x+1)(x^2 - 8x + 20)$
$2x + 1 > 0$ when $x > -\frac{1}{2}$, so $2x + 1 > 0$ for all real positive values of x.
Completing the square for $x^2 - 8x + 20$ gives $(x-4)^2 + 4$. As $(x-4)^2 \geqslant 0$ for all real values of x, $(x-4)^2 + 4 \geqslant 4$ for all real values of x and $(2x+1)(x^2 - 8x + 20) > 0$ for all real positive values of x.

G **a** $a = 9, b = 12, c = 4, d = -2$

b $p(x) = (x - 2)(9x^2 + 12x + 4)$. One real root when $x = 2$. Using discriminant for $9x^2 + 12x + 4$, $b^2 - 4ac = 144 - 4 \times 9 \times 4 = 0$ indicates one real root, so $p(x) = 0$ has exactly two real roots.

c

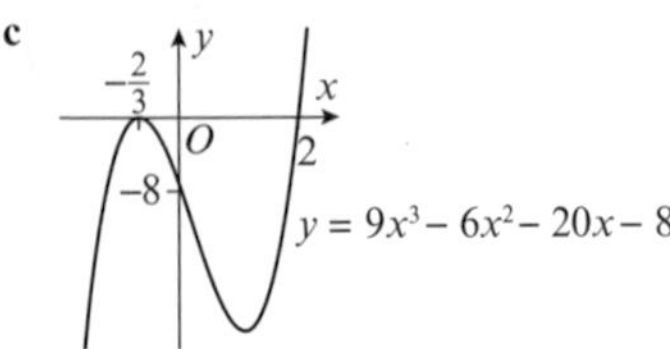

d $\{x : x < -\frac{2}{3}\} \cup \{x : -\frac{2}{3} < x < 2\}$

Problem solving: Set B

B **a** $f(-1) = -1 - 1 - p + q = 0$, which gives $q - p = 2$

b $f(-3) = -27 - 9 - 3p + q = 0$, which gives $q - 3p = 36$

c $p = -17$ and $q = -15$

d $f(x) = (x + 1)(x + 3)(x - 5)$

S **a** $f(-2) = -16 - 4 - 2p + q = 0$, gives $q - 2p - 20 = 0$

b $p = -13$ and $q = -6$

c $f(x) = (x + 2)(x - 3)(2x + 1)$

G **a** $f(2) = 8 + (p + 4) \times 4 + 16 + q = 0$, which gives $4p + q + 40 = 0$

b $f(-p) = -p^3 + p^3 + 4p^2 - 8p + q = 0$. Simplify to obtain $4p^2 - 8p + q = 0$

c $p = 5$ and $q = -60$

d $f(x) = (x - 2)(x + 5)(x + 6)$

CHAPTER 8

8.1 Pascal's triangle

1 **a** $243 - 405x + 270x^2 - 90x^3 + 15x^4 - x^5$

b $y^4 + 8xy^3 + 24x^2y^2 + 32x^3y + 16x^4$

c $243x^5 - 810x^4y + 1080x^3y^2 - 720x^2y^3 + 240xy^4 - 32y^5$

2 **a** 160 **b** −1080 **c** $-\frac{3}{2}$

3 $1 + 7x + 9x^2 - 27x^3 - 54x^4$

4 **a** $16 + 96x + 216x^2 + 216x^3 + 81x^4$

b $16 - 96x + 216x^2 - 216x^3 + 81x^4$

5 $k = \pm 5$

6 **a** $64 + 48k + 12k^2 + k^3$

b $64 - 48x + 60x^2 - 25x^3 + 15x^4 - 3x^5 + x^6$

7 $k = 1$ or $k = -\frac{3}{5}$

8.2 Factorial notation

1 **a** 120 **b** 1320 **c** 380

2 **a** 11 440 **b** 792

c 646 646 **d** 31 824

3 $p = {}^4C_2, q = {}^5C_3, r = {}^6C_3, s = {}^6C_4$

4 ${}^{15}C_7$

5 **a** 0.162 (3 s.f.) **b** 0.0130 (3 s.f.)

6 $p = 14$

7 $q = 22$

8 **a** ${}^{k+2}C_k = \frac{(k+2)!}{k!(k+2-k)!} = \frac{(k+1)(k+2)}{2}$

b $k = 7$ **c** k is a positive integer > 0

8.3 The binomial expansion

1 **a** $1 + 5x + 10x^2 + 10x^3 + 5x^4 + x^5$

b $64 + 192x + 240x^2 + 160x^3 + 60x^4 + 12x^5 + x^6$

c $10000 + 4000x + 600x^2 + 40x^3 + x^4$

2 **a** $81 - 216x + 216x^2 - 96x^3 + 16x^4$

b $1024 + 640x + 160x^2 + 20x^3 + \frac{5x^4}{4} + \frac{x^5}{32}$

c $64 - 192x + 240x^2 - 160x^3 + 60x^4 - 12x^5 + x^6$

3 **a** $1 + 10x + 45x^2 + 120x^3$

b $1 + 15x + 105x^2 + 455x^3$

c $512 + 2304x + 4608x^2 + 5376x^3$

4 **a** $1 - 5x + \frac{45x^4}{4} - 15x^3$

b $512 - 768x + 512x^2 - \frac{1792x^3}{9}$

c $6561 + 34992x + 81648x^2 + 108864x^3$

5 $32 - 240x + 720x^2 - 1080x^3$

6 $A = 531441, B = -2125764, C = 3897234$

7 $1 - 6x + 16x^2 - \frac{224x^3}{9}$

8 $x^5 - 10x^3 + 40x - \frac{80}{x} + \frac{80}{x^3} - \frac{32}{x^5}$

8.4 Solving binomial problems

1 **a** 364 **b** 15 360 **c** −21 875

2 **a** 489 888 **b** 354 375 **c** 7920

3 10 800

4 $\frac{160}{27}$

5 $n = 6$

6 $a = \pm 4$

7 0.5

8 **a** $256 - 3072x + 16128x^2 - 48384x^3$ **b** −40 320

9 **a** $729 + 1458px + 1215p^2x^2$ **b** $p = 6$

8.5 Binomial estimation

1 **a** $1 - \frac{9x}{5} + \frac{36x^2}{25} - \frac{84x^3}{125}$ **b** 0.8728

2 **a** $1 - 12x + 60x^2$

b Expand $(3 + x)(1 - 2x)^6$ up to terms in x^2

3 **a** $6561 + 4374x + \frac{5103x^2}{4} + \frac{1701x^3}{8}$

b Using $x = 0.2$, 7488.531

4 **a** $1 + \frac{9x}{4} + \frac{9x^2}{4} + \frac{21x^3}{16}$

b Using $x = 0.06$, $(1.015)^9 = 1.143$ (3 d.p.)

5 **a** $1 - 18x + 135x^2 - 540x^3$

b Expand $\left(1 + \frac{x}{2}\right)(1 - 3x)^6$ up to terms in x^2

6 **a** $128 - 224x + 168x^2$

b Substitute $x = 0.03$ in the expansion

7 **a** $59049 - 39366x + \frac{59049x^2}{5} - \frac{52488x^3}{25}$

b 58 072.198 320 **c** $1.64 \times 10^{-7}\%$ (3 s.f.)

Problem solving: Set A

B a $k = 2$ b 2160

S a $2048 - 11\,264px + 28\,160p^2x^2$

b $p = -4, q = 45\,056$

G $k = 2$

Problem solving: Set B

B a $1 - 20x + 150x^2$ b $A = 2$

c $k = 3$ d $B = 240$

S $A = 16, k = 6, B = -360$

G a $p = 12, q = \frac{1}{3}$ b $k = \frac{130}{9}$

CHAPTER 9

9.1 The cosine rule

1 13.6 cm (3 s.f.)

2 120° (3 s.f.)

3 a 2.75 (3 s.f.) b 4.08 (3 s.f.) c 5.08 (3 s.f.)

d 115 (3 s.f.) e 69.8 (3 s.f.) f 135 (3 s.f.)

4 24.1 km (3 s.f.)

5 $\cos\theta = \frac{5^2 + 6^2 - 8^2}{2 \times 5 \times 6} = -\frac{1}{20}$

6 313 m (3 s.f.)

7 Largest angle is $\angle BAC$.

$\cos A = \frac{10^2 + 7^2 - 12^2}{2 \times 10 \times 7} = \frac{5}{140} > 0$

So triangle ABC does **not** contain an obtuse angle.

8 34.4 cm (3 s.f.)

9 4

9.2 The sine rule

1 a 7.18 cm (3 s.f.) b 5.28 cm (3 s.f.) c 14.8 cm (3 s.f.)

2 a 44.2° (3 s.f.) b 42.8° (3 s.f.) c 47.6° (3 s.f.)

3 a 73.0° (3 s.f.) b 11.6 cm (3 s.f.)

4 a 74.6° and 105.4° (1 d.p.)

b

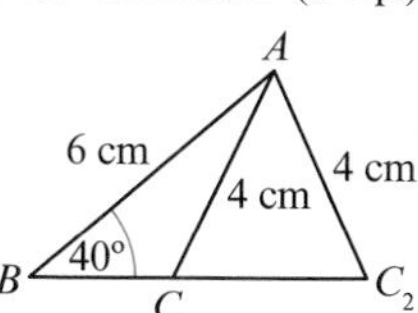

5 86.4° (1 d.p.)

6 a 0.475 b 28.39°, 151.61°

7 a 18.9 km (3 s.f.) b 057°

8 a 31.2° (3 s.f.) b 12.1 cm (3 s.f.)

c 5.62 cm (3 s.f.)

9 Using the sine rule, $\frac{2(4x - 3)}{\sqrt{3}} = \frac{4(2x + 1)}{3}$

Rearrange for x and rationalise the denominator.

9.3 Areas of triangles

1 a 11.5 cm² (3 s.f.) b 19.6 cm² (3 s.f.)

c 41.8 cm² (3 s.f.)

2 a 32.4° and 148° (3 s.f.) b 53.1° and 127° (3 s.f.)

c 8.81° and 171° (3 s.f.)

3 a $1500 = \frac{1}{2} \times 75 \times 80 \times \sin\theta \Rightarrow \sin\theta = \frac{1}{2}$

$\Rightarrow \theta = 30°$ or $150°$

b Perimeter = 305 m (3 s.f.)

4 a 20.1 cm b 132 cm²

5 a $A = \frac{18 + 3x - x^2}{5}$ cm²

b 4.05 cm², when $x = 1.5$ cm

6 5.94 cm

7 a 370 m² b Because the angles and side length are not given to 4 significant figures.

9.4 Solving triangle problems

1 a $x = 46.9, y = 68.1, z = 6.86$

b $x = 53.0, y = 12.4, z = 12.6$

c $x = 45.0, y = 10.8, z = 18.1$

2 a $x = 11.3$ b $x = 6.21$

c $x = 14.7$

3 3360 m² (3 s.f.)

4 72.1 cm² (3 s.f.)

5 a $\pm\frac{3}{5}$ b $\sqrt{433}$ cm

6 35.9 cm²

7 a 14.4 km (3 s.f.) b 057°

9.5 Graphs of sine, cosine and tangent

1 a $y = \sin\theta$ b 360°

c maximum 1, minimum −1

2 a b 360°

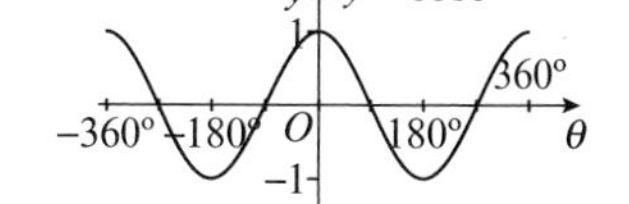

c maximum 1, minimum −1

3 a $y = \tan\theta$ b 180°

4 a −270° and 90° b −180° and 180°

c −360°, −180°, 0°, 180° and 360°

5 a $\theta = 30°$

b $y = \sin\theta$, $y = 0.5$

c $\theta = -330°, -210°, 30°, 150°$

6 a −45°

b $y = \tan\theta$, $y = -1$

c $\theta = -225°, -45°, 135°, 315°$

7 a 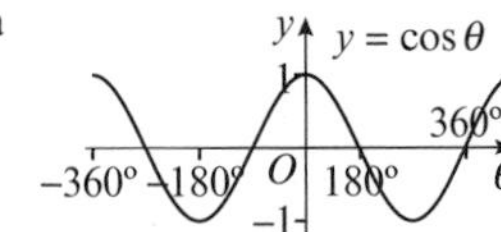

b i $\theta = \pm 315°, \pm 45°$

ii $\theta = \pm 225°, \pm 135°$

9.6 Transforming trigonometric graphs

1 a 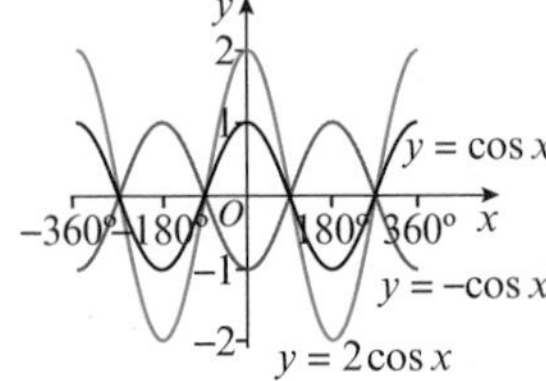

b i Stretch with scale factor 2 parallel to the y-axis

ii Reflection in the x-axis

2 a 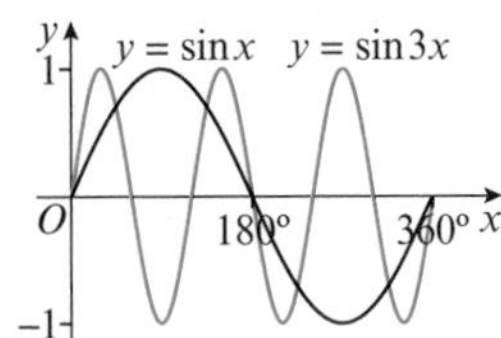

b 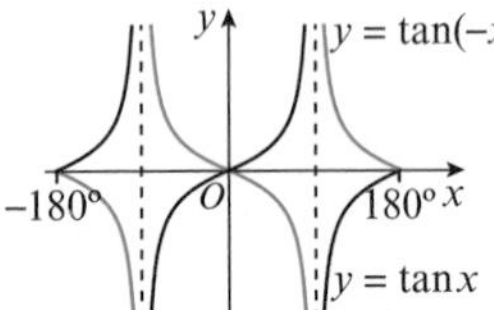

c i Stretch with scale factor $\frac{1}{3}$ parallel to the x-axis

ii Reflection in the y-axis

3 a 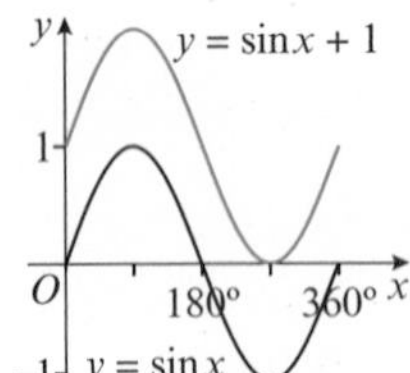

b Translation by $\begin{pmatrix} 0 \\ 1 \end{pmatrix}$

4 a 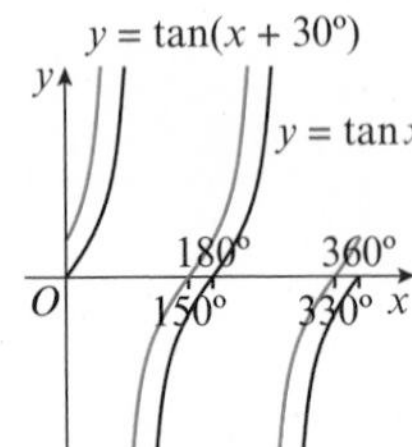

b Translation by vector $\begin{pmatrix} -30° \\ 0 \end{pmatrix}$

5 a 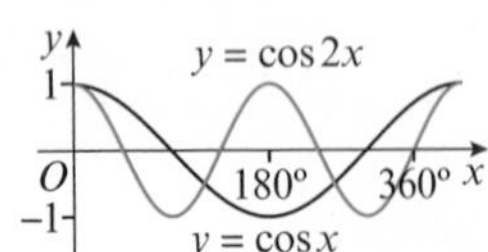

b Stretch with scale factor $\frac{1}{2}$ parallel to the x-axis

c 180°

d 45°, 135°, 225°, 315°

6 a 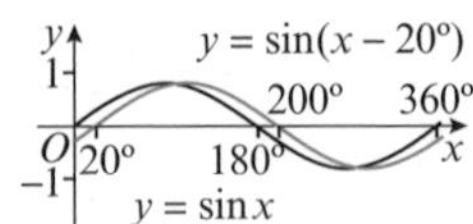

b Translation by vector $\begin{pmatrix} 20° \\ 0 \end{pmatrix}$

c i 110° ii 290°

7 a 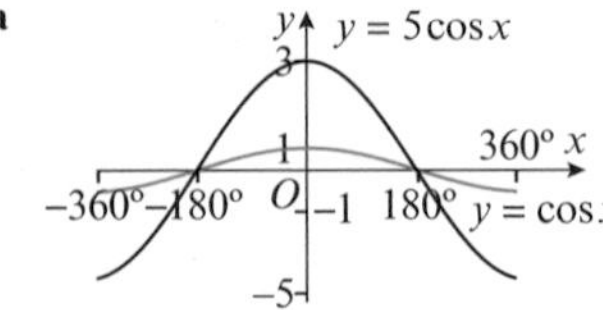

b Stretch with scale factor 5 parallel to the y-axis.

c maximum 5, minimum −5

d $-\theta$

8 a 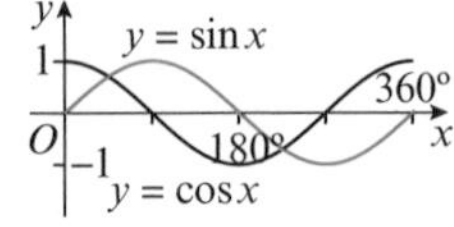

b 2

c $x = 45°$ and $x = 225°$

d i e.g. $k = -90°$

ii e.g. $k = 90°$

Problem solving: Set A

B a 31.3 cm b 269 cm^2 c 78.4 cm

S a 560 m (3 s.f.) b 541 m (3 s.f.)

G a 403 m (3 s.f.) b 269° (nearest degree)

Problem solving: Set B

B a 3 b (−180°, 2) and (180°, 2)

c $(0, A + 1)$

S a $k = 40$ b $x = -130°, 50°, 230°$

c E.g. $k = 400$. As the period of the graph is 360°, k can take any value of the form $40 \pm 360n$

G a $2k$

b 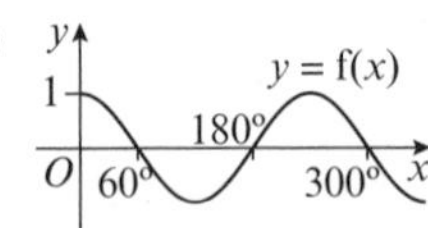

c $\frac{7}{6} \leqslant k < \frac{11}{6}$

CHAPTER 10

10.1 Angles in all four quadrants

1 a 1st and 2nd b 1st and 4th

c 1st and 3rd d 1st

2 a 2nd, negative b 3rd, positive c 1st, positive

d 3rd, positive e 4th, negative f 4th, positive

3 a 0 b −1 c 0

d −1 e −1 f −1

g $-\frac{1}{\sqrt{2}}$ h $\frac{1}{\sqrt{2}}$ i undefined

4 a $-\sin\theta$ b $\cos\theta$ c $-\tan\theta$
d $-\sin\theta$ e $\tan\theta$ f $-\cos\theta$
g $\sin\theta$ h $-\cos\theta$ i $\tan\theta$

5 a $-\sin 40°$ b $-\cos 20°$ c $\tan 30°$
d $-\sin 80°$ e $-\tan 50°$ f $\cos 80°$
g $\sin 85°$ h $-\cos 20°$ i $\tan 65°$

10.2 Exact values of trigonometric ratios

1 a $\sin 30° = \frac{1}{2}$ b $-\sin 45° = -\frac{1}{\sqrt{2}}$
c $-\sin 45° = -\frac{1}{\sqrt{2}}$ d $-\sin 60° = -\frac{\sqrt{3}}{2}$
e $-\cos 30° = -\frac{\sqrt{3}}{2}$ f $\cos 60° = \frac{1}{2}$
g $-\cos 45° = -\frac{1}{\sqrt{2}}$ h $-\cos 60° = -\frac{1}{2}$
i $\tan 45° = 1$ j $\tan 30° = \frac{\sqrt{3}}{3}$
k $\tan 45° = 1$ l $\tan 60° = \sqrt{3}$

2 $3 + \sqrt{2}$

3 $10^2 = x^2 + (2x)^2 - 2 \times x \times 2x \cos 30°$

$\Rightarrow 100 = 5x^2 - 4x^2 \times \frac{\sqrt{3}}{2} \Rightarrow 100 = x^2(5 - 2\sqrt{3})$

$\Rightarrow x^2 = \frac{100}{5 - 2\sqrt{3}} = \frac{100(5 + 2\sqrt{3})}{13}$

10.3 Trigonometric identities

1 a 1 b 6 c 4

2 a $2\tan x$ b $3\tan x$ c $\tan^2 x$

3 $\text{LHS} \equiv \frac{\sin^2 x - \cos^2 x}{1 - \sin^2 x} \equiv \frac{\sin^2 x - \cos^2 x}{\cos^2 x}$

$\equiv \frac{\sin^2 x}{\cos^2 x} - \frac{\cos^2 x}{\cos^2 x} \equiv \tan^2 x - 1 = \text{RHS}$

4 a $-\frac{\sqrt{55}}{8}$ b $-\frac{3}{\sqrt{55}}$

5 a 6 b $(2 - 3\sin\theta)(2 + 3\sin\theta)$ or $4 - 9\sin^2\theta$

6 Using $\sin^2 x \equiv 1 - \cos^2 x$,
$4\cos^2 x - 3(1 - \cos^2 x) = 2 \Rightarrow 7\cos^2 x = 5$

7 $\text{LHS} \equiv \frac{\cos^2 x - \sin^2 x}{\cos^2 x - \sin x \cos x} \equiv \frac{\cos x + \sin x}{\cos x}$

$\equiv \frac{\cos x}{\cos x} + \frac{\sin x}{\cos x} \equiv \text{RHS}$

8 a $\frac{2}{3}$ b $\frac{\sqrt{5}}{3}$

9 a Use $\sin 45° = \frac{1}{\sqrt{2}}$ and the sine rule, $\frac{\sin P}{10} = \frac{\sin 45°}{9}$
b $-\frac{\sqrt{31}}{9}$

10.4 Simple trigonometric equations

1 a $-36.9°$ (1 d.p.) b $x = 216.9°, x = 323.1°$ (1 d.p.)

2 $\theta = 45°, 315°$

3 $x = 48.6°, 131.4°$ (1 d.p.)

4 $\theta = 95.7°, 275.7°$ (1 d.p.)

5 a Line 3: Divided both sides by $\sin x$ and lost the solution(s) for $\sin x = 0$.
Line 6: Has not identified the negative solution for $\cos x = \frac{5}{9}$
b $x = -56.3°, 0.0°, 56.3°$ (1 d.p.)

6 a $\frac{5}{3}$ b $59.0°, 239.0°$

7 $\pm 30°, \pm 150°$

8 a Use $\sin^2 x \equiv 1 - \cos^2 x$: $4\cos^2 x - 3(1 - \cos^2 x) = 1$
$\Rightarrow 4\cos^2 x - 3 + 3\cos^2 x = 1$
b $x = 40.9°, 139.1°, 220.9°$ or $319.1°$

9 a $x = -63.4°, 14.5°, 116.6°, 165.5°$ (1 d.p.)
b $x = 80.4°$ or $279.6°$ (1 d.p.)

10.5 Harder trigonometric equations

1 a $10°, 110°, 130°, 230°, 250°, 350°$
b $90°, 270°$ c $60°, 150°, 240°, 330°$

2 a $\theta = 138.5°, 341.5°$ (1 d.p.)
b $\theta = -53.6°, 173.6°$ (1 d.p.)
c $\theta = 103.7°, 283.7°$ (1 d.p.)

3 a $\theta = -170°, -110°, -50°, 10°, 70°, 130°$
b $\theta = 17.0°, 62.0°, 107.0°, 152.0°, 197.0°, 242.0°, 287.0°, 332.0°$ (1 d.p.)
c $\theta = -98°, 82°$ (nearest degree)

4 a $x = 39.3°, 80.7°, 219.3°, 260.7°$ (1 d.p.)
b $x = 31.2°, 75.5°, 151.2°$ (1 d.p.)
c $x = 27.1°, 72.1°, 117.1°, 162.1°$ (1 d.p.)
d $x = -156.8°, -66.8°, 23.2°, 113.2°$ (1 d.p.)

5 a $x = 40°, 80°, 160°, 200°, 280°, 320°$
b $x = 245°, 335°$

6 a $x = 15.5°, 105.5°, 195.5°, 285.5°$
b $x = 20°, 50°, 140°, 170°$

7 a $23.1°, 96.9°$ b $152°, 208°$

8 $x = -54.4°, -18.4°, 17.6°, 53.6°$ or $89.6°$ (1 d.p.)

9 a $\frac{\sin 2x}{\cos 2x} = 4\sin 2x \Rightarrow \sin 2x = 4\sin 2x\cos 2x$
$\Rightarrow \sin 2x - 4\sin 2x\cos 2x = 0$
b $\sin 2x = 0$ gives $x = 0°, 90°, 180°$
$\cos 2x = \frac{1}{4}$ gives $x = 37.8°, 142.2°$

10.6 Equations and identities

1 a $\theta = 19.5°, 161°, 199°, 341°$
b $\theta = 60°, 120°, 240°, 300°$
c $\theta = 0°, 11.5°, 168°, 180°, 360°$
d $\theta = 71.6°, 104°, 252°, 284°$
e $\theta = 0°, 70.5°, 289°, 360°$
f $\theta = 21.1°, 38.9°, 81.1°, 98.9°, 141°, 159°, 201°, 219°, 261°, 279°, 321°, 339°$

2 a $\theta = 0°, 60°, 120°, 180°, 240°, 300°, 360°$
b $\theta = 0°, 78.7°, 180°, 259°, 360°$
c $\theta = 270°$
d $\theta = 45.5°, 150°, 225°, 330°$

3 a $2\cos^2 x - (1 - \cos^2 x) - 3\cos x + 2 = 0$
$\Rightarrow 3\cos^2 x - 4\cos x + 1 = 0$
b $x = 0°, 70.5°, 289°, 360°$

4 a $\theta = 11.5°, 90°, 168.5°$
b $\theta = 48.2°, 60°, 300°, 311.8°$
c $\theta = 30°$ or $150°$

5 a $2\sin^2 x - 3(1 - \sin^2 x) = 1 \Rightarrow 5\sin^2 x = 4$

b $x = 63.4°, 116.6°, 243.4°, 296.6°$

6 $x = 70.5°, 109.5°$

7 $x = 30°, 150°$

8 a $\frac{\sin\theta\sin\theta}{\cos\theta} = 2\cos\theta + 3 \Rightarrow \sin^2\theta = 2\cos^2\theta + 3\cos\theta$

$\Rightarrow 3\cos^2\theta + 3\cos\theta - 1 = 0$

b 74.7°, 285.3°

9 $\theta = 430.5°$ or $438.5°$

Problem solving: Set A

B a $\frac{5}{2}$ b $\theta = 48.2°, 228.2°$

S $x = 6.5°, 53.5°, 126.5°, 173.5°$

G $x = -103.2°, -61.8°, 76.8°, 118.2°$

Problem solving: Set B

B a $\frac{3\sin^2 x + \cos^2 x}{\cos^2 x} = \frac{3\sin^2 x}{\cos^2 x} + \frac{\cos^2 x}{\cos^2 x}$.

Use $\tan^2 x \equiv \frac{\sin^2 x}{\cos^2 x}$ and rearrange equation.

b $x = 49.1°, 130.9°, 229.1°, 310.9°$

S a $4(1 - \sin^2 x) + 9\sin x - 6 = 0 \Rightarrow 4\sin^2 x - 9\sin x + 2 = 0$

b $x = 14.5°, 165.5°, 374.5°, 525.5°$

G $x = 0°, 37.3°, 82.7°$

CHAPTER 11

11.1 Vectors

1 a 2**b** b −2**a** c −**b**
d **a** + **b** e **a** − **b** f −2**a** + 2**b**
g 2**a** − **b** h 2**a** − **b**

2 a −**a** + **b** b **a** − **b** + **d**
c −**b** + **c** d **a** − **c** + **d**

3 a **a** + **b** b $\frac{1}{2}\mathbf{a} + \frac{1}{2}\mathbf{b}$ c $\frac{1}{2}\mathbf{a} - \frac{1}{2}\mathbf{b}$

4 $-\frac{4}{5}$

5 $-\frac{7}{20}\mathbf{a} - \frac{2}{5}\mathbf{b}$

6 $\overrightarrow{AC} = \mathbf{a} + \mathbf{b}$, $\overrightarrow{DE} = \frac{1}{2}\mathbf{a} + \frac{1}{2}\mathbf{b} = \frac{1}{2}\overrightarrow{AC}$, so the two lines are parallel as one vector is a scalar multiple of the other.

7 a $\overrightarrow{MN} = \frac{1}{2}\mathbf{b} + k\mathbf{a} - \frac{1}{2}\mathbf{c}$ and $\overrightarrow{MN} = -\frac{1}{2}\mathbf{b} + \mathbf{a} + \frac{1}{2}\mathbf{c}$

Adding gives $2\overrightarrow{MN} = k\mathbf{a} + \mathbf{a}$, so $\overrightarrow{MN} = \left(\frac{1+k}{2}\right)\mathbf{a}$

b $\overrightarrow{EH}$, $\overrightarrow{FG}$ and $\overrightarrow{MN}$ are all scalar multiples of **a**, so are parallel to each other.

11.2 Representing vectors

1 a −21**i** + 9**j** b −11**i** + 13**j** c 13**i** − 18**j**

2 a $\begin{pmatrix}-1\\2\end{pmatrix}$ b $\begin{pmatrix}1\\-2\end{pmatrix}$ c $\begin{pmatrix}12\\-13\end{pmatrix}$

3 −3**i** + 2**j**

4 19**i** − 8**j**

5 $p = -19, q = -8$

6 $2a + 5b = 11\mathbf{i} + 4\mathbf{j}$, $2d - c = 5\mathbf{i} + 2\mathbf{j}$, so the vectors are not parallel.

7 $p = 7, q = -1$

11.3 Magnitude and direction

1 a 10 b $\sqrt{41}$ c $3\sqrt{5}$ d $8\sqrt{2}$

2 a 63.4° above (1 d.p) b 171.9° below (1 d.p)
c 90° above d 45° below

3 $-\frac{5}{13}\mathbf{i} + \frac{12}{13}\mathbf{j}$

4 $25\sqrt{2}\mathbf{i} - 25\sqrt{2}\mathbf{j}$

5 $p = 3.76, q = 10.3$ (3 s.f.)

6 $p = \pm 7$

7 a 38.7° (1 d.p.) b 74.1° (1 d.p.) c 112.7° (1 d.p.)

8 $p = -6$ and $q = \pm 2.5$

11.4 Position vectors

1 a −2**i** + 3**j** b 7**i** c 9**i** − 3**j**

2 a 6**i** − 2**j** b 2**i** + 4**j** c 5**i** − 5**j**

3 a 10 b $3\sqrt{10}$ c $\sqrt{34}$

4 a $\begin{pmatrix}-4\\5\end{pmatrix}$ b $\sqrt{41}$

5 $\frac{22}{5}\mathbf{i} + \frac{29}{5}\mathbf{j}$ and −2**i** − 7**j**

11.5 Solving geometric problems

1 a **q** − **p** b $\overrightarrow{PN} = \frac{1}{4}\overrightarrow{PQ}$
c $\overrightarrow{ON} = \overrightarrow{OP} + \overrightarrow{PN}$ d $\frac{3}{4}\mathbf{p} + \frac{1}{4}\mathbf{q}$

2 $\frac{3}{5}\mathbf{a} + \frac{2}{5}\mathbf{b}$

3 $\overrightarrow{XY} = -\frac{3}{10}\mathbf{a} + \frac{2}{5}\mathbf{b}$

4 78.7°

5 a $\overrightarrow{KL} = \mathbf{b} - \mathbf{a}$, $\overrightarrow{PR} = \frac{1}{2}\mathbf{b} - \frac{1}{2}\mathbf{a}$, $\overrightarrow{PR} = \frac{1}{2}\overrightarrow{KL}$

b Any two vectors, where one is a scalar multiple of the other, are parallel.

6 a i $\frac{1}{3}\mathbf{a} + \mathbf{b}$ ii −**a**

b $\overrightarrow{AM} = \overrightarrow{AW} + \overrightarrow{WM} = \overrightarrow{AW} + \lambda\overrightarrow{WY} = \frac{1}{3}\mathbf{a} + \lambda\left(\frac{1}{3}\mathbf{a} + \mathbf{b}\right)$
$= \frac{1}{3}(1 + \lambda)\mathbf{a} + \lambda\mathbf{b}$

c $\overrightarrow{AM} = \overrightarrow{AB} + \overrightarrow{BX} + \overrightarrow{XM} = \mathbf{a} + \frac{1}{2}\mathbf{b} - \mu\mathbf{a} = (1 - \mu)\mathbf{a} + \frac{1}{2}\mathbf{b}$

d Equating coefficients of **a** and **b**, $\mu + \frac{1}{3}\lambda = \frac{2}{3}$ and $\lambda = \frac{1}{2}$
So $\mu = \frac{2}{3} - \frac{1}{6} = \frac{1}{2} \Rightarrow M$ is the midpoint of WY and XZ.

7 a $\overrightarrow{PQ} = \frac{k}{1+k}\overrightarrow{OA} - \frac{k}{1+k}\overrightarrow{BA} = \frac{k}{1+k}\overrightarrow{OB}$. So $\overrightarrow{PQ}$ is a scalar multiple of $\overrightarrow{OB}$ and these vectors are parallel.

b $k = \frac{1}{2}$

11.6 Modelling with vectors

1 a $13\,\text{m s}^{-1}$ b $10\,\text{km s}^{-1}$ c $\sqrt{37}\,\text{mm s}^{-1}$ d $14\,\text{cm s}^{-1}$

2 a $5\,\text{m s}^{-1}$ b 20 m c 5.5 seconds

3 $\left(\frac{3}{2}\mathbf{i} - \frac{9}{2}\mathbf{j}\right)\,\text{km s}^{-2}$

4 a $(9\mathbf{i} - 16\mathbf{j})\,\text{m s}^{-1}$ b $\sqrt{337}\,\text{m s}^{-1}$

5 a (−3**i** + 6**j**) km b $\sqrt{137}$ km c 110.0° (1 d.p.)

6 a 56.3° (1 d.p) below the positive x-axis

b Solve $3 + p = 2k$ and $q - 7 = -3k$ by eliminating k.

c $3\sqrt{13}$ N

7 16.5 km²

8 a 76.0° b $a = 11, b = 25$ c $c = -8$

Problem solving: Set A

B a $7\mathbf{i} - 7\mathbf{j}$ b $AB = 13, AC = 13, BC = 7\sqrt{2}$ c 44.8°

S a $7\mathbf{i} + \mathbf{j}$ b 44.0° c 13.5

G 26

Problem solving: Set B

B a $\overrightarrow{OR} = \overrightarrow{OA} + \overrightarrow{AR} = \mathbf{a} + \frac{3}{5}(-\mathbf{a} + \mathbf{b}) = \frac{2}{5}\mathbf{a} + \frac{3}{5}\mathbf{b}$

$\overrightarrow{OP} = 5\overrightarrow{OR} = 5\left(\frac{2}{5}\mathbf{a} + \frac{3}{5}\mathbf{b}\right) = 2\mathbf{a} + 3\mathbf{b}$

b $\overrightarrow{QB} = \mathbf{a} + \mathbf{b}$, $\overrightarrow{BP} = 2\mathbf{a} + 2\mathbf{b}$, $\overrightarrow{QB} = \frac{1}{2}\overrightarrow{BP}$ so lines are parallel. As B is a shared point, Q, B and P are collinear.

S a $\frac{3}{5}$ b $\overrightarrow{OC} = 3\mathbf{b}$

G Let $\overrightarrow{AB} = \mathbf{a}$ and $\overrightarrow{AD} = \mathbf{b}$. Let point of intersection of PQ and MC be R, and let $MR = kMC$ and $PR = jPQ$. So going via M,

$\overrightarrow{AR} = \frac{1}{2}\mathbf{b} + k\left(\frac{1}{2}\mathbf{b} + \mathbf{a}\right) = k\mathbf{a} + \left(\frac{1}{2}k + \frac{1}{2}\right)\mathbf{b}$

And going via P,

$\overrightarrow{AR} = \frac{4}{5}\mathbf{a} + j\left(\mathbf{b} - \frac{2}{5}\mathbf{a}\right) = \left(\frac{4}{5} - \frac{2}{5}j\right)\mathbf{a} + j\mathbf{b}$

Equating coefficients of $\mathbf{a}$ and $\mathbf{b}$, $k = \frac{4}{5} - \frac{2}{5}j$ and $j = \frac{1}{2}k + \frac{1}{2}$ which gives $k = \frac{1}{2}$

CHAPTER 12

12.1 Gradients of curves

1 a For example:

x-coordinate	0	1	1.5	2
Estimate for gradient of curve	−1	2	5.75	11

b Gradient $= 3p^2 - 1$ c −0.25

2 a 4

b i 5 ii 4.5 iii 4.1 iv 4.01 v $4 + h$

12.2 Finding the derivative

1 a $f'(4) = \lim_{h\to 0}\frac{f(4+h) - f(4)}{h} = \lim_{h\to 0}\frac{12(4+h) - 12(4)}{h}$

$= \lim_{h\to 0}\frac{12h}{h} = 12$

b $f'(-2) = \lim_{h\to 0}\frac{f(-2+h) - f(-2)}{h}$

$= \lim_{h\to 0}\frac{12(-2+h) - 12(-2)}{h} = \lim_{h\to 0}\frac{12h}{h} = 12$

c $f'(0) = \lim_{h\to 0}\frac{f(0+h) - f(0)}{h} = \lim_{h\to 0}\frac{12(0+h) - 12(0)}{h}$

$= \lim_{h\to 0}\frac{12h}{h} = 12$

2 a $f'(3) = \lim_{h\to 0}\frac{f(3+h) - f(3)}{h} = \lim_{h\to 0}\frac{2(3+h)^2 - 2(3)^2}{h}$

$= \lim_{h\to 0}\frac{12h + 2h^2}{h} = \lim_{h\to 0}(12 + h) = 12$

b $f'(-1) = \lim_{h\to 0}\frac{f(-1+h) - f(-1)}{h}$

$= \lim_{h\to 0}\frac{2(-1+h)^2 - 2(-1)^2}{h}$

$= \lim_{h\to 0}\frac{-4h + 2h^2}{h} = \lim_{h\to 0}(-4 + h) = -4$

c $f'(0) = \lim_{h\to 0}\frac{f(0+h) - f(0)}{h} = \lim_{h\to 0}\frac{2(0+h)^2 - 2(0)^2}{h}$

$= \lim_{h\to 0}\frac{2h^2}{h} = \lim_{h\to 0}(2h) = 0$

3 $\frac{dy}{dx} = \lim_{h\to 0}\frac{8(x+h) - 8x}{h} = \lim_{h\to 0}\frac{8h}{h} = 8$

4 $\frac{dy}{dx} = \lim_{h\to 0}\frac{3(x+h)^2 - 3x^2}{h} = \lim_{h\to 0}\frac{6xh + 3h^2}{h}$

$= \lim_{h\to 0}(6x + 3h) = 6x$

5 $\frac{dy}{dx} = \lim_{h\to 0}\frac{(x+h)^3 - x^3}{h} = \lim_{h\to 0}\frac{x^3 + 3x^2h + 3xh^2 + h^3 - x^3}{h}$

$= \lim_{h\to 0}\frac{3x^2h + 3xh^2 + h^3}{h} = \lim_{h\to 0}(3x^2 + 3xh + h^2) = 3x^2$

6 $\frac{dy}{dx} = \lim_{h\to 0}\frac{(6 - 3(x+h)^2) - (6 - 3x^2)}{h}$

$= \lim_{h\to 0}\frac{6 - 3(x^2 + 2xh + h^2) - 6 + 3x^2}{h} = \lim_{h\to 0}\frac{-6xh - 3h^2}{h}$

$= \lim_{h\to 0}(-6x - 3h) = -6x$

At $x = 1$, $\frac{dy}{dx} = -6(1) = -6$

12.3 Differentiating x^n

1 a $5x^4$ b $-6x^{-7}$ c $\frac{1}{2}x^{-\frac{1}{2}}$ d $-\frac{1}{4}x^{-\frac{5}{4}}$

2 a $20x^3$ b $-10x^{-3}$ c $-\frac{3}{4}x^{-4}$ d $-6x^{-\frac{5}{3}}$

3 a 54 b 150 c 1.5 d $\frac{27}{8}$

4 $-\frac{15}{16}$

5 $\frac{1}{16}$

6 $\frac{3}{2}$

7 $\frac{8}{9}$

12.4 Differentiating quadratics

1 a $6x - 7$ b $\frac{3}{2}x + \frac{1}{2}$ c $-10x$

2 a −5 b 24 c $-\frac{1}{2}$

3 (3, 13)

4 −10

5 −2

6 a (2, −1) and (3, 1) b 1 and 3

7 a $\left(-\frac{4}{5}, 7\right)$ and (2, 7) b −14 and 14

12.5 Differentiating functions with two or more terms

1 a $5 - 14x$ b $20x^{\frac{3}{2}} - 24x^3$

c $20x^{\frac{3}{2}} - 8x - 2x^{-\frac{2}{3}}$

2 a $\frac{1}{2}x^{-\frac{1}{2}} - \frac{1}{2}x^{-\frac{3}{2}}$ b $-\frac{2}{7}x^{-3} + \frac{1}{4}x^{-\frac{3}{4}}$

3 a $6x^2 - 12x + 5$ b $3x^2 + \frac{4}{x^2}$ c $12x - 8$

4 a 16 b −10

5 −3

6 $p = \frac{5}{2}$

7 $p = 9, q = -4$

8 a $1 + 12x + 60x^2$

b i $12 + 120x$ ii $-12 + 120x$

12.6 Gradients, tangents and normals

1 a $y = -4x$ b $y = 11x + 8$
c $y = 17x - 28$ d $y = -\frac{7}{6}x - 4$

2 a $y = -\frac{3}{7}x + \frac{109}{14}$ b $y = \frac{4}{97}x - \frac{4575}{97}$

3 a $y = -4x + 11$ b $y = \frac{1}{4}x + \frac{27}{4}$

4 a $y = 5x - 21$ b $y = -x + 1$ c $\left(\frac{11}{3}, -\frac{8}{3}\right)$

5 a $y = -x + 2$ b $(2, 0)$
c $x = 2 \Rightarrow g(x) = 8 - 4 \times 2 = 0$, so A is on the curve $y = g(x)$

6 $\left(11, -\frac{43}{2}\right)$

7 a $3x - 16y + 28 = 0$ b $32x + 6y - 143 = 0$
c $\frac{265}{6}$

12.7 Increasing and decreasing functions

1 a $\left(-1, -\frac{1}{2}\right)$, $(0, 0)$ and $\left(1, -\frac{1}{2}\right)$
b i Decreasing ii Increasing iii Decreasing
iv Neither v Increasing vi Neither

2 a $x \geqslant \frac{4}{5}$ b $x \leqslant -\frac{3}{2}$
c $x \leqslant -2$ or $x \geqslant 2$ d $-\sqrt{3} \leqslant x \leqslant \sqrt{3}$

3 a $x \leqslant -\frac{7}{6}$ b $x \geqslant -\frac{3}{5}$
c All values of x d $x \geqslant \frac{3}{2}$

4 $f'(x) = 5x^4 + 6x^2 + 8$, $5x^4 \geqslant 0$ and $6x^2 \geqslant 0$ for all values of x, therefore $5x^4 + 6x^2 + 8 \geqslant 8 > 0$ for all values of x and therefore $f(x)$ is increasing for all real values of x.

5 $f'(x) = -6x^2 - 36x - 54 = -6(x + 3)^2$
$-6(x + 3)^2 \leqslant 0$ for all values of x.

6 $q \geqslant 0$

7 a $9x^2 + 4x - 5$ b $(-\infty, -1] \cup \left[\frac{5}{9}, \infty\right)$

12.8 Second order derivatives

1 a $18x - 6$, 18
b $-3x^2 - 4x^{-2}$, $-6x + 8x^{-3}$
c $-2x^{-\frac{3}{2}} + 12x^{\frac{1}{2}}$, $3x^{-\frac{5}{2}} + 6x^{-\frac{1}{2}}$

2 a $-18x^{-4} - 2x^3 - 9$, $72x^{-5} - 6x^2$
b $-2x^{-\frac{5}{4}} - \frac{1}{6}x^{-\frac{5}{6}}$, $\frac{5}{2}x^{-\frac{9}{4}} + \frac{5}{36}x^{-\frac{11}{6}}$
c $-5x^4 + 9x^2 + 10x$, $-20x^3 + 18x + 10$

3 a $16x^3 - \frac{1}{3}x^{-\frac{2}{3}} - 3x^{-4} - 2x$
b $48x^2 + \frac{2}{9}x^{-\frac{5}{3}} + 12x^{-5} - 2$

4 $-\frac{121}{4}$

5 a $1 - 12x + 48x - 64x^3$ b $96 - 384x$

6 $a = -3$, $b = 5$

7 a $p = -\frac{2}{5}$, $q = \frac{3}{10}$ b $-\frac{12}{5}x + \frac{3}{5}$

12.9 Stationary points

1 a

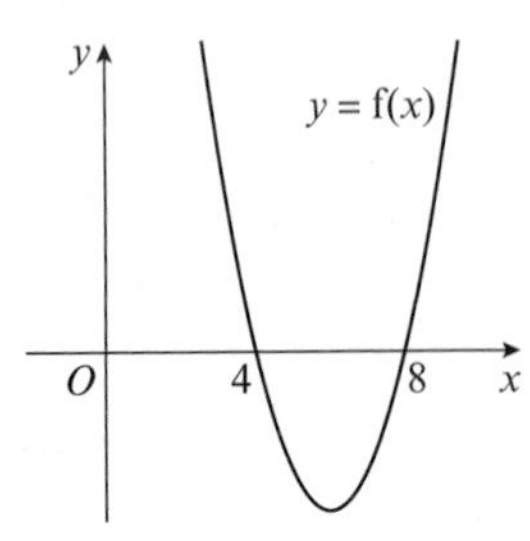

b The curve $y = f(x)$ changes from being a decreasing function to an increasing function between $x = 4$ and $x = 8$, so the curve has a local minimum in this interval.
c $2x - 12$ d $x = 6$ e $(6, -4)$

2 a $(3, -1)$ b $\left(\frac{3}{2}, -\frac{53}{4}\right)$ c $\left(-\frac{1}{3}, \frac{2}{3}\right)$

3 a $(-8, 72)$ b $\left(\frac{3}{2}, 9\right)$ c $\left(-\frac{15}{7}, \frac{351}{7}\right)$

4 a $(-1, -11)$ and $(2, 43)$ b $-24x + 12$
c $x = -1 \Rightarrow \frac{d^2y}{dx^2} > 0$, so $(-1, -11)$ is a local minimum.
$x = 2 \Rightarrow \frac{d^2y}{dx^2} < 0$, so $(2, 43)$ is a local maximum.

5 a $(-2, 2)$
b $q''(x) = 0$ and the gradient of the curve at $x = -2.1$ and $x = -1.9$ are both negative.
c

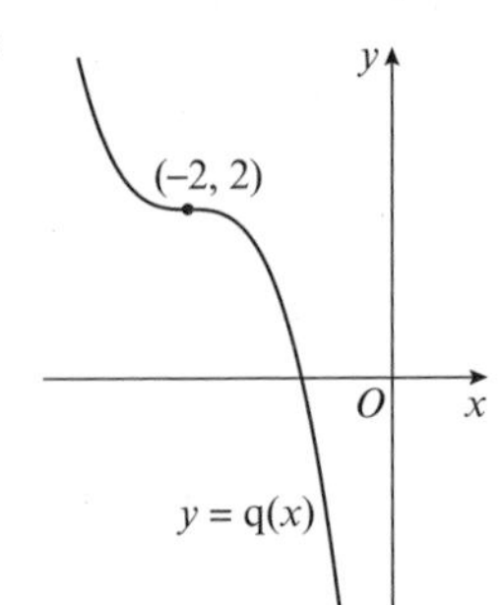

6 $\left(-\frac{1}{3}, \frac{119}{54}\right)$ and $\left(\frac{1}{2}, \frac{13}{8}\right)$

7 a $(9, 18)$ b $-x^{-\frac{3}{2}} + 108x^{-3}$
c $g''(9) = \frac{1}{9} > 0$, so it is a local minimum.

8 $(2, 6)$

9 a $(1, 5)$; point of inflection
b

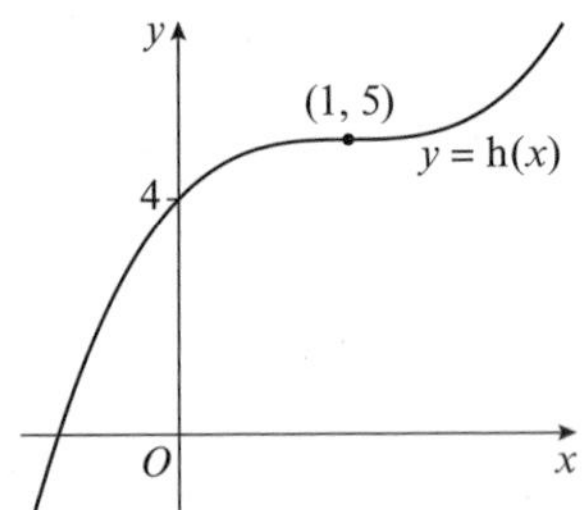

12.10 Sketching gradient functions

1 a $(1, 0)$ and $\left(\frac{7}{2}, 5\right)$ b Positive gradient
c Negative gradient d Negative gradient
e

y
y = f'(x)
O
1
7/2
x

2 a The gradient is negative for all values of x.
As x increases, the gradient moves closer to zero.

b

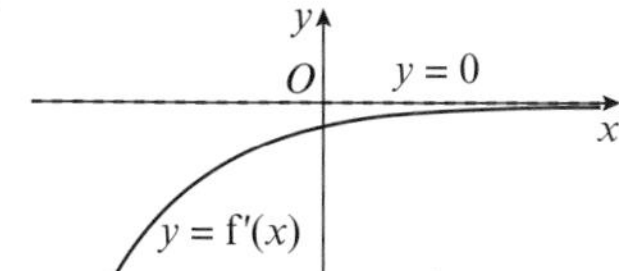

3 a

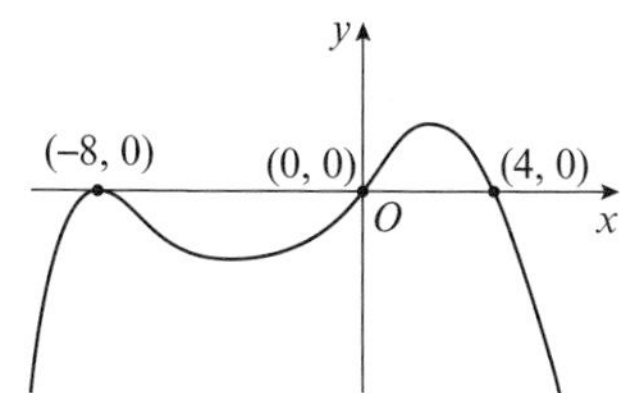

b

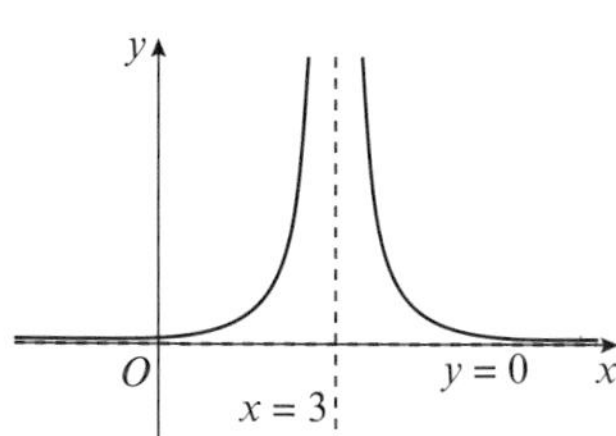

c

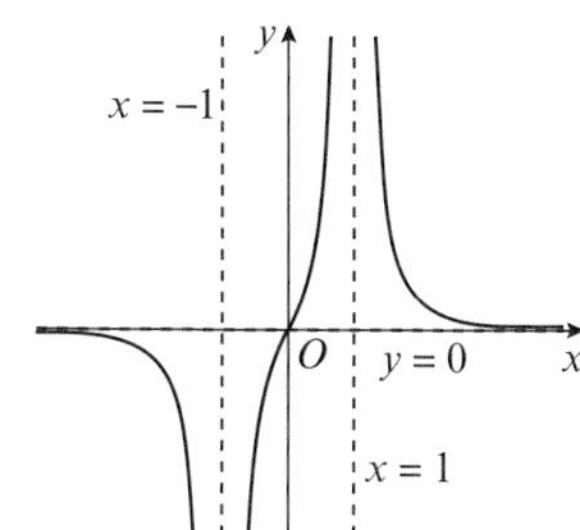

4

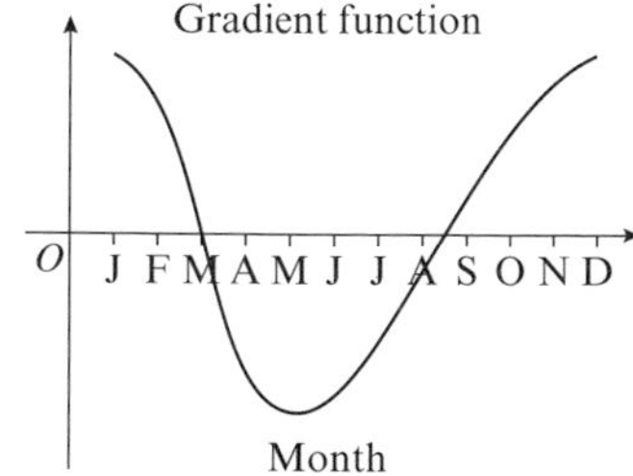

12.11 Modelling with differentiation

1 $12x$

2 a $-40\,000t^{-3}$ b -40

3 a 64 seconds b $-\frac{1}{2}t^{-\frac{1}{2}}$ c $-\frac{1}{6}\text{m s}^{-2}$

4 a $4\pi r^2$ b 64π cm³ per cm

5 a $y = 40 - x(1 + \sqrt{2})$

b $A = 2xy + x^2 = 2x(40 - x(1 + \sqrt{2})) + x^2 = 80x - x^2(1 + 2\sqrt{2})$ c $\frac{40}{1 + 2\sqrt{2}}$

6 a Substitute $r = \sqrt{\frac{1024}{h}}$ into $S = 2\pi r^2 + 2\pi rh$ and simplify.

b 384π cm³

7 a Use $2x^2 + 3xy = 608$ and $V = x^2y$ to write V in terms of x.

b $V = 1360.1$ cm³

c Draw a sketch of the graph or find $\frac{d^2V}{dx^2} = -4x$. Interpret sketch or conclude that as $x > 0$, $\frac{d^2V}{dx^2} < 0$, so it is a maximum.

8 a $\pi r^2 + 2\pi rh + 2\pi r^2 = 800\pi \Rightarrow h = \frac{800 - 3r^2}{2r}$

$V = \pi r^2 h + \frac{2}{3}\pi r^3 = \pi r^2\left(\frac{800 - 3r^2}{2r}\right) + \frac{2}{3}\pi r^3$

$= \frac{\pi r(2400 - 5r^2)}{6}$

b 10 600 cm³

Problem solving: Set A

B a (3, 23) b Local maximum

S a (−2, 0), (0, 4) b $x = -2$ and $x = 1$

c (−2, 0) is a minimum and (0, 4) is a maximum

G a $\left(1, -\frac{52}{3}\right), \left(4, -\frac{56}{3}\right)$

b $\left(1, -\frac{52}{3}\right)$ is a maximum and $\left(4, -\frac{56}{3}\right)$ is a minimum

Problem solving: Set B

B a $D(-2, 0)$, $E(1, 0)$, $F(2, 0)$

b $h'(x) = 3x^2 - 2x - 4$

c i $y = -3x + 3$ ii $y = 4x - 8$

d $\left(\frac{11}{7}, -\frac{12}{7}\right)$

S a $g'(x) = 3x^2 - 2x - 2$

b i $y = -\frac{1}{3}x - \frac{1}{3}$ ii $y = -\frac{1}{6}x + \frac{1}{3}$

c (−4, 1)

G 9

CHAPTER 13

13.1 Integrating x^n

1 a $3x^3 + c$ b $-6x^{-2} + c$

c $\frac{12}{7}x^{\frac{7}{2}} + c$ d $25x + c$

2 a $\frac{2}{3}x^{\frac{3}{2}} + c$ b $\frac{9}{2}x^{\frac{2}{3}} + c$

c $4x^{\frac{5}{4}} + c$ d $-\frac{1}{3}x^{\frac{1}{2}} + c$

3 a $\frac{1}{5}x^5 + \frac{1}{4}x^4 - \frac{5}{2}x^2 + 6x + c$

b $\frac{4}{3}x^3 - \frac{1}{3}x^{-1} + \frac{5}{2}x^{-2} + c$ c $4x^{\frac{3}{4}} - 8x^{\frac{1}{4}}$

4 a $16x^2 + 24x + 9$ b $\frac{16}{3}x^3 + 12x^2 + 9x + c$

5 $3x^3 - 15x^2 + 25x + c$

6 $x - 3x^2 + 3x^3 + c$

7 a 2 b $16x - 80x^2 + 200x^3 - 250x^4 + 125x^5 + c$

8 a $16 - 16\sqrt{x} + 4x$ b $16x - \frac{32}{3}x^{\frac{3}{2}} + 2x^2 + c$

9 a $6561 - 34\,992x + 81\,648x^2$

b $6561x - 17\,496x^2 + 27\,216x^3 + c$

13.2 Indefinite integrals

1 a $\frac{1}{5}x^5 + c$ b $-x^7 + c$ c $-4x^{-1} + c$

2 a $\frac{5}{3}x^3 - \frac{9}{2}x^{-2} + c$ b $\frac{1}{8}x^6 + x^{-4} + 4x^2 + c$

c $\frac{8}{5}x^{\frac{5}{2}} + 20x^{-\frac{1}{2}} + c$

3 a $36x - 40x^{\frac{3}{2}} + \frac{25}{2}x^2 + c$ b $\frac{18}{5}x^{\frac{5}{2}} - 8x^{\frac{3}{2}} + 8x^{\frac{1}{2}} + c$

c $-5x^{-1} - \frac{7}{2}x^2 - \frac{8}{5}x^{-\frac{5}{2}} + c$

4 **a** $2Px^{\frac{1}{2}} - \frac{2}{5}Rx^{\frac{5}{2}} + c$ **b** $-\frac{1}{2}Ax^{-2} - \frac{1}{4}Bx^{-4} + c$

c $\frac{3}{4}Mx^{\frac{4}{3}} - \frac{4}{7}Nx^{\frac{7}{4}} + c$

5 $\frac{180}{7}x^{\frac{7}{5}} - 50x^{\frac{6}{5}} + 25x + c$

6 $\frac{3}{4}x^4 - \frac{14}{3}x^{\frac{3}{2}} + 6x^{\frac{2}{3}} + 2x^{-1} + c$

7 $a = 3, b = 2$

8 **a** $65\,536 - 262\,144x + 245\,760x^2$

b $65\,536x - 131\,072x^2 + 81\,920x^3 + c$

13.3 Finding functions

1 **a** $y = x^4 - \frac{10}{3}x^3 + 15$ **b** $y = 10x^{\frac{1}{2}} + \frac{1}{6}x^3 - \frac{3946}{6}$

2 **a** $y = -x^{-3} + x^{-1} + 4$ **b** $y = \frac{1}{3}x^3 - 5x^2 + 25x - \frac{44}{3}$

3 $16x^{\frac{1}{2}} - \frac{8}{3}x^{\frac{3}{2}} - 6$

4 $y = 10x - 3x^2 - \frac{4}{3}x^3 + 18$

5 $f(x) = -10x^{-\frac{1}{2}} - 6x^{\frac{1}{2}} + 24$

6 $y = 3x^2 - 6x^{\frac{2}{3}} - 56$

7 **a** $p'(x) = \frac{1 - 15x + 75x^2 - 125x^3}{x^{\frac{3}{2}}}$

$= x^{-\frac{3}{2}} - 15x^{-\frac{1}{2}} + 75x^{\frac{1}{2}} - 125x^{\frac{3}{2}}$

b $p(x) = -2x^{-\frac{1}{2}} - 30x^{\frac{1}{2}} + 50x^{\frac{3}{2}} - 50x^{\frac{5}{2}} + 8$

8 $-37 - \frac{1}{2}\sqrt{2}$

13.4 Definite integrals

1 **a** $\frac{992}{3}$ **b** $\frac{135}{2}$ **c** $\frac{3}{8}$ **d** 10

2 **a** 25.5 **b** 446 **c** 2385

3 $\frac{13}{8}$

4 $\frac{1359}{8}$

5 161.5

6 25

7 **a** $y = \left(2x^{\frac{1}{3}} - \frac{1}{2}x^{-\frac{1}{3}}\right)^2 = 4x^{\frac{2}{3}} + \frac{1}{4}x^{-\frac{2}{3}} - 2$ **b** $\frac{1223}{20}$

8 -9

13.5 Areas under curves

1 $\frac{4}{3}$

2 **a** (2, 0) **b** $\frac{45}{2}$

3 $\frac{75}{8}$

4 **a** $A(-2, 0), B(6, 0)$ **b** $\frac{256}{3}$

5 $\frac{27}{4}$

6 **a** $A(4, 0)$ **b** $\frac{64}{3}$

7 3

13.6 Areas under the x-axis

1 $\frac{125}{6}$

2 **a** 2 and 4 **b** $\frac{16}{3}$

3 **a** $\frac{8}{3}$ **b** $\frac{5}{12}$ **c** $\frac{37}{12}$

4 **a** (4, 0) **b** $\frac{64}{3}$

5 **a** $A(3, 0), B(10, 0)$ **b** $\frac{343}{12}$

6 **a** $A(-4, 0), B(1, 0)$ **b** $\frac{131}{4}$

7 **a** $f(2) = 8 + 24 - 8 - 24 = 0$

b $f(x) = (x - 2)(x + 2)(x + 6)$ **c** 128

13.7 Areas between curves and lines

1 **a** $A(-4, 6), B(4, 6)$ **b** $\frac{80}{3}$

c 48 **d** $\frac{64}{3}$

2 **a** $P(1, 3), Q(4, 0)$ **b** 9

c $\frac{9}{2}$ **d** $\frac{9}{2}$

3 **a** $\frac{554}{15}$ **b** $\frac{194}{15}$

4 **a** $A(3, 5), B(7, 3)$ **b** $\frac{16}{3}$

5 $\frac{31}{30}$

6 **a** $5x + \frac{16}{x^2} - 21 = 0$, then multiply by x^2 to get

$5x^3 - 21x^2 + 16 = 0$

b $\left(4, \frac{1}{4}\right)$ **c** $\frac{27}{8}$

7 **a** $-18 - 9x = -x^3 - 2x^2 + 16x + 32$

$\Rightarrow x^3 + 2x^2 - 25x - 50 = 0$

b $(5, -21)$ **c** 123.86 (2 d.p.)

Problem solving: Set A

B **a** $f'(x) = -5x^{\frac{3}{2}} + 7x^{\frac{1}{6}}$

b $f(x) = -2x^{\frac{5}{2}} + 6x^{\frac{7}{6}} - 14$

S 113.3

G $f(x) = x^3 - \frac{5}{2}x^2 - 8x^{-1} + 17$

Problem solving: Set B

B **a** $A(-4, 9), B(4, 9)$ **b** $\frac{128}{3}$

S $\frac{32}{3}$

G **a** $3x + 6 = -x^3 - 2x^2 + 4x + 8 \Rightarrow x^3 + 2x^2 - x - 2 = 0$

b $\frac{37}{12}$

CHAPTER 14

14.1 Exponential functions

1 **a** $\frac{8}{27}$ **b** $\frac{1}{16}$ **c** 25 **d** $\frac{27}{64}$

2 **a**

x	−3	−2	−1	0	1	2	3
y	0.58	0.69	0.83	1	1.2	1.44	1.73

b

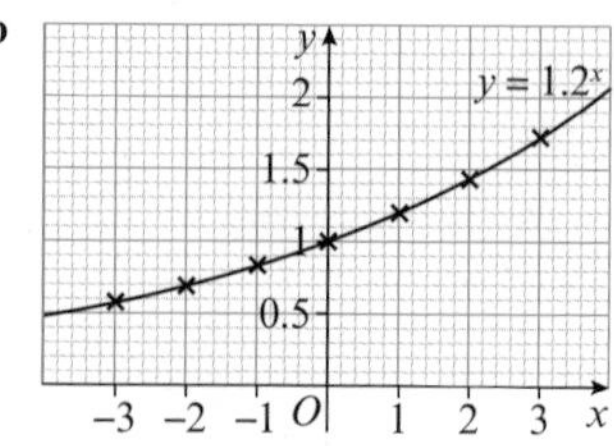

3 **a**

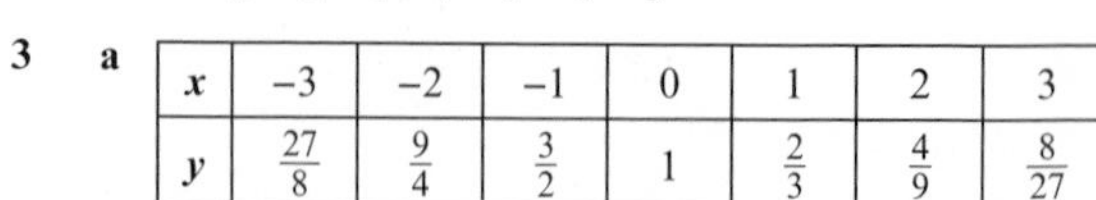

x	−3	−2	−1	0	1	2	3
y	$\frac{27}{8}$	$\frac{9}{4}$	$\frac{3}{2}$	1	$\frac{2}{3}$	$\frac{4}{9}$	$\frac{8}{27}$

b

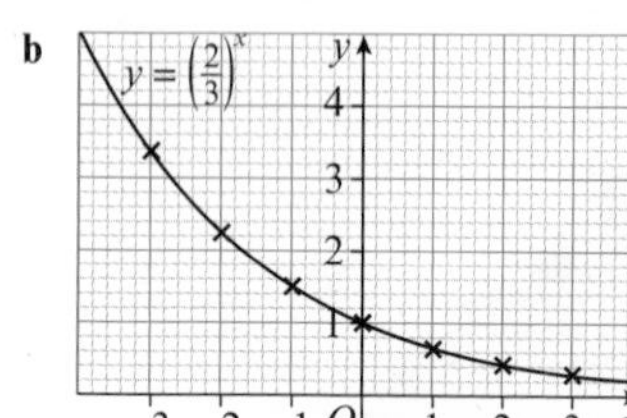

c $x = -1.7$

4 a
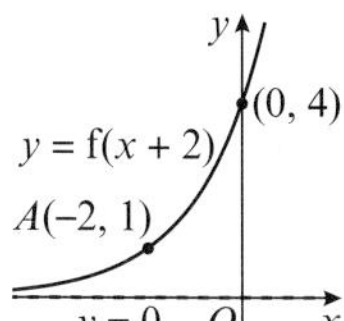

b
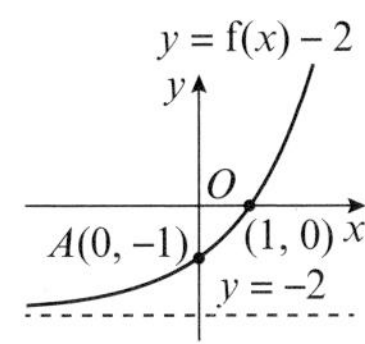

c
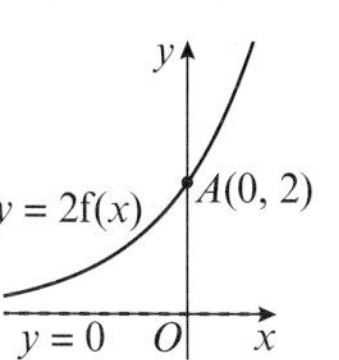

d
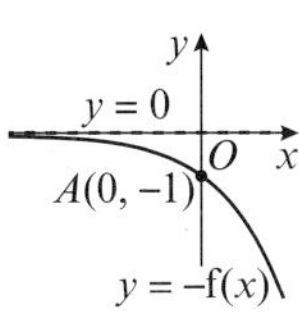

5 a
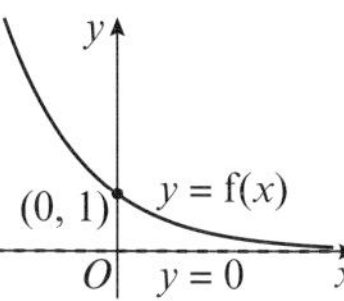

b i $y = f(x + 3) - 4$
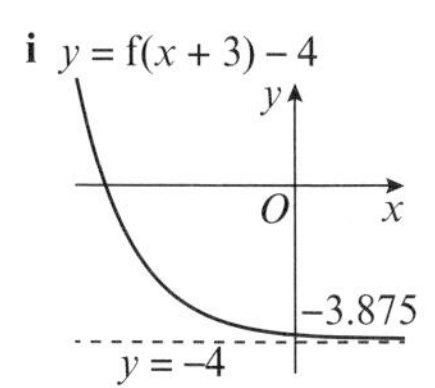

ii
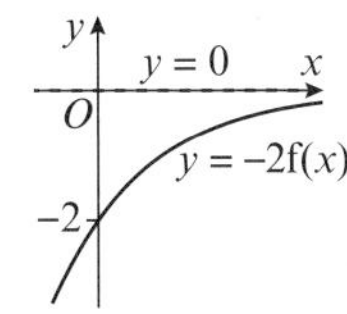

6 a $y = g(-x)$
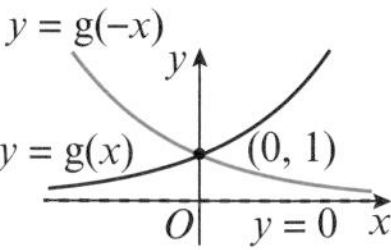

b
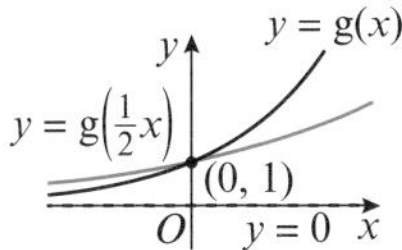

7 $k = 8, a = \frac{1}{4}$

14.2 $y = e^x$

1 a
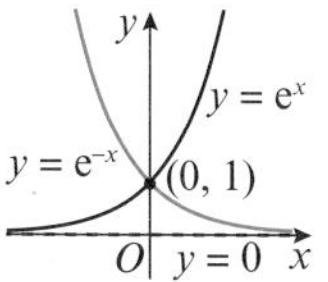

b
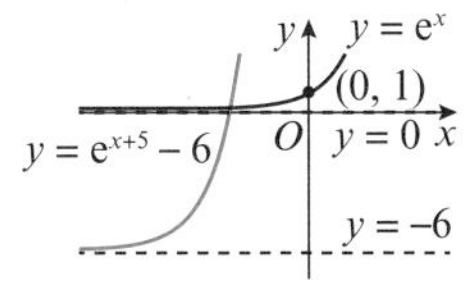

c
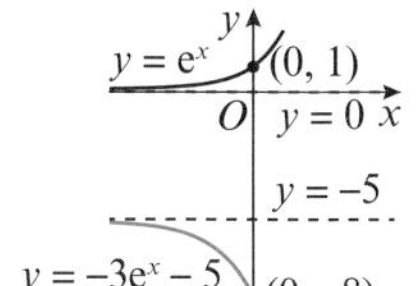

d
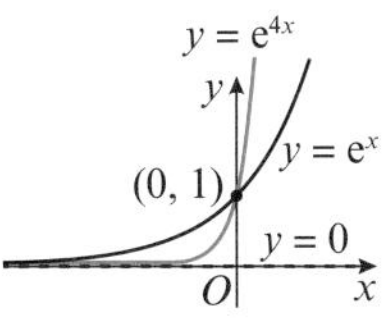

e
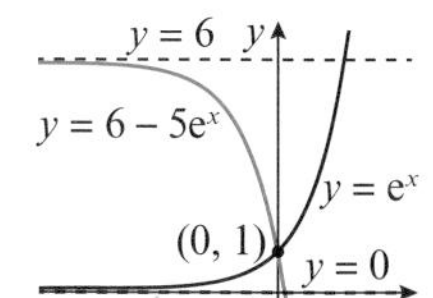

f
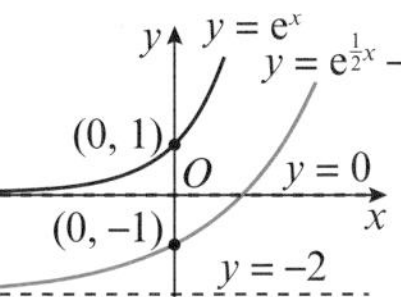

2 a $\frac{1}{2}e^{\frac{1}{2}x}$ b $-5e^{-5x}$ c $-24e^{6x}$ d $\frac{7}{3}e^{\frac{1}{3}x}$

3 a $6e^{2x} + 4e^{4x}$ b $-12e^{-2x} - 5e^{-x}$

c $e^x - 8e^{2x}$ d $-6e^{-3x} - 2e^{2x}$

4 a $A = e^4, b = 5$ b $A = e^{-1}, b = 7$

c $A = e^3, b = -\frac{1}{2}$

5 a $16x + \frac{6}{e^{6x}}$ b $32e^{4x} + 6e^{-x}$

6 $\frac{5}{2\sqrt{x}} - \frac{20}{x^3} + 8e^{-2x}$

7 a $e^{2(3x-1)} = e^{6x-2} = e^{6x} \times e^{-2} = e^{-2}(e^{6x})$,

so $A = e^{-2}, b = 6$

b
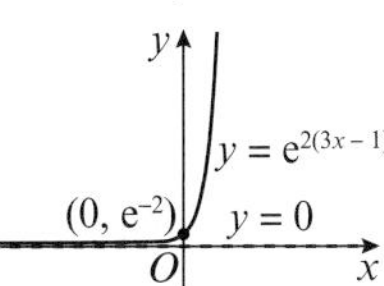

8 $y = 4e^2x - e^2$

14.3 Exponential modelling

1 a $\frac{dP}{dt} = 200e^{0.01t}$ b 221.0 (1 d.p.)

2 a
32000 V
$V = 32000e^{-0.08t}$
O t

b $\frac{dV}{dt} = -2560e^{-0.08t}$

c −516.9 (1 d.p.)

d The slope is negative, so the value of the car is decreasing.

3 a i 1300 ii 1669 iii 1776

b It is the initial deer population. c $\frac{dD}{dt} = 162.5e^{\frac{1}{8}t}$

d Substitute $t = 8$ into $\frac{dD}{dt} = 162.5e^{\frac{1}{8}t}$

$162.5e^1 = 441.72\ldots = 440$ (2 s.f.)

e $\frac{dD}{dt} \approx 3\,579\,300$ which is not a reasonable value.

4 a £14 115.82

b i −2737.46 ii −1693.90

c The value decreases over the first 6 years of ownership. The decrease happens more quickly at first and then more slowly.

5 a $P_0 = 18000$ b 594

c For large values of t, the population will be increasing at an increasingly rapid rate, making it very unlikely the model is valid.

6 a 101.6 mg (1 d.p.) b $\frac{dP}{dt} = -0.15P$, so $k = -0.15$

c k is negative, so the amount of paracetamol in the body is decreasing.

7 a £150 000 b £228 294 c $9000e^{0.06t}$

d $t = 8$, $\frac{dV}{dt} = 14\,544.67$ and $t = 9$, $\frac{dV}{dt} = 15\,444.06$. During this time the property prices are increasing by £15 000 per year, so the conditions are right for a crash.

14.4 Logarithms

1 a $\log_3 1000 = 10$ b $2^3 = 8$ c $\log_2 49 = 7$

d $\log_6 \frac{1}{216} = -3$ e $4^{-4} = \frac{1}{256}$ f $9^1 = 9$

g $\log_{25} 5 = \frac{1}{2}$ h $27^{\frac{1}{3}} = 3$

2 a 2 b −4 c 1

d $\frac{1}{3}$ e −2 f 0

3 a 81 b 42 c 9 d 4

4 a 1.7959 b −0.5108 c 2.1133

5 $\frac{33}{2}$

6 256

7 $\frac{1}{625}$

8 $\frac{2}{3}$

14.5 Laws of logarithms

1 a $\log_7 16$ b $\log_3 8$ c $\log_5 42$
d $\log_4 8000$ e $\log_7\left(\frac{64}{81}\right)$ f $\log_6\left(\frac{12}{5}\right)$

2 a 2 b 2 c 3
d −1 e $\frac{2}{3}$

3 a 3 b $2\log_a x + 3\log_a y$
c $3\log_a x - 4\log_a z$ d $\log_a x + \frac{1}{2}\log_a y - 2\log_a z$
e $\frac{1}{3} + \frac{2}{3}\log_a x + \frac{1}{3}\log_a y + \frac{4}{3}\log_a z$
f $3 + 4\log_a x + \log_a y - \frac{1}{2}\log_a z$

4 a 9 b 288 c 12 d 7

5 $\frac{5}{8}$

6 $5\sqrt[3]{5}$

7 $\frac{7}{17}$

8 a $\log_2\frac{x+5}{(x-1)^2} = 1$ b $x = 3$
$x + 5 = 2x^2 - 4x + 2$
$2x^2 - 5x - 3 = 0$

14.6 Solving equations using logarithms

1 a 1.92 b 3.87 c 2.01

2 a −3.75 b 4.10 c −0.0581

3 a 1, 1.11 b 1, 1.95 c 0.431, 0.365

4 a $(3u + 4)(3u - 5) = 0$
$u = -\frac{4}{3}, u = \frac{5}{3}$
b $3^{2x+2} - 3(3^x) - 20 = 0$
$3^2(3^{2x}) - 3(3^x) - 20 = 0$
$9(3^x)^2 - 3(3^x) - 20 = 0$
$9u^2 - 3u - 20 = 0$
c $u = \frac{5}{3} \Rightarrow 3^x = \frac{5}{3} \Rightarrow x = \log_3\frac{5}{3} = 0.465$
$u = -\frac{4}{3} \Rightarrow 3^x = -\frac{5}{3}$
This has no real solution as you cannot take the logarithm of a negative number

5 a 10.3 b 1.41 c 1.89

6 a 4.04 b 0.613

7 a 19.6 b 0.257

8 a b 3.70

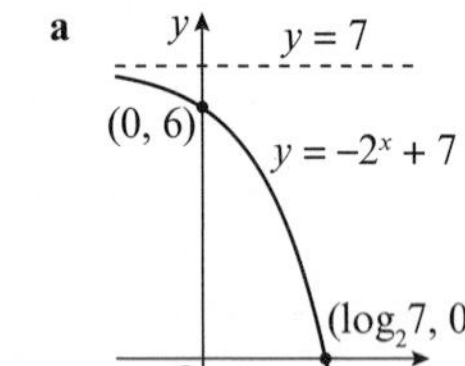

9 a $9^x - 3^{x+1} - 10 = 0$ b 1.46
$(3^2)^x - 3(3^x) - 10 = 0$
$(3^x)^2 - 3(3^x) - 10 = 0$
$u^2 - 3u - 10 = 0$

14.7 Working with natural logarithms

1 a $\ln 8$ b $2\ln 6$ or $\ln 36$ c $-5 + \ln 81$
d $\frac{1}{5} + \frac{1}{5}\ln 23$ e $1 - \ln\frac{29}{4}$ f $\frac{1}{3}\ln\frac{21}{2}$

2 a e^{11} b $\frac{1}{6}e^3$ c $\frac{3e - 1}{5e}$
d $\frac{5 - e^{\frac{2}{3}}}{2}$ e $\frac{5 + e^{\frac{7}{2}}}{3}$ f 4, 5

3 $0, \ln 4$

4 $\frac{1}{2}\ln\frac{7}{2}, \frac{1}{2}\ln 2$

5 $\ln 16$

6 $\frac{1}{4}\ln\frac{3}{2}, \frac{1}{4}\ln 5$

7 $\frac{1 - \ln 4}{6 + \ln 12}$

8 a $(0, 3 - \ln 4)$ b $\left(-2 + e^{\frac{3}{2}}, 0\right)$

9 a
T
(0, 38)
$T = 31e^{-0.05x} + 7$
$T = 7$
O
t

b The minimum temperature of the bath water.
c 10.9 minutes

14.8 Logarithms and non-linear data

1 a $V = 5x^3$
$\log V = \log 5x^3$
$\log V = \log 5 + 3\log x$
b c 3
log V
log 5
O
log x

2 a $H = 3\left(\frac{3}{2}\right)^t$ b
$\log H = \log\left(3\left(\frac{3}{2}\right)^t\right)$
$\log H = \log 3 + \log\left(\frac{3}{2}\right)^t$
$\log H = \log 3 + t\log\frac{3}{2}$
log H
log 3
O
t
c $\log\frac{3}{2}$

3 a $y = ab^x$
$\log y = \log(ab^x)$
$\log y = \log a + \log b^x$
$\log y = \log a + x\log b$

b

x	0.5	2	5	7
log y	2.93	3.27	3.97	4.43

c

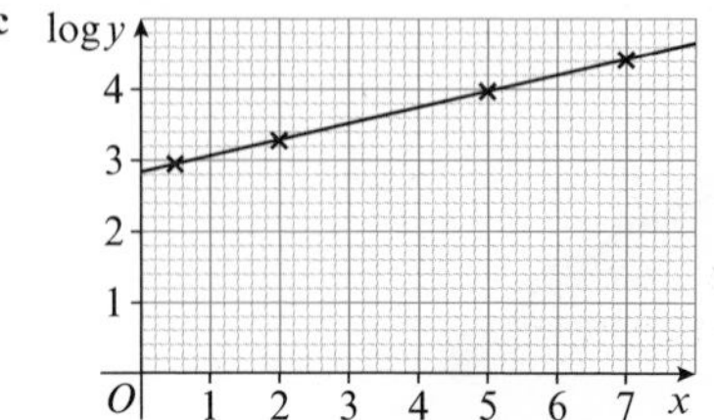

d $a = 645.7$ (answers may vary), $b = 1.7$

4 a

Time, t	0	2	4	6	8	10
$\log_{10} V$	4.40	4.17	3.94	3.72	3.49	3.26

b $V = ab^t$

$\log V = \log(ab^t)$

$\log V = \log a + \log b^t$

$\log V = \log a + t\log b$

c 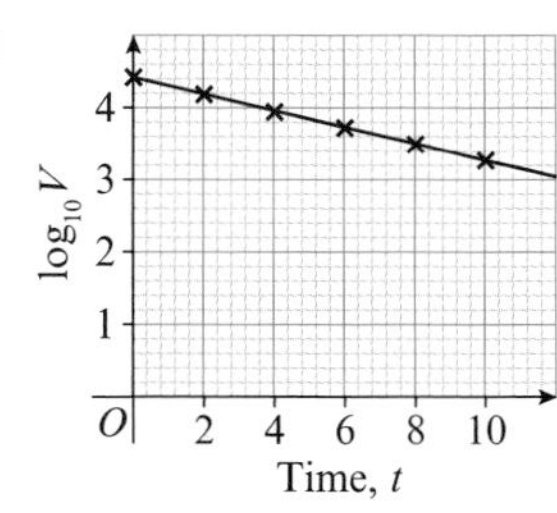

d $\frac{3.26 - 4.4}{10 - 0} = -0.114$ e $a = 25\,100, b = 0.769$

5 a

Time in days since 12/3/18, t	0	1	2	3	4
log P	1.44	1.47	1.51	1.54	1.58

b $P = ab^t$

$\log P = \log(ab^t)$

$\log P = \log a + \log b^t$

$\log P = \log a + t\log b$

c 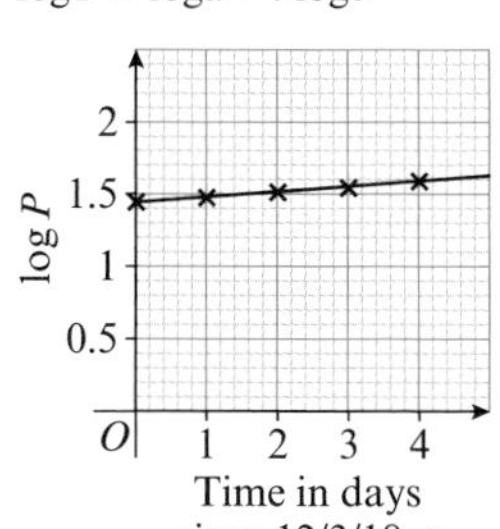

d $a = 27.43, b = 1.084$

6 a $\log_4 P = -0.075t + 5.2$ b 1350

c $a = 1350, b = 0.901$ d 59 people

7 a $\log_2 N = 0.26t + 9.3$ b $a = 630.3, b = 1.2$

c The initial number of bacteria in the petri dish.

Problem solving: Set A

B 1.25

S a $9^x - 3^{x+2} + 20 = 0$ b $\log_3 4, \log_3 5$

$(3^2)^x - (3^x)(3^2) + 20 = 0$

$(3^x)^2 - 9(3^x) + 20 = 0$

$u^2 - 9u + 20 = 0$

G $\frac{1}{2}, \frac{\log 6}{2\log 2}$

Problem solving: Set B

B a 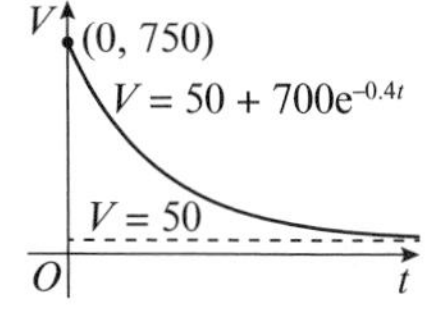

b £144.73

c $-\frac{5}{2}\ln\frac{1}{14}$ or $\frac{5}{2}\ln 14$

S a 14643 computers   b $\frac{dC}{dt} = \frac{1}{4}e^{\frac{1}{4}t}$

c $4 \ln 48$

G a Temperature of soup will decrease, so k must be negative.

b $a = 45$ c $\frac{1}{5}\ln\frac{2}{3}$ d 10.96 minutes

Exam Question Bank

1 a 243   b $8x^{-\frac{1}{2}}$

2 $\frac{13}{23}\sqrt{3} - \frac{1}{23}$

3 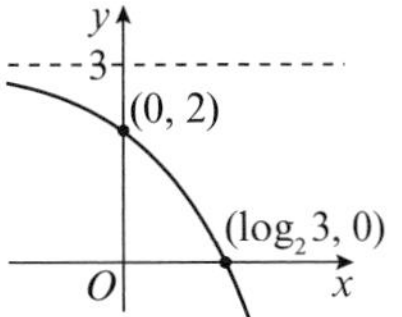

4 a $2\sqrt{2}$ b $4\sqrt{6}$

5 a $\frac{2\sqrt{5}}{5}\mathbf{i} - \frac{\sqrt{5}}{5}\mathbf{j}$ b 26.6° (1 d.p.) below the x-axis

6 $\frac{27}{5}x^{\frac{5}{3}} - \frac{4}{x} - \frac{3}{4}x^4 + 2x + c$

7 $\text{LHS} \equiv \frac{2\sin^2 x - (\sin^2 x + \cos^2 x)}{\cos^2 x} \equiv \frac{\sin^2 x - \cos^2 x}{\cos^2 x}$

$\equiv \frac{\sin^2 x}{\cos^2 x} - \frac{\cos^2 x}{\cos^2 x} \equiv \tan^2 x - 1 \equiv \text{RHS}$

8 $\frac{203}{64}$

9 $n = 5$

10 a $A(5 - 2\sqrt{3}, 15 - 4\sqrt{3}), B(5 + 2\sqrt{3}, 15 + 4\sqrt{3})$

b $x^2 - 8x + 18 \leqslant y \leqslant 2x + 5$

11 $x = \frac{1}{16}, x = \frac{81}{16}$

12 a i 18   ii $c = \frac{3}{2}$

b $f(x) = 4x^3 - 4x^2 - 15x + 18 \Rightarrow f'(x) = 12x^2 - 8x - 15$

13 a $k^2 - 10k + 4$ b $k = 5 \pm \sqrt{21}$

c Substituting $k = 8$ into f(x) gives

$f(x) = x^2 + 4x + 7 = (x + 2)^2 + 3$

So $f(x) > 0$ for all values of x

14 a $p = \frac{5}{3}$ b $\frac{38}{3}$

15 a $f'(x) = \frac{1}{6\sqrt{x}} - \frac{3}{2\sqrt{x^3}}$ b (9, 0)

c $f''(x) = -\frac{1}{12}x^{-\frac{3}{2}} + \frac{9}{4}x^{-\frac{5}{2}}$

$f''(9) = \frac{1}{162}$ therefore local minimum

16 a Ship: $-2\mathbf{i} + 6\mathbf{j}$, boat: $12\mathbf{i} - 4\mathbf{j}$ b $2\sqrt{74}$

17 a e.g. $k = -15°, k = 165°$

b e.g. $x = 105°, x = -75°$

18 $-\frac{7}{3}$

19 $8\mathbf{i} - 6\mathbf{j}$

20 a $1 + \frac{4}{x} + \frac{4}{x^2}$ b $1 + \frac{12x}{5} + \frac{12x^2}{5} + \frac{32x^3}{25}$

c $\frac{428}{25}$

21 a $2x + 5y - 66 = 0$ b 208.8

22 a $x = \frac{3}{2}, x \leqslant -1$ b $x = \frac{3}{4}, x = -\frac{1}{2}$

23 $25k^2 - 20k < 0 \Rightarrow 5k(5k - 4) < 0 \Rightarrow 0 < k < \frac{4}{5}$

24 a $x(x + 6)(x + 6)$

b $y = x^3 + 12x^2 + 36x$

25 $\frac{1}{3}\ln\frac{4}{3}, \frac{1}{3}\ln 5$

26 25

27 $a = \frac{8}{9}, b = -\frac{5}{3}, c = 1$

28 **a** C_1: radius 4, centre $(0, -2)$; C_2: radius 4, centre $(0, -5)$

b

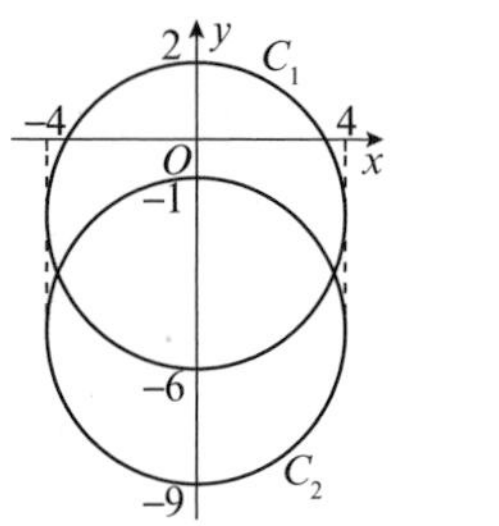

c $\sqrt{55}$

29 **a** $q^{100} + 100q^{99}p + 4950q^{98}p^2 + 161\,700q^{97}p^3$

b 0.858 96

30 **a** Consider $(p - q)^2 = p^2 + q^2 - 2pq = p^2 - q^2 + 2q^2 - 2pq$
If $p > q > 0$ then $2pq > 2q^2$, so $2q^2 - 2pq$ is negative, and $(p - q)^2 < p^2 - q^2$, so $\sqrt{p^2 - q^2} > p - q$ as required.

b $p = 0, q = 0$ would give $0 > 0$ which is untrue.

31 $-\frac{8}{\sqrt{17}}$

32 **a** $x > 1$ **b** $-3 \leqslant x \leqslant 7$ **c** $1 < x \leqslant 7$

33 **a** $(5, -4.5)$ **b** 10

c $x \geqslant 5$ **d** $k \leqslant 1.5$

34 **a** When $x = -1$, $5^x = 5^{-1} = \frac{1}{5}$, $2^x = 2^{-1} = \frac{1}{2}$ and $\frac{1}{5} < \frac{1}{2}$, so not true for all values of x.

b Write the consecutive odd numbers as $2n - 1$ and $2n + 1$, where n is an integer. Then:

$$\frac{1}{2}(2n-1)^2 + \frac{1}{2}(2n+1)^2 = \frac{1}{2}(4n^2 - 4n + 1) + \frac{1}{2}(4n^2 + 4n + 1)$$
$$= 2n^2 - 2n + \frac{1}{2} + 2n^2 + 2n + \frac{1}{2}$$
$$= 4n^2 + 1$$

which is one more than an even number, so must be odd.

35 $x = 0, y = 2$ or $x = 6, y = -4$

36 $x = \frac{3}{2}$ or $x = \frac{5}{2}$

37 **a** 34.8° and 145.2° (1 d.p.) **b** 19.2 cm (3 s.f.)

38 **a** $p = 3, q = 25$ **b** 100

c $y = 16 - 6x - x^2$; $(-3, 25)$, $(0, 16)$, $(-8, 0)$, $(2, 0)$

39 **a** $y = \cos 3x$; $y = \cos x$

b Stretch with scale factor $\frac{1}{3}$ parallel to the x-axis.

c $x = 20°, 100°, 140°, 220°, 260°$ or $340°$

40 **a** $1024 - \frac{1024}{3}x + \frac{256}{5}x^2$ **b** $p = \frac{1}{2}$ **c** $q = \frac{1}{12}$

41 **a** $(x + 1)(2x + 1)(3x - 2)$

b $\frac{6x^3 + 5x^2 - 3x - 2}{2x^3 + 3x^2 + x} = \frac{(x+1)(2x+1)(3x-2)}{x(2x+1)(x+1)} = 3 - \frac{2}{x}$

42 **a** $(x - 2)^2 + (y + 1)^2 = 72$ **b** $8\sqrt{17} - 4$

43 **a** $f(x) = (x - 3)^2 - 9 + 16 = (x - 3)^2 + 7$

b $y = f(x)$; 16; $(3, 7)$ **c** $k > 7$

44 **a** $A(-1, 1), B(2, 4)$ **b** $\frac{9}{2}$

45 15.4 cm² (3 s.f.)

46 **a**

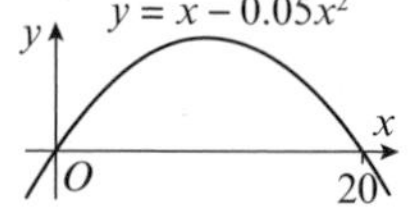

b 20 m

c Model gives maximum height of 5 m

47 $x = 68.2°, 146.3°, 248.2°$ or $326.3°$

48 **a** $3x^2 + 6x - 9$ **b** $\{x : x < -3\} \cup \{x : x > 1\}$

49 **a** $x(6 + 5x)(6 - 5x)$

b

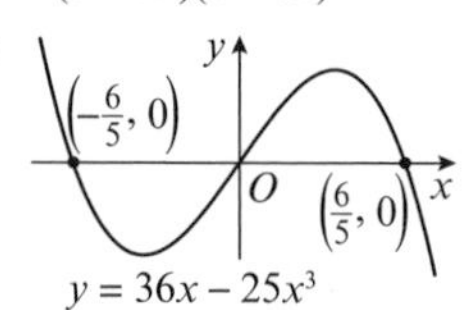

c $A(-1, -11)$ and $B(1, 11)$
Distance is $\sqrt{(-1-1)^2 + (-11-11)^2} = 2\sqrt{122}$

50 **a** $(1, 2.5)$ **b** $y = -\frac{2}{7}x + \frac{64}{7}$

c 128

51 **a** $1 + 5x + 10x^2 + 10x^3 + 5x^4 + x^5$

b **i** $(1 + \sqrt{5})^5 = 1 + 5\sqrt{5} + 10(5) + 10(5\sqrt{5}) + 5(25) + (25\sqrt{5})$
$= 176 + 80\sqrt{5}$

ii $\log_2(1 + \sqrt{5})^5 = \log_2(176 + 80\sqrt{5}) = \log_2 16(11 + 5\sqrt{5})$
$= \log_2 16 + \log_2(11 + 5\sqrt{5}) = 4 + \log_2(11 + 5\sqrt{5})$

52 **a** $(x - 3)^2 + 4$

b C; $P(0, 13)$; $Q(3, 4)$ **c** $3\sqrt{10}$

53 **a** 4 **b** $\frac{208}{15}$

54 **a** $\log_3 \frac{7x - 18}{(x-2)^2} = 1$

$7x - 18 = 3x^2 - 12x + 12$

$3x^2 - 19x + 30 = 0$

b $x = 3, x = \frac{10}{3}$

55 **a** $2x - 4 = x^2 + px - 2p \Rightarrow x^2 + (p - 2)x + 4 - 2p = 0$
So $(p - 2)^2 - 4(4 - 2p) < 0 \Rightarrow p^2 + 4p - 12 < 0$

b $-6 < p < 2$

56 **a** $|\overrightarrow{DE}| = \sqrt{58}$, $|\overrightarrow{EF}| = \sqrt{89}$ and $|\overrightarrow{DF}| = \sqrt{65}$, so triangle DEF is scalene.

b 50.9°

57 **a** Factor theorem states if $(x + 3)$ is a factor of g(x) then g(−3) = 0.

$g(-3) = 4 \times (-3)^3 - 16 \times (-3)^2 - 35 \times (-3) + 147$
$= -108 - 144 + 105 + 147 = 0$

so $(x + 3)$ is a factor of g(x).

b By division $g(x) = (x + 3)(4x^2 - 28x + 49)$
$= (x + 3)(2x - 7)^2$

c **i** $g(x) \leqslant 0$ when $x \leqslant -3$

ii $g(2x) \leqslant 0$ when $x \leqslant -\frac{3}{2}$

58 **a** **i** (4, −6) **ii** 6 **b** $0 < k < 2.4$

59 **a** £805.19 **b** $24.5e^{0.035t}$ **c** 19.8 years

60 **a** Use $\sin^2 x \equiv 1 - \cos^2 x$ and rearrange

b $x = 36.42°, 97.94°, 262.06°$ or $323.58°$

61 **a** $1 + 12x + 54x^2 + 108x^3 + 81x^4$

b Expand $(1 - 3x)^4$ in a similar way to $(1 + 3x)^4$ and add the two expansions. The x and x^3 terms cancel out.

c (0, 2)

62 $\frac{1}{2}$ and −2

63 **a** Using cosine rule, $\frac{3}{4} = \frac{6^2 + (2x-5)^2 - (x+2)^2}{2 \times 6 \times (2x - 5)}$

b $x = 7 \pm \sqrt{15}$

64 **a** $1 + x^2 < 1 + x^2 + 2x \Rightarrow 0 < 2x \Rightarrow x > 0$

b For $x = -1$, $1 + (-1)^2 < (1 + (-1))^2 \Rightarrow 2 < 0$ (not true)

65 $k = -11$

66 **a** $x = 45°$ or $105°$

b $x = 67.5°, 112.5°, 247.5°$ or $292.5°$

67 **a** **i** $P = 150 - 6.25(x - 15)^2$ **ii** 10.1 and 19.9

iii

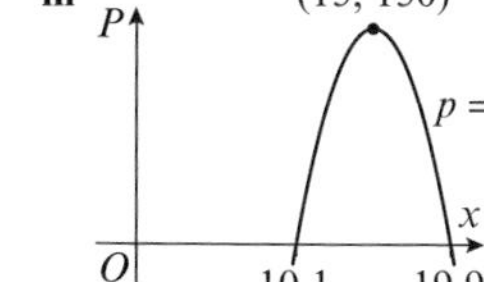

b The company would make an annual loss of £475 000.

c **i** £150 000 **ii** £15

68 **a** $C = ab^t$

$\log C = \log(ab^t)$

$\log C = \log a + \log b^t$

$\log C = \log a + t \log b$

b

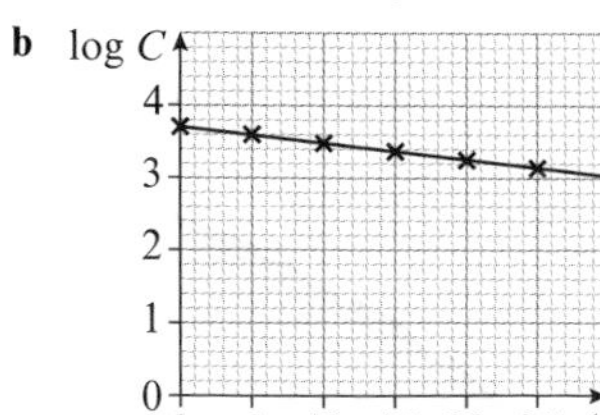

c $a = 5100$, $b = 0.95$

69 **a** 7

b $f(1) = (1 + k)(2 + 5) + 7 = 0$, $7 + 7k + 7 = 0$ so $7k = -14$ and $k = -2$

c $x = -\frac{1}{2}$, $x = -3$ and $x = 1$

70 **a** **i** (5, −3) **ii** $3\sqrt{5}$

b $(x - 5)^2 + y^2 = 36$

71 **a** $x^2 - 8x + 17 = (x - 4)^2 + 1$. Minimum value is $y = 1$.

b Sometimes true; e.g. substitute in −5 and 5 to obtain different results.

72 **a** (−3, 0), (5, 0) **b** $\frac{863}{6}$

73 **a** 258.6 m (1 d.p) **b** 087° (nearest degree)

74 **a** Use $P = 300 = 2\pi r + 2y$ to find $y = 150 - \frac{\pi}{2}x$

Substitute into $A = \frac{\pi x^2}{4} + xy$

b $\frac{22\,500}{\pi}$ m²

75 **a** $f(5) = 250 + 25k - 120 + 45 = 0$ so $25k = -175$ and $k = -7$

b $f(x) = (x - 5)(2x - 3)(x + 3)$

c $5^y = 5$ so $y = 1.5^y = \frac{3}{2} \Rightarrow y\ln 5 = \ln \frac{3}{2}$ so $y = 0.25$

($5^y = -3$ has no solutions)

76 $x = 49.0°, 84.4°$ or $169.0°$

77 **a** $y = 3x$ **b** $\left(\frac{6}{5}, \frac{18}{5}\right)$

c $\sqrt{\left(\frac{6}{5}\right)^2 + \left(\frac{18}{5}\right)^2} = \sqrt{\frac{360}{25}} = \frac{6\sqrt{10}}{5}$ **d** $2\sqrt{10}$

e 12

78 **a**

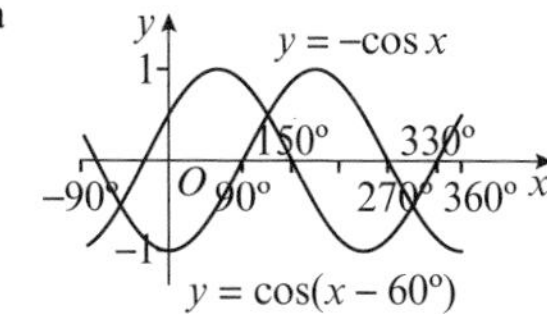

b (0, 0.5), (−30°, 0), (150°, 0), (330°, 0)

c $x = -60°, 120°$ or $300°$

79 **a** $(\sqrt[3]{16}, 8)$

b $4x^{\frac{3}{2}} - \frac{1}{2}x^3 = 0 \Rightarrow x^3 - 8x^{\frac{3}{2}} = 0$

$\Rightarrow x^{\frac{3}{2}}(x^{\frac{3}{2}} - 8) = 0 \Rightarrow x^{\frac{3}{2}} = 0$ or $x^{\frac{3}{2}} = 8$

$\Rightarrow x = 0$ or $x = 4$, so B is the point (4, 0).

c $\frac{96}{5}$

80 **a** $\frac{3}{5}\mathbf{a} - \frac{2}{5}\mathbf{b}$

b $\overrightarrow{CA} = \mathbf{a} - \mathbf{b}$. As $\overrightarrow{CA}$ and $\overrightarrow{CZ}$ are not parallel, $\overrightarrow{CZ}$ will not pass through the point A.

c Create a vector equation using $\overrightarrow{OZ} = \overrightarrow{OA} + \overrightarrow{AP} + \overrightarrow{PZ}$ and equate **a** and **b** parts.

81 **a** (2, 9) **b** 40 **c** $3 < k < 15$

82 **a** $(x - 5)^2 + (y - 3)^2 = 100$

b $4x + 3y - 79 = 0$ **c** $10\sqrt{5}$

83 **a** $f(-2) = 0$

b $f(x) = (x + 2)(x^2 - 4x + 9)$; using discriminant on quadratic, $b^2 - 4ac = 16 - 36 = -20 < 0$ so no real roots. Only one root when $x + 2 = 0$, i.e. $x = -2$

c $x^3 - 2x^2 + x + 18 \geqslant 18$, means $x^3 - 2x^2 + x \geqslant 0$. Factorising $x(x^2 - 2x + 1) = x(x - 1)^2$, $(x - 1)^2 \geqslant 0$ for all values of x. If $x \geqslant 0$ then $x(x - 1)^2 \geqslant 0$ and $f(x) = x^3 - 2x^2 + x + 18 = x(x - 1)^2 + 18 \geqslant 18$ for all $x \geqslant 0$.

84 **a** $f(x) = (x + 2)(2x - 1)(3x + 1)$

b $y = 30°, 150°, 199.5°$ or $340.5°$

85 **a** $\theta = 33.4°, 176.6°, 213.4°$ or $356.6°$

b $x = 379.47°$ or $520.53°$

86 **a** $4x + 3y - 13 = 0$ **b** $3x - 4y + 34 = 0$

c -10 **d** $(x + 3)^2 + y^2 = 50$

87 **a** $3x + 4y = 19$

b $3 \times 13 + 4y = 19 \Rightarrow y = -5$

c $(x - 13)^2 + (y + 5)^2 = 125$

d $50\sqrt{5}$

88 **a** **i**

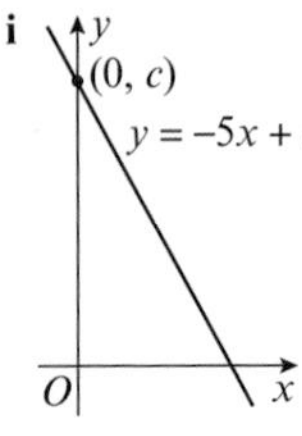

ii

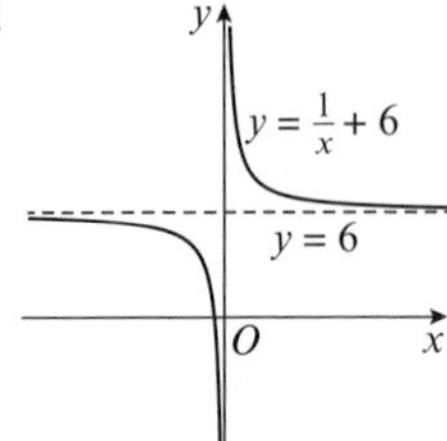

b $\frac{1}{x} + 6 = -5x + c$ rearranges to $5x^2 + (6 - c)x + 1 = 0$

Using $b^2 - 4ac > 0$, $(6 - c)^2 - 20 > 0 \Rightarrow (6 - c)^2 > 20$

c $c < 6 - 2\sqrt{5}$ or $c > 6 + 2\sqrt{5}$

89 **a** 4 and 8 **b** $(7, 5)$

90 **a** Multiply dimensions: x, $36 - 2x$, $60 - 2x$

b $16 - 2\sqrt{19}$ **c** 7092.4 cm^3

91 **a** Perimeter $= 500 = 2\pi r + 2x \Rightarrow x = 250 - \pi r$

Area $= \pi r^2 + 2rx = \pi r^2 + 2r(250 - \pi r) = 500r - \pi r^2$

b $A = \frac{62\,500}{\pi}$, $r = \frac{250}{\pi}$

92 **a** -2 **b** $(x - 3)^2 + (y + 2)^2 = 169$

c 78

93 **a** Gradient of $PQ = \frac{15}{25} = \frac{3}{5}$; gradient of $SR = \frac{9}{15} = \frac{3}{5}$

So PQ and SR are parallel.

b Length of $PS = \sqrt{194}$; length of $QR = \sqrt{74}$

So trapezium is not isosceles.

c $\left(\frac{5}{2}, \frac{31}{4}\right)$

d Midpoint of PR is $(0, 5)$; midpoint of SQ is $(5, 8)$, so neither diagonal of the trapezium bisects the other.